Materials Technology for the Energy and Environmental Nexus, Volume 2

Online at: https://doi.org/10.1088/978-0-7503-5729-6

Materials Technology for the Energy and Environmental Nexus, Volume 2

Edited by
R V Mangalaraja
Faculty of Engineering and Sciences, Universidad Adolfo Ibáñez, Diagonal las Torres 2640, Peñalolén, Santiago, Chile

Arunachalam Arulraj
Departamento de Electricidad, Facultad de Ingeniería, Universidad Tecnológica Metropolitana, Santiago, Chile

Radhamanohar Aepuru
Departamento de Ingeniería Mecánica, Facultad de Ciencias Físicas y Matemáticas, Universidad de Chile, Santiago, Chile

Departmento de Mecánica, Facultad de Ingeniería, Universidad Tecnológica Metropolitana, Santiago, Chile

Chan Siew Hwa
School of Mechanical & Aerospace Engineering, Nanyang Technological University, Singapore

IOP Publishing, Bristol, UK

ISBN 978-0-7503-5729-6 (ebook)
ISBN 978-0-7503-5727-2 (print)
ISBN 978-0-7503-5730-2 (myPrint)
ISBN 978-0-7503-5728-9 (mobi)

DOI 10.1088/978-0-7503-5729-6

Version: 20231101

IOP ebooks

British Library Cataloguing-in-Publication Data: A catalogue record for this book is available from the British Library.

Published by IOP Publishing, wholly owned by The Institute of Physics, London

IOP Publishing, No.2 The Distillery, Glassfields, Avon Street, Bristol, BS2 0GR, UK

US Office: IOP Publishing, Inc., 190 North Independence Mall West, Suite 601, Philadelphia, PA 19106, USA

The editor Prof. R.V. Mangalaraja would like to dedicate this book on the remembrance and honor of (Late) Prof. Manohar Paramasivam, Head (Retired) of the Department of Ceramic Technology, Anna University, Chennai, India.

Contents

Acknowledgements

The editors gratefully acknowledge all the contributors for their sincere efforts and kind support for the successful completion of the book. The editor Prof. R.V. Mangalaraja and Dr. A. Arulraj sincerely acknowledges the Universidad Adolfo Ibáñez, and Universidad Tecnológica Metropolitana, Santiago, Chile for their kind support and management.

Editor biographies

R V Mangalaraja

Prof. R V Mangalaraja is a Full Professor in the Faculty of Engineering and Sciences, Universidad Adolfo Ibáñez, Peñalolén, Santiago, Chile. He worked as the Director of International Relations at the University of Concepción for the years 2018 and 2019 and as the Director for the Department of Materials Engineering at University of Concepcion for six years from 2010 to 2016. Further, he also served as the President of the Chilean Metallurgy and Materials Society (SOCHIMM) for the period from 2015 to 2017. Prof. Mangalaraja is well recognized in the scientific community, and he also serves in many professional bodies like Life-Member of Indian Ceramic Society (ICS), Fellow of Indian Institute of Ceramics (IIC), Member of The Mineral, Metals and Materials Society (TMS), Singapore Materials Research Society (MRS-S) during 2017-2018; member of American (ACerS) and PAN-American Ceramic Societies.

Prof. Mangalaraja earned his B.Sc. degree in Physics from Madurai Kamaraj University and M.Sc. degree in Materials Science, M.Tech. in Ceramic Technology. Later, he continued his Ph.D. in Materials Technology (Advanced Ceramics) from Anna University-Chennai, India. Prior to joining Universidad Adolfo Ibáñez, he served as Professor for 17 years at the University of Concepcion, he was a Postdoctoral researcher at AIST-Nagoya, Japan, Lulea University of Technology-Lulea, Sweden and Anna University-Chennai, India. His primary research focuses on the areas of nanotechnology- driven advanced ceramics (functional ceramics for energy and environmental applications) and micro- nano-structure-property relationship of functional and structural ceramics and metallic composite materials.

Prof. Mangalaraja has published over 280 refereed international journal publications, 40 book chapters, three patents, edited two Books, and more than 200 research presentations/invited talks in several International Conferences. Further, he has participated as Responsible and Co-Investigator of 40 research projects funded by ANID-National Agency for Research Development (formerly CONICYT - National Commission of Scientific and Technological Research), Government of Chile and two projects funded by Universidad Adolfo Ibáñez, Peñalolén, Santiago, Chile. For his credits, he has guided 30 Postdoctoral researchers and 13 doctoral candidates in the area of materials science and technology. Also, he had been acted as a member in the expert committee of FONDECYT-ANID for two times (four years) the area of Materials Science and Engineering, and now extending his expertise as a Reviewer of various projects in the area of Materials Science and Engineering.

Arunachalam Arulraj

Dr. Arulraj Arunachalam is an Assistant Professor in the Department of Electricidad, Universidad Tecnológica Metropolitana, Santiago, Chile. Priorly, he worked as a FONDECYT Postdoctoral Researcher in the Faculty of Engineering and Sciences at the Universidad Adolfo Ibáñez, Peñalolén, Santiago, Chile. Also, he worked as a Research Fellow, in the Graphene and Advanced 2D Materials Research Group, at Sunway University, Malaysia, and Institute of Physics, National Taiwan University, Taiwan. He graduated his B. E (Electrical and Electronics Engineering), M. Tech (Nanoscience and Nanotechnology), and Ph.D. (Technology) degrees from Anna University, India. For his academic achievement, he was awarded the meritorious TEQIP-II fellowship, India for the period of 2013 and 2016.

Dr. Arulraj also serves in many professional bodies like the American Ceramics Society (Associate Member), Royal Society of Chemistry (Affiliated Member), American Chemical Society (Member), International Society for Development & Sustainability (Associate Member). To his credit, he has published 30 research articles including 7 book chapters in various peer-reviewed international journals. His research interests are focused on designing and studying various two-dimensional materials which include graphene, chalcogenides, and MXenes for energy conversion and storage.

Radhamanohar Aepuru

Dr. RADHAMANOHAR AEPURU is currently working as the Assistant Professor in the Department of Mechanical Engineering, Facultad de Ciencias Físicas y Matemáticas, Universidad de Chile, Santiago, Chile. Earlier, he worked as an Assistant Professor in Department of Mechanical Engineering, Facultad de Ingeniería, Universidad Tecnologica Metropolitana, Santiago, Chile during April 2020- July 2022. He worked as a Principal Investigator for the FONDECYT Postdoctoral Project in the Universidad de Concepcion and Universidad Tecnologica Metropolitana during the years 2018-2021 funded by ANID National Agency for Research Development, Government of Chile. Dr. Aepuru served as an Invited researcher (2017-2020) in the Department of Microelectronics, Technical University of Sofia, Bulgaria for a project funded by the Bulgarian National Scientific Fund, Ministry of Education and Science, Bulgaria. He worked as a National Postdoctoral Researcher as a Principal Investigator during the years 2017 and 2018 in the Institute of Nanoscience and Technology (INST), Mohali, India funded by the Science and Engineering Research Board (SERB), Department of Science and Technology (DST), Govt. of India. He is a member in professional bodies like Life-Member of Electron Microscope Society of India (EMSI).

Chan Siew Hwa

Prof. CHAN Siew Hwa is a Fellow of Academy of Engineering, Singapore (SAEng), a Fellow of ASEAN Academy of Engineering and Technology (AAET), President's Chair in Energy and a Professor in Nanyang Technological University (NTU). He joined NTU in 1991 after obtaining his PhD and subsequently working as a post-doctoral researcher at Imperial College London. He leads the hydrogen and fuel cell research at Energy Research Institute @ NTU. Since 2017, he has been appointed as Senior Vice President of China-Singapore International Joint Research Institute in China- Singapore Guangzhou Knowledge City, focusing on incubation and commercialization of technologies. Prof. Chan has been active in serving academic & research communities and industries such as Singapore's covering focal point for the Sub- committee on Sustainable Energy Research (SCSER) under the ASEAN Committee on Science, Technology, and Innovation (COSTI), Ministry of Trade and Industry (MTI) Singapore's Future Energy Technology (Hydrogen) Watch Group, NUS's Governing Board of Centre for Hydrogen Innovation, Advisory Board of Total SA and Hydrogen Energy, etc. He is also active in commercialization of technologies including 12 technology licensing. Dr Chan was a Director of Maz Energy Pte Ltd (2004 – 2022) with core business in nitroparaffin-based fuel additives and founded Xin Xiang (Guangzhou) Hydrogen Technologies Co., Ltd., which is a tech-company manufacturing key components of PEMFC. He has published ~300 refereed journal papers with total citations of >15K and h-index of 64. He was a recipient of NTU's Teacher-of-the-year, Nanyang Award (Research Excellence), Nanyang Award (Innovation Entrepreneurship), George-Stephenson Medal from IMechE, UK, Outstanding Scientific Achievement from International Association of Hydrogen Energy (IAHE), USA, Star of Innovation Talent award from Guangzhou government, World's Most Influential Scientific Minds from Thomson- Reuters, etc.

List of contributors

Radhamanohar Aepuru
Departamento de Ingeniería Mecánica, Facultad de Ciencias Físicas y Matemáticas, Universidad de Chile, Santiago, Chile
Departmento de Mecánica, Facultad de Ingeniería, Universidad Tecnológica Metropolitana, Santiago, Chile

Ali Akbari-Fakhrabadi
Advanced Materials Laboratory (AmLab), Department of Mechanical Engineering, University of Chile, Santiago, Chile

Viswanathan Alagan
Department of Physics, University College of Engineering-BIT Campus, Anna University, Tiruchirappalli, Tamil Nadu, India

Md Abdullah Al Bari
Department of Chemical and Petroleum Engineering, University of Calgary, 2500 University Drive, NW Calgary, Alberta, Canada T2N 1N4

Maria Aleksandrova
Department of Microelectronics, Technical University of Sofia, Sofia 1000, Bulgaria

Padmanaban Annamalai
Facultad de Ingeniería, Universidad Católica de la Santísima Concepción, Concepción, Chile
Department of Physiology, Saveetha Institute of Medical and Technical Sciences, Saveetha University, Chennai 600077, India

Yamunadevi Balakrishnan
Department of Biotechnology, Sri Venkateswara College of Engineering, Sriperumbudur-602 117, Tamil Nadu, India

V Balasubramanian
Department of Physics, Francis Xavier Engineering College (Autonomous), Tirunelveli 627003, Tamil Nadu, India

B V Bhaskara Rao
Departamento de Ingeniería Química, Biotecnología y Materiales, Facultad de Ciencias Físicas y Matemáticas, Universidad de Chile, Santiago, Chile

Cristian H Campos
Department of Physical-Chemistry, Faculty of Chemical Sciences, University of Concepcion, Edmundo Larenas 129, Concepción, Chile

P V Chandrasekar
School of Science, Optoelectronic Research Centre, Minzu University of China, Beijing 100081, People's Republic of China

Rajasekaran Chandrasekaran
School of Bio Sciences and Technology, Vellore Institute of Technology, Vellore 632014, Tamil Nadu, India

Seung-Cheol Chang
Department of Cogno-Mechatronics Engineering, College of Nanoscience and Technology, Pusan National University, Busan 46241, Republic of Korea

Chang Tang Chang
Department of Environmental Engineering, National Ilan University, Yilan, Taiwan

Kuan-Chung Chen
Department of Environmental Science and Engineering, National Pingtung University of Science and Technology, Taiwan

Bavatharani Chokkiah
Department of Chemistry, Saveetha Engineering College (Autonomous), Chennai 602105, Tamil Nadu, India

Kavery Elangovan
Department of Chemistry, Padmavani Arts and Science College for Women, Salem 636011, India

Rodrigo Espinoza-Gonzáleza
Departamento de Ingeniería Química, Biotecnología y Materiales, Facultad de Ciencias Físicas y Matemáticas, Universidad de Chile, Santiago, Chile

Muthusankar Eswaran
Division of Systems and Synthetic Biology, Department of Biology and Biological Engineering, Chalmers University of Technology, SE-412 96 Göteborg, Sweden
School of Electronic Systems and Automation, Kerala University of Digital Sciences, Innovation and Technology (Digital University Kerala), Technocity Campus Mangalapuram, Thonnakkal PO, Kerala 695317, India

Vishwajit M Gaikwad
Department of Physics, Amolakchand Mahavidyalaya, Yavatmal, Maharashtra 445001, India

Mano Ganapathy
Department of Physics, Bharathiyar Institute of Engineering for Women, Salem, Tamil Nadu, India

Karthick Kannimuthu
Department of Chemical and Petroleum Engineering, University of Calgary, 2500 University Drive, NW Calgary, Alberta, Canada T2N 1N4

Md Golam Kibria
Department of Chemical and Petroleum Engineering, University of Calgary, 2500 University Drive, NW Calgary, Alberta, Canada T2N 1N4

Abirami Krishnan
Department of Biotechnology, Sri Venkateswara College of Engineering, Sriperumbudur 602117, Tamil Nadu, India

Pawan Kumar
Department of Chemical and Petroleum Engineering, University of Calgary, 2500 University Drive, NW Calgary, Alberta, Canada T2N 1N4

R V Mangalaraja
Faculty of Engineering and Sciences, Universidad Adolfo Ibáñez, Diagonal las Torres 2640, Peñalolén, Santiago, Chile

Shankar Mani
Department of Physics Knowledge Institute of Technology, KIOT-Campus, Salem, Tamil Nadu, India

Mudaliar Mahesh Margoni
Advanced Materials Laboratory (AmLab), Department of Mechanical Engineering, University of Chile, Santiago, Chile

Ivan Mijakovic
Division of Systems and Synthetic Biology, Department of Biology and Biological Engineering, Chalmers University of Technology, SE-412 96 Göteborg, Sweden
Novo Nordisk Foundation Center for Biosustainability, Technical University of Denmark, DK-2800 Lyngby, Denmark

Kanagam Nachiappan
School of Bio Sciences and Technology, Vellore Institute of Technology, Vellore 632014, Tamil Nadu, India
Department of Biotechnology, Sri Venkateswara College of Engineering, Sriperumbudur 602117, Tamil Nadu, India

Durga Prasad Pabba
Departamento de Ingeniería Mecánica, Facultad de Ingeniería, Universidad Tecnológica Metropolitana, Santiago, Chile

Sathishkumar Panneerselvam
Department of Chemistry, School of Advanced Sciences, Vellore Institute of Technology, Vellore 632014, Tamil Nadu, India

Vinoth Kumar Ponnusamy
Department of Medicinal and Applied Chemistry, Kaohsiung Medical University (KMU), Kaohsiung City 807, Taiwan
Research Center for Precision Environmental Medicine, Kaohsiung Medical University (KMU), Kaohsiung City 807, Taiwan

Franck Quero
Laboratorio de Nanocelulosa y Biomateriales, Departamento de Ingeniería Química, Biotecnología y Materiales, Facultad de Ciencias Físicas y Matemáticas, Universidad de Chile, Avenida Beauchef 851, Santiago 8370456, Chile

Kazi Hasibur Rahman
Faculty of Physics, Department of Humanities and Sciences, Lords Institute of Engineering and Technology, Hyderabad 500091, Telangana, India

Govindan Rajivgandhi
Laboratorio de Nanocelulosa y Biomateriales, Departamento de Ingeniería Química, Biotecnología y Materiales, Facultad de Ciencias Físicas y Matemáticas, Universidad de Chile, Avenida Beauchef 851, Santiago 8370456, Chile

C Rajkumar
The Institute of Scientific and Industrial Research (ISIR: SANKEN), Osaka University, Osaka, Japan

Manikandan Ramalingam
Department of Cogno-Mechatronics Engineering, College of Nanoscience and Nanotechnology, Pusan National University, Busan 46241, Republic of Korea.
Engineering Research Center for Color-Modulated Extra-Sensory Perception Technology, Pusan National University, Busan 46241, Republic of Korea.

Kiruthiga Ramakrishnan
Departamento de Ingeniería Mecánica, Facultad de Ingeniería, Universidad Tecnológica Metropolitana, Santiago, Chile

Prasanta Kumar Sahoo
Department of Mechanical Engineering, Siksha 'O' Anusandhan, Deemed to Be University, Bhubaneswar 751030, India

Felipe Sanhueza-Gómez
Electrochemical Technologies and Green Energies Laboratory, Department of Materials Engineering, Faculty of Engineering, University of Concepcion, Concepción, Chile

Prabhu Saravanan
Department of Physical-Chemistry, Faculty of Chemical Sciences, University of Concepcion, Edmundo Larenas 129, Concepción, Chile

Manda Sathish
Centro de Investigación de Estudios Avanzados del Maule (CIEAM), Vicerrectoría de Investigación y Postgrado, Universidad Católica del Maule, Talca 3460000, Chile

A Sathiya Priya
Department of Physics, Sri Sai Ram Engineering College, Chennai, India

Gopinath Sekar
Department of Biotechnology, Sri Venkateswara College of Engineering, Sriperumbudur 602117, Tamil Nadu, India

B Shunmugapriya
Futuristic Materials Research Centre for Planetary Exploration, Department of Physics and Nanotechnology, SRM Institute of Science and Technology, Kattankulathur 603203, Chengalpattu, Tamil Nadu, India

Krishnamoorthy Shanmugaraj
Faculty of Engineering and Sciences, Universidad Adolfo Ibáñez, Diagonal las Torres 2640, Peñalolén, Santiago, Chile

Bharathkumar Sivakumar
Facultad de Ingeniería, Universidad Católica de la Santísima Concepción, Concepción, Chile
Department of Physics, K L University, University in Guntur, Andhra Pradesh 522302, India

Balaji Sundaramoorthy
School of Bio Sciences and Technology, Vellore Institute of Technology, Vellore 632014, Tamil Nadu, India

Kalaivani Thiagarajan
School of Bio Sciences and Technology, Vellore Institute of Technology, Vellore 632014, Tamil Nadu, India

Arun Thirumurugan
Sede Vallenar, Universidad de Atacama, Costanera #105, Vallenar 1612178, Chile

Hector Valdes
Facultad de Ingeniería, Universidad Católica de la Santísima Concepción, Concepción, Chile

Harini Gurumurthy Vedavalli
Department of Biotechnology, Sri Venkateswara College of Engineering, Sriperumbudur 602117, Tamil Nadu, India

Kaaviya Velmurugan
Department of Biotechnology, Sri Venkateswara College of Engineering, Sriperumbudur 602117, Tamil Nadu, India

Chapter 1

An introduction to the role of materials in the energy–environment nexus

Radhamanohar Aepuru, Maria Aleksandrova, Vishwajit M Gaikwad, A Sathiya Priya, Prasanta Kumar Sahoo, Krishnamoorthy Shanmugaraj and R V Mangalaraja

1.1 Introduction

Energy is vital to society and is essential for promoting development, and the global demand for energy is increasing continuously. The non-renewable energy resources, such as coal, oil, and natural gas, are currently the major sources of energy, which are being depleted drastically [1]. This is resulting in increased energy costs and energy shortages, which are affected by issues such as technological improvements, population growth, and industrialization. In addition to the energy crisis, these energy sources are highly polluting and responsible for environmental damage such as global warming, climate change, natural resource depletion, improper waste disposal, loss of biodiversity, deforestation, ocean acidification, ozone layer depletion, acid rain, water pollution, and urban sprawl [2]. Therefore, the production of energy and the protection of the environment are the foremost challenges to realize a sustainable future. Considering the energy problems and the increase in energy consumption across the globe, efficient clean energy systems need to be explored to minimize the potential environmental problems which affect the growth of a sustainable environment. Thus sustainability is a vital and integral aspect in meeting the future global demand for energy and environmental resources, which can balance the ecosystem by providing a greener pathway to achieve renewable energy production and positive socio-economic impacts. In the last two decades a multidisciplinary field of research centered around the concept of the energy–environment nexus has carried out extensive studies in order to establish clean renewable energy for sustainable development, focusing on the generation, storage, and efficient utilization of energy and natural resources [3]. The development of a low-carbon

doi:10.1088/978-0-7503-5729-6ch1

system needs a shift toward technologies such as solar, wind power, electromobility, and energy efficiency. Climate change goes hand-in-hand with other significant aspects, among which energy security is central and can be affected by bottlenecks in the supply of material. The critical materials with high supply risks and scarce geological occurrence but of major economic importance, need to be studied. The development of the importance of each material in the future is difficult to predict, as it will depend on a wide range of factors such as technological developments in the energy sector and in other sectors. Research needs to focus on understanding the role of materials in connecting energy and the environment, paving the way for the development of greener energy and sustainability. The development of the material nexus is important and highlights the need for and role of the materials for greener energy and sustainable environment systems. The requirements vary given the integration of the newly developed technologies for energy systems such as solar energy conversion and storage. Energy–environment material nexus challenges do not only have implications for the energy supply but also have a significant influence on other sectors such as sustainable development. In fact, energy is not a replaceable aspect of our lives and it is mostly correlated with the environmental society, where conventional energy systems are the main drivers of an environmental pollution. Therefore, energy production should be oriented in an environmentally friendly manner to provide a path for smarter and cleaner systems which are the foremost challenges to realize sustainability.

Recently, great progress has been made in the carbon-based materials and metal oxides for energy and environmental applications because of their stability, efficiency, optical and electrical characteristics, and suitability for creating a nexus for the development of next-generation sustainable materials. Figure 1.1 shows the emerging attributes of carbon and metal oxide-based materials for energy and environmental applications. In particular, the carbon-based materials such as carbon nanotubes, graphene, graphene oxide, reduced graphene oxide, carbon dots, and graphitic carbon nitride materials have chemical stability, diverse

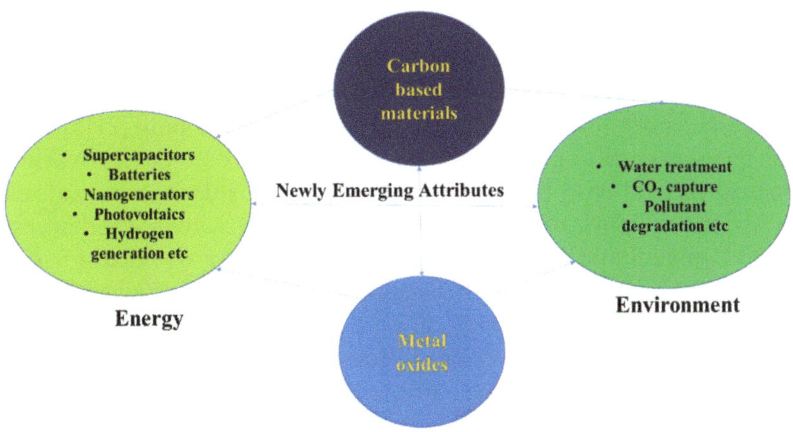

Figure 1.1. Overview of materials and their newly emerging attributes for energy and environment applications.

structures, electron transport properties, and catalytic, redox, fluorescence, and luminescence properties [4–6]. The metal oxide materials, such as TiO_2, ZnO, and Al_2O_3, have wide bandgaps, electron–hole pair separation, transfer efficiency, and electron transport properties [7, 8].

1.2 The role of carbon-based materials in the energy–environment nexus

The carbon-based nanomaterials offer incredible potential in the sectors of energy and environment due to their unique structural features, such as a large surface area with a wide range of pores, excellent electrical properties, high mechanical strength, intrinsic hydrophobic nature, good thermal stability in an inert atmosphere, the presence of a vast number of functional groups, and fast mass transfer performance [9]. The carbon-based nanomaterials are classified as zero-dimensional (0D) (Buckminster fullerenes and carbon dots), one-dimensional (1D) (carbon nanotubes (CNTs) and carbon nanofibers (CNFs)), two-dimensional (2D) (graphene), and three-dimensional (3D) (carbon sponges) carbon nanostructures based on their size, shape, and dimensionality, as shown in figure 1.2(a) [10]. Generally, the chemical and physical properties of these carbon-based nanomaterials depend strongly on their structural and interfacial interactions with neighboring bulk materials. Due to their incredible characteristics, these carbon-based materials have inspired innovative solutions to address the energy and environmental challenges in solar energy storage [11], batteries and fuel cells [12, 13] supercapacitors [14], electrochemical energy storage and conversion [15], protection against electromagnetic waves [16], catalysis [17], gas storage [18], detection and separation [19], pollutant elimination in gases as well as liquids [19], and photocatalysis [20]. Figure 1.2(b) shows the role of carbon-based materials in energy and environmental applications.

1.3 The role of metal oxides in the energy–environment nexus

Metal oxides play an important role in key applications to overcome the energy crisis, including energy production, conversion, and storage. Metal oxides are multifunctional compounds that show a large variety of properties such as optical, electrical, magnetic, catalytic, and chemical properties, which can be tuned using the crystal structure, morphology, defects, composition, and doping [21]. In the energy sector, for next-generation portable and wearable electronic devices various piezoelectric nanostructures have been reported. The metal oxides can be realized as self-powered devices by fabricating nanogenerators (NGs), which can harvest energy from the environment and show a greener path as a materials nexus for sustainable energy and environment. Numerous materials have been employed to obtain the desirable electrical and mechanical properties required for energy storage and conversion devices, such as a superior dielectric constant, high piezoelectric charge constant, and high elastic modulus. For example, lead zirconate titanate (PZT) [22, 23], lithium niobate (LN type) [24, 25], ZnO based [26], and potassium sodium niobate (KNN type) [27] materials are being developed for energy harvesters due to their prominent electrical properties and high Curie temperature (T_C). Although

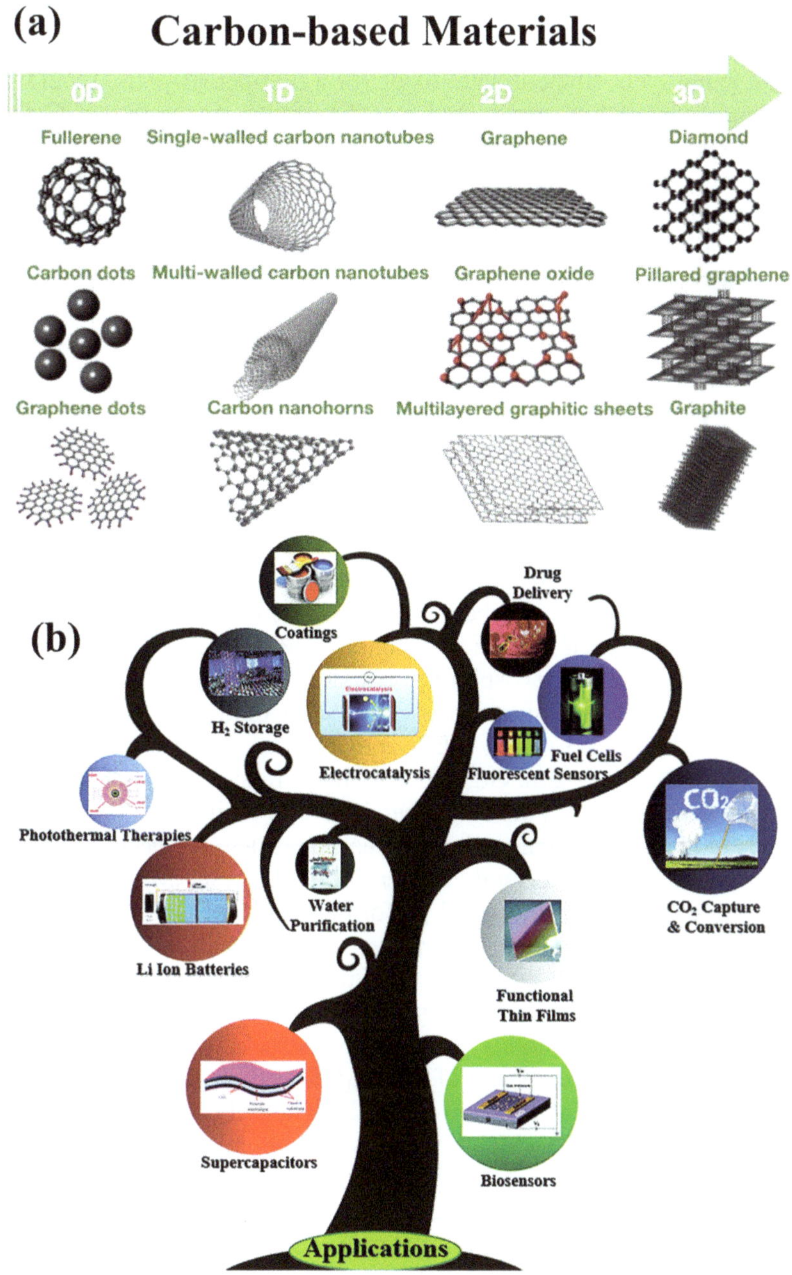

Figure 1.2. (a) The classification and (b) role of carbon-based nanomaterials for energy and environment applications. ((b) Reproduced with permission from [53]. Copyright 2021 the Royal Society of Chemistry.)

they have a superior energy harvesting performance, PZT based compounds are toxic in nature due to the lead (Pb) content and may cause severe issues in the human body and the environment. Also, the PZT based compounds are not viable for the energy and environmental applications. Currently, the green synthesis techniques are being employed to prepare metal oxide-based materials. Cupric oxide (CuO) has been an extensively studied metal oxide for a number of years. Its 2D structure has been implemented for energy devices. The optical properties of CuO are the important factors to decide its use for optoelectronic devices. The optical features of CuO indicate good absorption in the visible region of the solar spectrum, which is useful for the absorber layer in photovoltaics [28]. Another important metal oxide, ZnO, is an eco-friendly and biosafe material. Its wide band gap is helpful for versatile applications. ZnO is easy to synthesize and is a cost-effective material. It shows piezoelectricity and pyroelectricity, near UV emission, and transparent conductivity and biocompatibility. ZnO has both the piezoelectric and semiconducting properties required for energy harvesting devices. It also plays an important role in contributing to carbon dioxide (CO_2) emission free energy [29]. Cobalt oxide (Co_3O_4) is a p-type semiconducting magnetic material which is kinetically stable at room temperature. The nanostructural form of Co_3O_4 shows excellent electrochemical performance due to its features, such as a high surface area, a short path length for ion transport, and an easily tunable surface [30]. The effective redox charge transfer makes it a suitable candidate for the electrode material in supercapacitors.

Perovskite type oxides are another important class of oxides that exhibit versatile applications in the oxygen reduction reaction (ORR), oxygen evolution reaction (OER), electrochemical water splitting reaction, metal–air batteries, solid-state batteries, oxygen separation membranes, and solid oxide fuel cells [31] as shown schematically in figure 1.3. High OER activity was observed in the perovskites $Ln_{0.5}Ba_{0.5}CoO_{3-\delta}$ (Ln = Pr, Sm, Gd, and Ho) that is related to O (p-band center) [32]. In the perovskites, the B sites are redox catalysts and their 8 coordinating sites with oxygen leads to metallic or semiconducting characteristics and electronic conductivity at high temperatures. Nanocrystals of perovskite $La_{0.5}Sr_{0.5}Mn_{1-x}Ni_xO_{3-\delta}$ exhibit high electronic conductivity and good performance toward the ORR [33]. A triple perovskite $Nd_{1.5}Ba_{1.5}CoFeMnO_{9-\delta}$ composite with N-doped reduced graphene oxide exhibited enhanced electrocatalytic activity, better than Pt/C and Ir/C [34]. Perovskite oxides can be used as potential electrodes in electrochemical water splitting because of their good electrochemical stability, composition, structural flexibility, and low cost. Shao *et al* described the development of $SrNb_{0.1}Co_{0.7}Fe_{0.2}O_{3-\delta}$ oxide as a bifunctional OER/HER electrocatalyst which showed excellent activity for overall water splitting [35]. Perovskite $La_{0.6}Sr_{0.4}Co_{0.8}Mn_{0.2}O_3$ has a porous microstructure and large O_2 diffusion and demonstrated good catalytic behavior in Li–O_2 batteries [36]. $LaMnO_{3+\delta}$ delivers a high power density, even better than the Pt/C catalyst, as an alternative to noble metal catalysts for the positive electrode of Zn–air batteries [37]. Among the various solid electrolytes, the perovskite $Li_{0.05-3x}La_{0.5+x}TiO_3$ (LLTO) shows excellent Li-ion conductivity and has promising characteristics for energy storage devices such as lithium batteries [38]. The complex perovskite oxide $Ba_{0.5}Sr_{0.5}Co_{0.8}Fe_{0.2}O_{3-\delta}$ has a dense structure, extraordinary thermal stability, and high oxygen permeation flux. It is also considered an excellent ceramic-based membrane for oxygen separation [39]. The cathode is an important electrode component in solid oxide fuel cells (SOFCs).

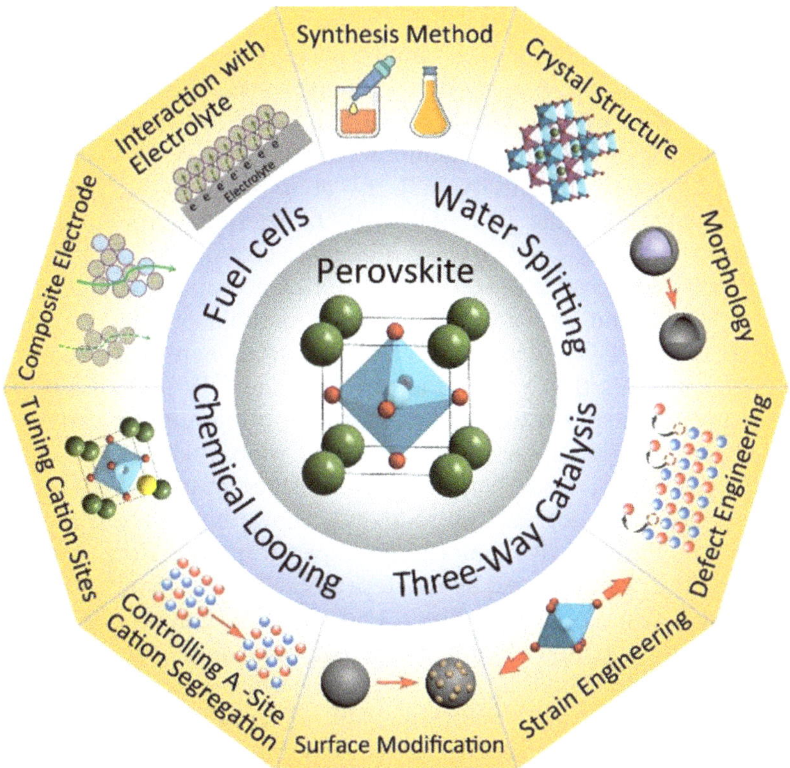

Figure 1.3. A schematic overview of enhancing the performance of perovskite oxide-based materials for energy and the environment. (Reproduced from [51]. CC BY 4.0.)

High electron density states are found near the Fermi surface in lanthanum cobaltite ($LaCoO_3$). Thus, $LaCoO_3$ displays improved performance as a cathode for intermediate temperature solid oxide fuel cells (IT-SOFCs) [40]. Another important perovskite $Ba_{0.9}Co_{0.7}Fe_{0.2}Nb_{0.1}O_{3-\delta}$ exhibits oxygen ion diffusion in its structure via a vacancy mechanism that is required for the operation of cathode materials in the IT-SOFCs [41].

The current research focus has concentrated on achieving environmental sustainability. In line with this, it is mandatory to provide a clean environment for human survival in the coming decades. Solar energy is an alternative source of energy since the Sun is the ultimate and abundant source of energy. However, sunlight is not available 24 h a day on Earth and it also depends on seasonal conditions. Thus it is important to transform sunlight into energy that can be stored. The photothermal, photovoltaic, and photocatalysis methods are three different approaches for solar energy harvesting [42]. Among them, photocatalysis is a promising approach for the transformation of earth-abundant H_2O, carbon dioxide (CO_2), and nitrogen (N_2) into energy-rich products such as H_2, CH_4, CH_3OH, and NH_3, by exploiting solar energy through processes such as water splitting for hydrogen (H_2) generation, organic pollutant degradation, CO_2 reduction, and N_2 fixation. The layered perovskites are popular due to their semiconducting properties and are used for environmental applications. Among the layered perovskites, Bi_2WO_6 is a fascinating semiconducting oxide with an energy band gap of 2.8 eV, which is required for visible light harvesting. The heterojunction formation of Bi_2WO_6 with BiOI showed the highest photocatalytic activity [43]. Li *et al* demonstrated the significant photocatalytic activity of the $Bi_2Fe_4O_9/Bi_2WO_4$ heterojunction for the complete degradation of rhodamine B (RhB) [44]. Titanate perovskites $ATiO_3$ (A = Ca, Sr, and Ba) and their composites are studied for different photocatalytic applications. They have a wide band gap and are therefore capable of harvesting UV light. Calcium titanate ($CaTiO_3$) decorated with gold (Au) nanoparticles showed the highest activity towards the degradation of RhB under UV and visible irradiation [45]. The improved activity is due to the separation of photogenerated charges and the localized plasmon resonance effect of the Au nanoparticles. The niobate family of perovskites ($ANbO_3$, A = Na, K) are also the most studied materials for photocatalytic pollutant degradation. $NaNbO_3$ shows high optical absorbance in visible light to obtain high photocatalytic performance. Fore example, 95% degradation of RhB was achieved by Ru doped $NaNbO_3$ in 45 min [46]. The mixed perovskite oxides such as $La_{0.7}Sr_{0.3}MnO_3$ and $La_{0.5}Ca_{0.5}CoO_3$ have been explored for organic pollutant degradation under different irradiation sources [47, 48]. It was found that the perovskite $SrTi_{0.1}Fe_{0.9}O_{3-\delta}$ exhibited magnetic properties along with the capability of visible light harvesting [49]. The other promising perovskite materials, such as $BaZrO_3$, $ASnO_3$ (Ca, Sr and Ba), $LaCoO_3$, and $LaNiO_3$ are also investigated for the photocatalytic degradation of pollutants [42]. A binary heterojunction photocatalyst $LaNiO_3/TiO_2$ sample showed the degradation of methyl orange (MO) dye which was found to be higher than for pristine $LaNiO_3$ and TiO_2 [50].

Recently, multifunctional materials have been developed which can perform multiple tasks with combined effects to improve the efficiency beyond their individual materials for various applications [52]. Figure 1.4 shows the application of multifunctional nanomaterials in energy and environmental applications. These multifunctional materials can be a potential materials nexus for energy and environment applications because of their special optical, magnetic, chemical, and electrical properties. In addition to the role of materials in the energy sector, these materials are highly beneficial in environmental applications, in particular as nanoadditives, porous membranes, and protective coatings for the reduction of emissions from fuels, CO_2 capture, green hydrogen production, and corrosion protection inhibitors, etc.

1.4 Summary

Sustainable materials have been investigated toward the development of energy conversion, storage, and environmental applications in the academic and industrial fields. Various carbon-based nanostructures and inorganic metal oxides have been reported using green synthesis methods for energy storage and conversion devices,

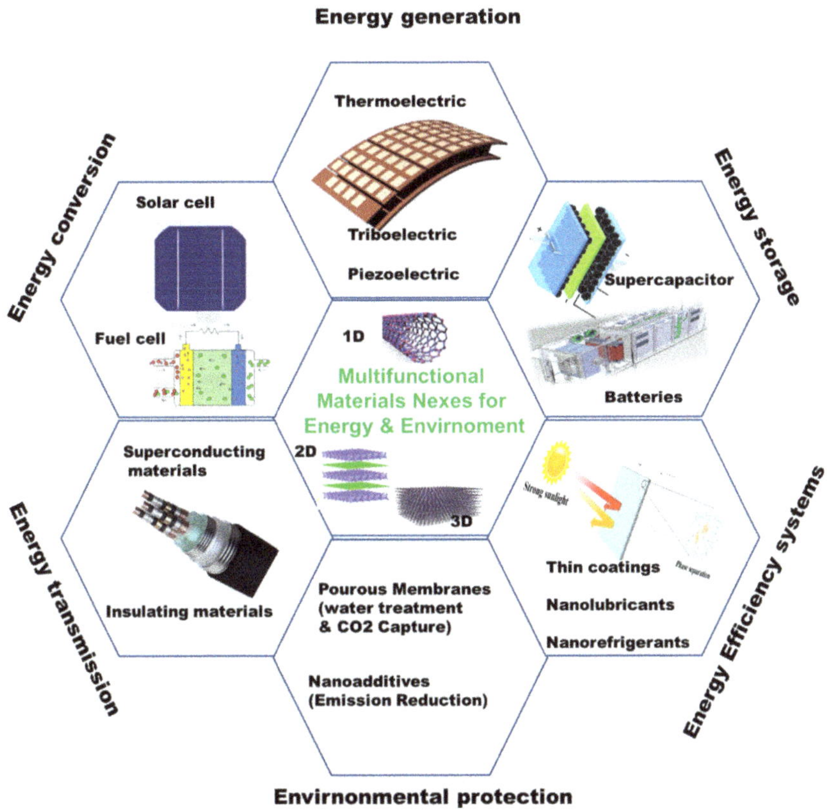

Figure 1.4. Applications of multifunctional materials in the energy and environmental sector.

such as supercapacitors, batteries, and nanogenerator based self-powered sensors. In particular, for sustainable engineering, nanostructures for water treatment, green hydrogen generation using electrochemical based water splitting, and CO_2 capture technologies are being realized. In addition, the development of multifunctional materials could allow multiple tasks to be performed with combined effects and are potential materials for the energy and environment nexus. Moreover, integrating different functions in a single system is a fundamental challenge and needs to be carefully designed to perform the synergetic effects which could potentially have great impacts on the energy and environment applications in the near future.

Acknowledgments

RA acknowledges the project supported by the Competition for Research Regular Projects, year 2021, code LPR21-03, Universidad Tecnologica Metropolitana, Santiago, Chile. RA also acknowledges Direccion Academia y de Investigacion, FCFM, Universidad de Chile for the Fondos de Instalación-Apoyo a la Inserción Academia.

References

[1] Arulraj A, Thachnatharen N, Rajkumar C and Mangalaraja R V 2022 Engineering 2D chalcogenides for energy and environmental remediation, Nanostructured Materials for Sustainable Energy and Environmental Remediation (Bristol: IOP Publishing) pp 2-1–2-28

[2] Sadhu S D, Garg M and Kumar A 2018 Major environmental issues and new materials ed C M Hussain and A K Mishra *New Polymer Nanocomposites for Environmental Remediation* (Amsterdam: Elsevier) pp 77–97

[3] Wan R and Ni M 2021 Sustainable water–energy–environment nexus *Environ. Sci. Pollut. Res.* **28** 40049–52

[4] Iijima S 1991 Helical microtubules of graphitic carbon *Nature* **1991** 56–8

[5] Geim A K 2009 Graphene: status and prospects *Science* **324** 1530–4

[6] Xu X, Ray R, Gu Y, Ploehn H J, Gearheart L, Raker K and Scrivens W A 2004 Electrophoretic analysis and purification of fluorescent single-walled carbon nanotube fragments *J. Am. Chem. Soc.* **126** 12736–7

[7] Kumaravel V, Mathew S, Bartlett J and Pillai S C 2019 Photocatalytic hydrogen production using metal doped TiO_2: a review of recent advances *Appl. Catal.* B **244** 1021–64

[8] Sang L, Zhao Y and Burda C 2014 TiO_2 nanoparticles as functional building blocks *Chem. Rev.* **114** 9283–318

[9] Xin W and Song Y 2015 Mesoporous carbons: recent advances in synthesis and typical applications *RSC Adv.* **5** 83239–85

[10] Sahoo P K, Tseng C A, Huang Y J and Lee C P 2021 *Carbon-Based Nanocomposite Materials for High-Performance Supercapacitors Novel Nanomaterials* (London: IntechOpen)

[11] Kim A, Dash J K, Kumar P and Patel R 2022 Carbon-based quantum dots for photovoltaic devices: a review *ACS Appl. Electron. Mater.* **4** 27–58

[12] Hong M *et al* 2019 Scalable synthesis of γ-Fe_2O_3/CNT composite as high-performance anode material for lithium-ion batteries *J. Alloys Compd.* **770** 116–24

[13] Noh S H *et al* 2017 Self-assembled nitrogen-doped fullerenes and their catalysis for fuel cell and rechargeable metal-air battery applications *Nanoscale* **9** 7373–9

[14] Niraj Kumar P K, Sahoo H S and Panda H S 2018 Localize current burst in modified carbon nanotube/polyaniline composite fibers mat electrode miniaturized resistance and improved rate capability for solid-state supercapacitor *J. Mater. Sci., Mater. Electron.* **29** 7562–74

[15] Kumar S, Sahoo P K and Satpati A K 2020 Insight into the catalytic performance of HER catalysis of noble metal/3D-G nanocomposites *Electrochim. Acta* **333** 135467

[16] Aepuru R, Viswanathan M R, Bhaskara Rao B V, Panda H S, Sahu S and Sahoo P K 2020 Tailoring the performance of mechanically robust highly conducting silver/3D graphene aerogels with superior electromagnetic shielding effectiveness *Diam. Relat. Mater.* **109** 108043

[17] Memetova A *et al* 2022 Porous carbon-based material as a sustainable alternative for the storage of natural gas (methane) and biogas (biomethane): a review *Chem. Eng. J.* **446** 137373

[18] Kante K, Florent M, Temirgaliyeva A, Lesbayev B and Bandosz T J 2019 Exploring resistance changes of porous carbon upon physical adsorption of VOCs *Carbon* **146** 568–71

[19] Mohammad Mehdi S, Sahar M, Mehrorang G, Hadi H and Vellaisamy A L R 2021 Carbon based materials: a review of adsorbents for inorganic and organic compounds *Mater. Adv.* **2** 598–627

[20] Zhang J, Dai M, Zhang S, Dai M, Zhang P, Wang S and He Z 2022 Recent progress on carbon-nanotube-based materials for photocatalytic applications: a review *Sol. RRL* **6** 2200243

[21] Danish M M S, Bhattacharya A, Stepanava D, Mikhaylov A, Grilli M L, Khosravy M and Senjyu T 2020 A systematic review of metal oxide applications for energy and environmental sustainability *Metals* **10** 1604

[22] Fan F R, Tang W and Wang Z L 2016 Flexible nanogenerators for energy harvesting and self-powered electronics *Adv. Mater.* **28** 4283–305

[23] Lee H, Kim H, Kim D Y and Seo Y 2019 Pure piezoelectricity generation by a flexible nanogenerator based on lead zirconate titanate nanofibers *ACS Omega* **4** 2610–7

[24] Li X, Sun M, Wei X, Shan C and Chen Q 2018 1D piezoelectric material based nanogenerators: methods, materials and property optimization *Nanomaterials* **8** 188

[25] Zhang Y, Liang G, Tang S, Peng B and Zhang Q 2020 Phase-transition induced optimization of electrostrain, electrocaloric refrigeration and energy storage of $LiNbO_3$ doped BNT-BT ceramics *Ceram. Int.* **46** 1343–51

[26] Xi Y *et al* 2009 Growth of ZnO nanotube arrays and nanotube based piezoelectric nanogenerators *J. Mater. Chem.* **19** 9260–4

[27] Teka A, Bairagi S, Ziauddin S and Wazed A S 2018 Poly (vinylidene fluoride) (PVDF)/potassium sodium niobate (KNN)-based nanofibrous web: a unique nanogenerator for renewable energy harvesting and investigating the role of KNN nanorods *Polym. Adv. Technol.* **29** 2537–44

[28] Arregui F J 2008 *Sensors Based on Nanostructured Materials* (Berlin: Springer)

[29] Lide D R 2005 *CRC Handbook of Chemistry and Physics* (Boca Raton, FL: CRC Press)

[30] Tanaka M, Mukaia M, Fujimuri Y, Kondoh M, Tasaka Y, Baba H and Usami S 1996 Transition metal oxide films prepared by pulsed laser deposition for atomic beam detection *Thin Solid Films* **281–2** 453–6

[31] Sun C, Alonso J A and Bian J 2020 Recent advances in perovskite-type oxides for energy conversion and storage applications *Adv. Energy Mater.* **11** 2000459

[32] Grimaud A, May K J, Carlton C E, Lee Y L, Risch M, Hong W T, Zhou J G and Shao-Horn Y 2013 Double perovskites as a family of highly active catalysts for oxygen evolution in alkaline solution *Nat. Commun.* **4** 2439

[33] Sunarso J, Torriero A A J, Zhou W, Howlett P C and Forsyth M 2012 Oxygen reduction reaction activity of la-based perovskite oxides in alkaline medium: a thin-film rotating ring-disk electrode study *J. Phys. Chem.* C **116** 5827–34

[34] Kim N I *et al* 2018 Oxygen-deficient triple perovskites as highly active and durable bifunctional electrocatalysts for oxygen electrode reactions *Sci. Adv.* **4** eaap9360

[35] Zhu Y L, Zhou W, Zhong Y J, Bu Y F, Chen X Y, Zhong Q, Liu M L and Shao Z P 2017 A perovskite nanorod as bifunctional electrocatalyst for overall water splitting *Adv. Energy Mater.* **7** 1602122

[36] Zhang X L, Gong Y D, Li S Q and Sun C W 2017 Porous perovskite $La_{0.6}Sr_{0.4}Co_{0.8}Mn_{0.2}O_3$ nanofibers loaded with RuO_2 nanosheets as an efficient and durable bifunctional catalyst for rechargeable $Li–O_2$ batteries *ACS Catal.* **7** 7737

[37] Kuai L *et al* 2018 Mesoporous $LaMnO_{3+\delta}$ perovskite from spray−pyrolysis with superior performance for oxygen reduction reaction and Zn−air battery *Nano Energy* **43** 81

[38] Alonso J, Sanz J, Santamaria J, Leon C, Varez A and Fernandez-Diaz M 2000 On the location of Li^+ cations in the fast Li-cation conductor $La_{0.5}Li_{0.5}TiO_3$ perovskite *Angew. Chem. Int. Ed.* **39** 619

[39] Teroka Y, Zhang H M, Furukawa S and Yamazoe N 1985 Oxygen permeation through perovskite-type oxides *Chem. Lett.* **14** 1743
[40] Petrov A N, Kananchuk O F, Andreev A V, Cherepanov V A and Kofstad P 1995 Crystal structure, electrical and magnetic properties of $La_{1-x}Sr_xCoO_{3-y}$ *Solid State Ion.* **80** 189
[41] Gong Y D, Sun C W, Haung Q A, Alonso J A, Feranandez-Diaz M T and Chen L Q 2015 Insight into the structure and functional application of the $Sr_{0.95}Ce_{0.05}CoO_{3-\delta}$ cathode for solid oxide fuel cells *Inorg. Chem.* **54** 3477
[42] Kumar A, Kumar A and Krishnan V 2020 Perovskite oxide based materials for energy and environment-oriented photocatalysis *ACS Catal.* **10** 10253
[43] Wu R, Sung H, Luo N, Sheng Y and Ji G 2019 Microwave-assisted preparation and enhanced photocatalytic activity of Bi_2WO_6/BiOI heterojunction for organic pollutants degradation under visible-light irradiation *Solid State Sci.* **87** 101
[44] Li B *et al* 2018 Facile hydrothermal synthesis of Z-scheme $Bi_2Fe_4O_9/Bi_2WO_6$ heterojunction photocatalyst with enhanced visible light photocatalytic activity *ACS Appl. Mater. Interfaces* **10** 18824
[45] Yan Y, Yang H, Zhao X, Zhang H and Jiang J 2018 A hydrothermal route to the synthesis of $CaTiO_3$ nanocuboids using P25 as the titanium source *J. Electron. Mater.* **47** 3045
[46] Li G, Yi Z, Bai Y, Zhang W and Zhang H 2012 Anisotropy in photocatalytic oxidization activity of $NaNbO_3$ photocatalyst *Dalton Trans.* **41** 10194
[47] Ghiasi M and Malekzadeh A 2014 Solar photocatalytic degradation of methyl orange over $La_{0.7}Sr_{0.3}MnO_3$ nano-perovskite *Purif. Technol.* **134** 12
[48] Tavakkoli H, Beikenjad D and Tabari T 2014 Fabrication of perovskite-type oxide $La_{0.5}Ca_{0.5}CoO_{3-\delta}$ nanoparticles and its dye removal performance *Desalin. Water Treat.* **52** 7377
[49] Chen H X, Wei Z X, Wang Y, Zeng W W and Xio G M 2011 Preparation of $SrTi_{0.1}Fe_{0.9}O_{3-\delta}$ and its photocatalysis activity for degradation of methyl orange in water *Mater. Chem. Phys.* **130** 1387
[50] Chen C, Zhou J, Geng J, Bao R, Wang Z, Xia J and Li H 2020 Perovskite $LaNiO_3/TiO_2$ step-scheme heterojunction with enhanced photocatalytic activity *Appl. Surf. Sci.* **503** 144287
[51] Zhang M, Jeerh G, Zhou P, Lan R, Wang M, Wang H and Tao S 2021 Recent development of perovskite oxide-based electrocatalysts and their applications in low to intermediate temperature electrochemical devices *Mater. Today* **49** 35–377
[52] Wang H, Liang X, Wang J, Jiao S and Xue D 2020 Multifunctional inorganic nanomaterials for energy applications *Nanoscale* **12** 14–42
[53] Baig N, Kammakakam I and Falath W 2021 Nanomaterials: a review of synthesis methods, properties, recent progress, and challenges *Mater. Adv.* **2** 1821

Chapter 2

Flexible multifunctional polymer composites for energy harvesting applications

Durga Prasad Pabba, Radhamanohar Aepuru, B V Bhaskara Rao, Arun Thirumurugan and R V Mangalaraja

Over the decades, the energy shortage, environmental degradation, and greenhouse effect have mandated the necessity to discover and use alternative clean and sustainable energy sources, such as solar thermal energy, biomechanical energy, and wind energy, to replace the traditional fossil fuels, as they are sustainable and environmentally friendly. In recent decades, there has been a surge in the research and development efforts dedicated to power generation that transforms the waste resources that exist in the environment, such as mechanical vibration, light illumination, thermal heat, wind flow, radiations, and flow of water, into electricity for low-power gadgets. Much emphasis has been placed on ambient sources of energy that are appropriate for power generation, such as sunlight, radio waves, sound energy, thermal gradients, and kinetic mechanical energy. Kinetic and magnetic energy are the most abundant and versatile of these sources, manifesting themselves in our surroundings as mechanical movements and stray magnetic fields. This chapter focuses on the various fabrication processes used to enhance the piezoelectric and energy harvesting performance of flexible polymer nanogenerators. It gives a brief overview of the various fillers used in polyvinylidene fluoride (PVDF) to improve the performance of nanogenerators and discusses the mechanisms for understanding energy generation. Finally, we summarize the recent advances in this field and describe the future opportunities for improving the energy harvester performance.

2.1 Flexible energy harvesters

The destruction of the environment as well as resource scarcity have emerged as significant issues that must be tackled if humankind is to progress in a sustainable manner. It is indeed crucial to reduce the use of fossil energy and develop highly

doi:10.1088/978-0-7503-5729-6ch2

regarded alternative renewable energies [1–3]. According to a report, a growing percentage of electric cars is currently being used, and most European nations are aiming to phase out fossil fuel automobiles by 2050, causing an increase in the demand for electrical power [4]. There are different types of accessible energy sources in the environment that may be transformed into electric power, such as biomechanical, thermal, chemical, solar, thermal, nuclear energy, and so on, all of which are present in different forms. Harvesting power from the external environment that could be utilized instantaneously as well as stored for future use, has emerged as an appealing and effective technique in both the scientific and commercial disciplines to power numerous gadgets and establish self-powered devices, since mechanical energy sources are widespread within the ambient atmosphere, which include vibrations, mechanical affects, wind, automobile transport, etc [5]. A device used to convert mechanical energy to electricity is known as a piezoelectric nanogenerator (PENG). Wang originally suggested the piezoelectric nanogenerator in 2006 by scanning an atomic force microscope (AFM) tip over vertically produced ZnO nanowires [6].

Nanogenerators primarily use three types of approaches: triboelectric, piezoelectric, and pyroelectric. The transformation of biomechanical energy into electrical energy has been enhanced using nanogenerators. Additionally, some devices use the pyroelectric effect to convert thermal energy into electricity. The piezoelectric systems have been reported to perform better than the triboelectric ones [7]. The piezoelectric nanogenerators, based on the piezoelectric effect, were introduced in 2006, while triboelectric nanogenerators based on the triboelectric effect were proposed later, in 2012. When two materials of different electron affinities come in contact, electrostatic charges form on their surfaces. Subsequently, when two such materials are kept apart, a voltage is generated that causes electrons to flow between them, generating an alternating current for the triboelectric nanogenerator [8].

Over the years, numerous initiatives have been taken to enhance the efficiency of PENGs. A diverse variety of piezoelectric materials with large piezoelectric coefficients (d_{33}), such as lead magnesium niobate–lead titanate (PMN-PT), lead zirconium titanate (PZT), barium titanate ($BaTiO_3$), and zinc oxide (ZnO), have been studied comprehensively, with d_{33} values of 2500, 304, 100, and 23 pC N^{-1}, respectively. In general, they were synthesized using the hydrothermal process and the molten salt approach. Because of their limited power output and brittle nature, they were inefficient in large-scale applications. Piezoelectric polymers offer good mechanically flexible alternatives to inorganic polymers. Furthermore, the polymer-based nanogenerators are well-suited for low-weight applications [6, 9, 10].

Poly(vinylidene fluoride) (PVDF) and its copolymers are favourable piezoelectric polymers because of their outstanding flexibility, light weight, simple production, and high stability in strong external electric fields. Despite having a lower d_{33} than inorganic piezoelectric materials, PVDF's unique flexibility allows for good processability as fibres, nanowires, or thin films, as well as easy integration with functional materials to improve the performance, and it has sparked a lot of interest in the field of energy harvesting [6]. In general, PVDF is available in five various polymorphs α, β, γ, δ, and ε. Among these, the most significant are α and β and the

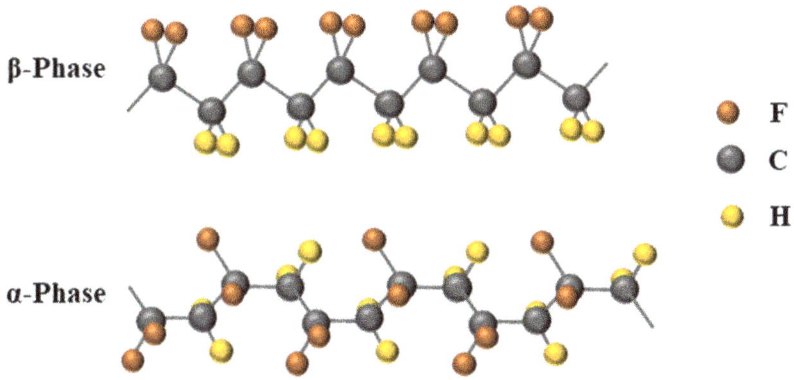

Figure 2.1. Molecular arrangement in the α- and β-phases of the PVDF chain.

have a chain structure of all trans (TTTT) and trans gauche–trans gauche (TGTG′) conformations, as shown in figure 2.1. The dipole orientation from the β-phase determines its piezoelectric activity.

Numerous paths, such as incorporating materials, structural design, and processing methods, have been examined over the years to optimize the PVDF-based energy harvesting devices for improved sensitivity and larger power generation, establishing the foundation for a wide range of uses in wearable electronics and implantable fields, manufacturing ambient, as well as other electronic gadgets, thanks to the work of researchers [11].

We review the significant advances of the polymer-based PENGs in the following areas, as depicted in figure 2.2, based on the relevant works reported in the different application fields, which include preparation techniques, performance enhancement with the incorporation of various materials, and pertinent energy harvesting applications in various fields. These topics will be examined and investigated individually in-depth. The problems and solutions in this sector are then detailed, and future approaches are suggested in the hope of facilitating the advancement of flexible PVDF-based PENG devices.

2.2 Fabrication techniques

The preparation method has a significant effect on the piezoelectric, dielectric, ferroelectric, and electroactive properties of PVDF and PVDF-based composites. According to numerous studies, the primary factors that influence the piezoelectric properties of PVDF and PVDF composites include crystalline phase content, dipole alignments, temperature treatment, polarization under electric fields, and so on. The fabrication procedure, however, represents the most significant factor that affects the output performance of PVDF-based PENGs. Some of the important procedures for the preparation include spin-coating, electrospinning, hot pressing, and 3D printing techniques. For example, the spin-coating process offers extendable and uniform thickness along with outstanding reproducibility. Electrospinning is an easy and adaptable method for the production of continuous long fibres within the nanometre

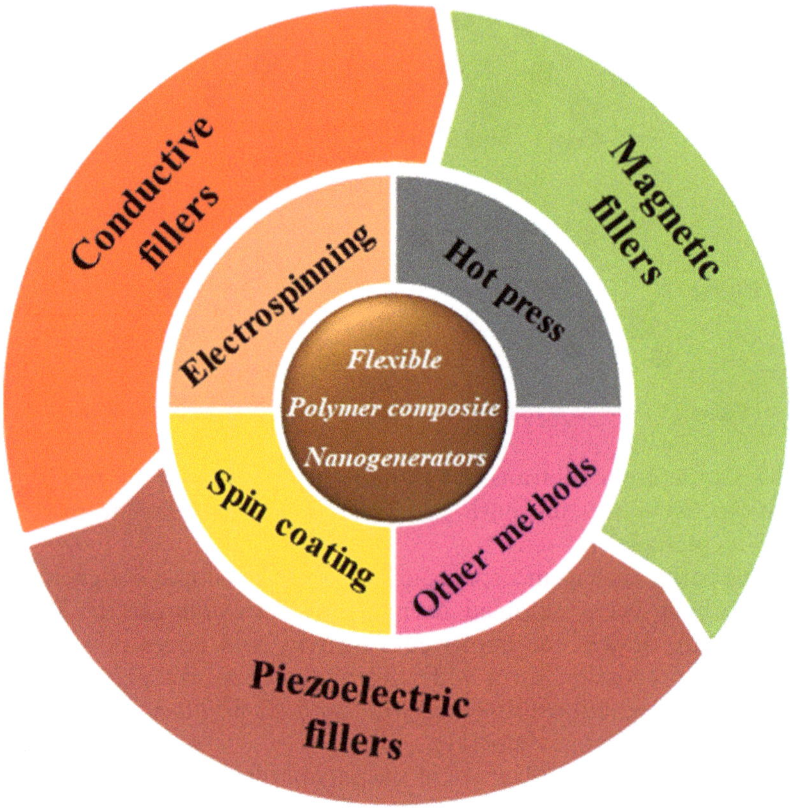

Figure 2.2. An overview of polymer-based nanogenerators, including the fabrication processes and materials incorporation.

to micrometre range. During the hot-pressing procedure, the PVDF is re-melted at elevated temperatures and pressures, which results in its high density. Other important techniques for improving the nanogenerator performance include 3D printing and the formation of interdigital electrodes [12, 13].

2.2.1 Spin coating

The spin-coating technique has been in use for over a century and it has emerged as the favoured candidate for making preparations of polymer composite films. The basic idea behind spin-coating is to pour drops of polymer solution into the centre of a substrate and then rapidly rotate the substrate. Due to the high centrifugal force, the solution is pushed toward the boundary and eventually out of the substrate, having left a layer of material on the surface for the subsequent formation of the film. By adjusting the parameters, such as the viscosity of the solution, the speed and time of rotation, and the amount of solution, could improve the properties of the resulting film [14–16]. Currently, despite a few deficiencies such as the lack of a pattern, this strategy has several benefits, including low cost, simplicity, and ease of operation, providing a significant motivation for the study and implementation of

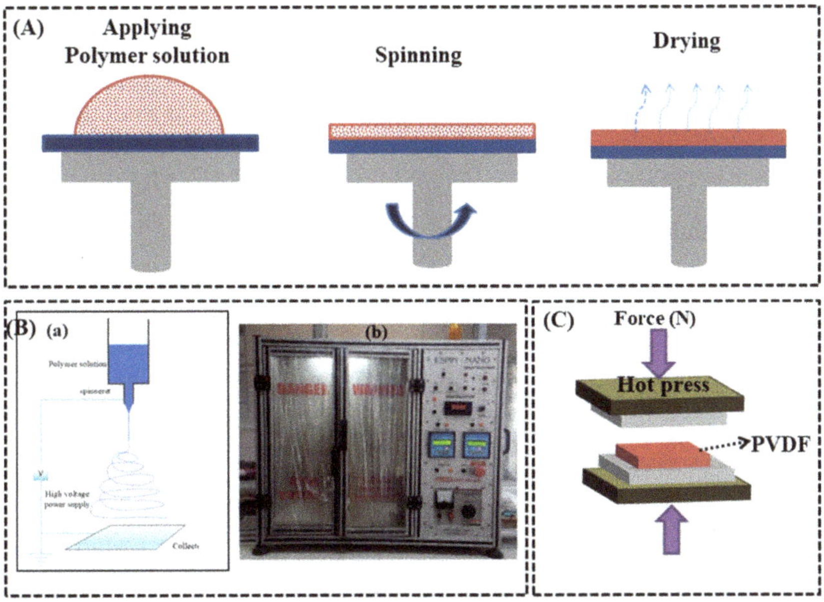

Figure 2.3. Schematics of the fabrication processes (A) spin-coating, (B) electrospinning, and (C) the hot press technique.

the spin-coating technique. Figure 2.3(A) depicts a schematic of the spin-coating process for preparing PVDF films. The critical procedures are spinning speed and drying, both of which have a direct impact on the ultimate thickness and piezoelectric β-phase content.

In general, PVDF and its copolymers are dissolvable in a broad range of organic polar solvents, such as dimethylsulphoxide (DMSO), *N*,*N*-dimethylmethanamide (DMF), hexamethylphosphoric triamide (HMPA), 1-methylpyrrolidin-2-one (NMP), *N*,*N*-dimethylacetamide (DMAc), and triethyl phosphate (TEP). Nishiyama *et al* [17] examined the impact of different polar solvents and the rate of evaporation of the solvent on crystallization in spin coated PVDF films. It was discovered that crystallization depends on the type of solvent used in the initial stage of the evaporation after spin-coating. As the dipole moment of the solvent increased, the crystalline phase altered in the order of α, γ, and β. The higher dipole moment HMPA prompted the formation of the β-phase PVDF, the lower dipole moment TEP induced α-phase PVDF, and the intermediate dipole moment DMAc and NMP resulted in the development of the γ-phase. Similarly, the crystallization of PVDF in the final stage of evaporation was predominately dependent on the evaporation rate and varied in the order of β, γ, and α with increasing evaporation rate of the solvent. A low evaporation rate resulted in the formation of β-phase crystallization.

RamaSundaram *et al* [18] studied the effects of various heat treatments on the crystal phase formation of PVDF thin films. The samples were prepared at various temperatures and it was noted that elevated temperatures such as 40 °C, 50 °C, 60 °C, and 70 °C aided β-phase crystallization, whereas room temperature and 30 °C encouraged α-phase crystallization. Cardoso *et al* [19] studied the effect of rotational velocity on the percentage of β-phase in the PVDF films and stated that a high rotational speed as well as low solvent concentration aided in the creation of β-phase in the films. This was due to the shear and elongation forces extending the polymer chains and helping to promote α -to β-phase transition during spinning.

2.2.2 Electrospinning

Electrospinning is a popular electrostatic fibre formation technology that uses an electrostatic force to produce fibres of diameters from nanometres to micrometres using both natural and synthetic polymers. This method enables the development of new natural nanofibers and fabrics with tunable pore structure. The electrospinning method has received attention in the last decade, not only because of its versatility in synthesizing a wide range of polymeric fibres, but also because of its ability to consistently generate fibres in the submicron range. A schematic and photograph of the electrospinning set-up are provided in figure 2.3(B). Electrospinning is done at a high voltage, and the PVDF is stretched and polarized during the spinning. Baii *et al* found that the electrospun PVDF fibres had outstanding piezoelectric properties with no further poling and could be used for PENG applications directly [20]. However, a few problems arose during the electrospinning of PVDF fibres. In the prepared PVDF fibre mats, the possibility of disordered fibre arrangement has an offsetting effect, minimizing the total piezoelectric property of the fibre mats. A rotating drum collector was used to obtain aligned fibres to solve this problem. Persano *et al* [21] fabricated PVDF-TrFE fibre arrays by using a rotating drum collector with a rotation speed of 4000 rpm and a linear speed 16 m s^{-1} at the collector surface. They obtained well-aligned fibres with a diameter of 260 nm, and more than 48% crystallinity has been obtained in the aligned PVDF-TrFE arrays compared to 40% in the random mat, and a percentage of phase near to 85%. Furthermore, the obtained PVDF-TrFE nanofibers were extremely sensitive to pressure measurements, even at incredibly low operating pressures of 0.1 Pa.

2.2.3 Hot press

Another important tool for preparing high quality piezoelectric composite films is the hot press. The polymeric composite materials are hot pressed near their melting point in this technique. The ferroelectric, dielectric, piezoelectric, and electroactive properties of the composites are related not only to the properties of the polymer matrix, but also to the temperature and pressure encountered during the pressing process [22]. The material rolls more uniformly as temperature and pressure increase, reducing the internal porosity and increasing the interface adhesion force. As a result, the crystal structure of the material is compacted and the dielectric loss is reduced, resulting in better piezoelectric and dielectric properties. Whenever the temperature goes up too high, the shrinkage of PVDF increases with the number of internal defects [23]. The hot press technique has the advantage of producing PVDF composite films through complete evaporation or without the use of organic solvents.

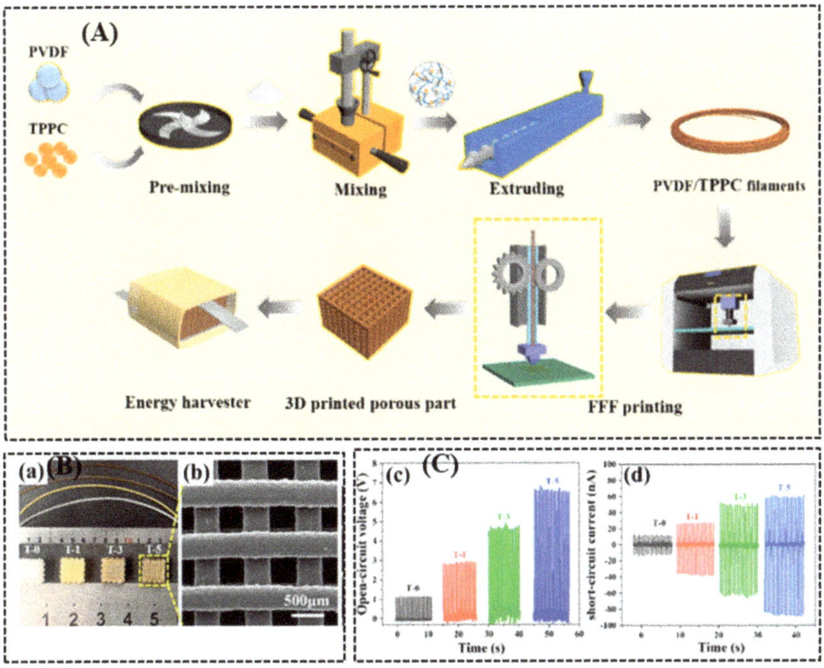

Figure 2.4. (A) A schematic diagram of the fabrication of PVDF/TPPC filaments and the 3D printed components for a nanogenerator. (B) (a) A digital photography of the filaments and 3D printed components and (b) an SEM micrograph of the top surface. (C) Output voltage and current of PVDF/TPPC nanogenerators. (Reproduced with permission from [27]. Copyright 2021 Elsevier.)

A typical hot press consists of high-temperature pressing followed by high-pressure cooling. A schematic of the hot press set-up is shown in figure 2.4(C). The important factors of the hot press are the temperature, pressure, and processing times. In most hot press processes, the temperature exceeds 150 °C and in some cases exceeds 200 °C [24]. The elevated temperature could cause the PVDF matrix to melt completely. The pressure in the hot press differs based on the experimental conditions. Following theoretical analysis, Fu *et al* [25] created a unique sandwich structure by hot pressing three stacked sheets at 200 °C and 10 MPa, one $FeTiNbO_6$/PVDF composite as an intermediate layer and the other two as the upper and lower PVDF layers. The sandwich structure improved the interfacial polarization and electroactive phase content in the PVDF, allowing the energy harvester to achieve an enhanced power and charge density of 110 $\mu W\ cm^{-3}$ and of 75 $\mu C\ cm^{-2}$, respectively. It ultimately showed outstanding applications in human movement and bicycle riding.

2.2.4 Other methods

3D printing has recently emerged as a method for producing PENGs with the required dimensions and geometry. It has been used extensively in conjunction with precise digital control to develop and manufacture polymer-based components with complicated structures and shapes. The fused filament fabrication (FFF) is the most widely used additive manufacturing and is capable of producing precise porous parts. Marandi *et al* [26] used the FFF 3D printing technique to produce almost only β-phase P(VDF-TrFE) and demonstrated it as a highly sensitive P(VDF-TrFE) sensor, with a sensitivity of 19 V at 2 mm displacement.

Using the FFF 3D printing, Pei *et al* [27] developed a high performance piezoelectric PVDF/tetraphenylphosphonium chloride (TPPC) composite for PENGs. Figure 2.4(A) depicts the fabrication procedure for PVDF/TPPC composite filaments and 3D printed parts. Figure 2.4(B)(a) depicts the PVDF/TPPC composite filaments used in FFF 3D printing as well as the 3D printed parts. The scanning electron microscopy (SEM) images (figure 2.4(B)(b)) of the FFF 3D printed parts show a clear observation of the 3D printed porous configuration and the dispersion of TPPC nanoparticles in the PVDF. Also, the relatively smooth surface of the FFF printed part with a 0°/90° cross deposition pattern and the stable multilayer grid configuration was found, indicating the successful printing of the defined porous structures with no defects. The energy harvester with a porous structure exhibited the highest output voltage and current values of 6.62 V and 87.6 nA, respectively.

2.3 Performance enhancement with the incorporation of fillers

The output signals of plain PVDF-based nanogenerators fall short of the requirements for real-world applications due to the low d_{33} values of PVDF and its copolymers, in particular in comparison to inorganic piezoelectric materials. Incorporating filler is one of the best ways to enhance the efficiency of PVDF based nanogenerators. The following section covers the different kinds of filler loading in PVDF that can be used to improve β-phase and nanogenerator performance.

2.3.1 Piezoelectric fillers

When piezoelectric additives are added to the PVDF or its copolymers, they not only deliver an extra piezoelectric effect, but also act as nucleation agents, stretching the PVDF molecules and increased the β-phase percentage via electrostatic interaction. The piezoelectric fillers also serve as stress reinforcement agents, improving the force transmission to the crystalline interior. To improve the energy harvesting of PVDF based PENGs, this section discusses the incorporation of piezoelectric materials such as PZT, PMN-PT, $BaTiO_3$ nanoparticles, $0.5(Ba_{0.7}Ca_{0.3})TiO_3$–$0.5Ba(Zr_{0.2}Ti_{0.8})O_3$ (BCT-BZT), and organic–inorganic lead halide perovskites. In general, a piezoelectric material with a large d_{33} value converts mechanical deformation more efficiently into electrical energy. The lead-based PZT and PMN-PT have the highest known piezoelectric coefficients, with PMN-PT nearly four times higher than PZT, 20 times more than $BaTiO_3$, and 90 times more than ZnO [28]. As a result, more goals have been set to improve the piezoelectric effect of the PVDF based composites in order to develop high performance flexible nanogenerators.

Chamankar *et al* [29] developed a highly flexible pressure sensor based on PVDF/PZT composite electrospun fibres with the incorporation of PZT particles in the

PVDF. The incorporation of PZT particles increased the dielectric constant and d_{33} values, as well as the PVDF molecular chain orientation, which was required to improve the dielectric and piezoelectric properties. For the pristine PVDF fibres, the output voltage and piezoelectric sensitivity were 22 mV and 29.490 mV N^{-1} μm^{-1}, respectively, and these values increased to 184 mV and 173.507 mV N^{-1} μm^{-1} for the composite fibres. The PVDF-PZT composite fibres at 0.37 PZT loading achieved an output power density of 30.69 μW cm^{-2}. Li *et al* [30] used a hydrothermal technique to produce novel $0.7Pb(Mg_{1/3}Nb_{2/3})O_3$–$0.3PbTiO_3$(PMN-PT) nanorods with a high yield of 78%. The PMN-PT nanorods were incorporated into PVDF to generate high piezoelectric composites. With a 20 wt% nanorod content, they obtained a highest Young's modulus of 1.71 GPa, and maximum output voltage and current values of 10.3 V and 46 nA, respectively, providing a 13-fold higher output voltage and 4.5-fold higher output current than for plain PVDF. The PVDF/PMN-PT composites exhibited a current density of 20 nA cm^{-2}. The PVDF/PMN-PT composites demonstrated significant promise for flexible self-powered sensing applications. Although PMN-PT and its composite materials possess excellent features, the high toxicity of the lead component precludes their use.

To develop eco-friendly nanogenerators, among the most studied piezoelectric materials with a considerably large piezoelectric coefficient and the possibility to replace lead inclusive counterparts are $BaTiO_3$ and BCT-BZT.

Despite the fact that nanocomposites composed of a flexible piezoelectric polymer and reinforcing fillers have shown a promising future for the creation of highly efficient PENGs, the poor interfacial interaction and lower dispersion capabilities of the reinforcing phase significantly hinder the dielectric and piezoelectric properties of the polymer composites, and thus ultimately reduce the PENG performance. To address the above mentioned problem, Shi *et al* [31] synthesized lead-free $BaTiO_3$ nanowires through the hydrothermal method and modified the surface with surface-initiated polymerization of a layer of poly(methyl methacrylate) (PMMA). The transmission electron microscopy (TEM) micrograph of PMMA modified $BaTiO_3$ nanowire in figure 2.5(A)(a) clearly shows a uniformly coated layer of PMMA with a thickness of 8–10 nm on the $BaTiO_3$ nanowire surface. The PMMA coated $BaTiO_3$ nanowires demonstrated a strong interfacial interaction with the polymer matrix in the PVDF-TRFE/$BaTiO_3$ composite fibres, and also effectively improved the dispersion of the $BaTiO_3$ and stress transfer at the interface, resulting in improved energy harvesting performance.

The performance of the composite fibre PENGs was measured under mechanical bending and releasing motions at a frequency of 2 Hz and a displacement of 4 mm, as shown in figure 2.5(A)(c). Enhanced output voltage and current values were exhibited in the PMMA coated $BaTiO_3$ nanowire based PENGs compared to the corresponding $BaTiO_3$ nanowire based PENGs, and obtained a maximum voltage 12.6 V and a current of 1.30 A at 10 wt% loading. The improved output performance of the PMMA coated $BaTiO_3$ based PENGs can be attributed to the interfacial PMMA, which improved the dispersion capability of the $BaTiO_3$ nanowires and the nanocomposite fibres' micromechanical properties. Furthermore, as shown in figures 2.5(A)(g) and (h), both the output voltage and current values increased

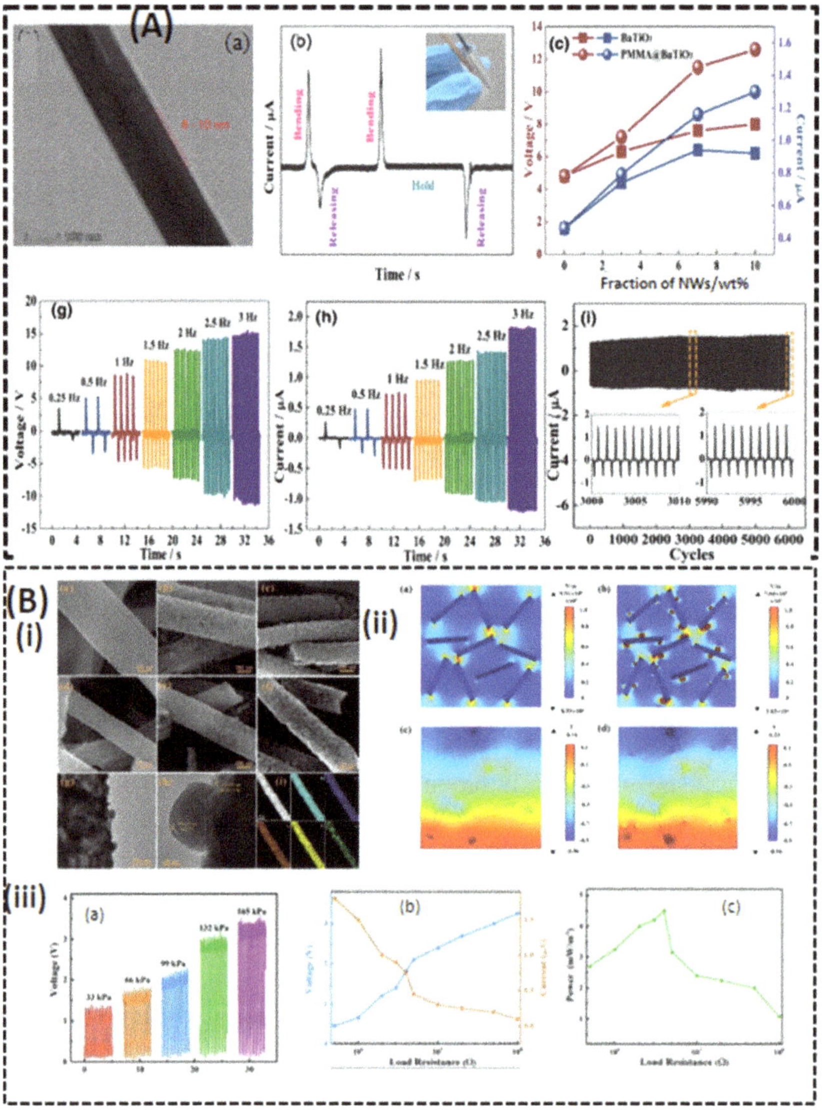

Figure 2.5. (A) (a) TEM image of PMMA coated $BaTiO_3$ nanowire. (b) Output current of PENG under finger impartation. (c) Output voltage and current values of the PENGs at various loadings of nanowires in PVDF-TrFE fibres. Output (g) voltages and (h) currents of the PENG based on 10 wt% PMMA coated $BaTiO_3$/PVDF-TrFE fibres at different frequencies; (i) the durability of the PENG. (Reproduced with permission from [31]. Copyright 2021 Elsevier.) (B) (i) (a)–(f) SEM micrograph of the BCZT nanowires with a range of Ag content, (g)–(h) HRTEM micrograph, and (i) EDS mapping of the 0.04Ag-BCZT nanowires; (B) (ii) electric field distribution of (a) BCZT/PVDF–TrFE, (b) Ag-BCZT/PVDF–TrFE; piezoelectric potentials of (c) BCZT NWs/PVDF–TrFE and (d) Ag-BCZT/PVDF–TrFE, respectively. (B) (iii) (a) Output voltage produced from the PENG. (b) and (c) Voltage, current, and power values at various load resistances. (Reproduced with permission from [33]. Copyright 2022 American Chemical Society.)

with an increase of mechanical bending frequency and also exhibited long term stability.

Another important lead-free piezoelectric material, barium calcium zirconate titanate, $x(Ba_{0.7}Ca_{0.3})TiO_3$–$(1-x)Ba(Zr_{0.2}Ti_{0.8})O_3$ (BCT-BZT) with a d_{33} value of 620 pC N^{-1}, comparable to PZT, is considered to be an outstanding candidate for a lead-free piezoelectric material. Lu *et al* [32] demonstrated a creative 3D core/shell structure of 3D-Tb-BCZT-coated 3D-Tb-BCZT/PVDF composite fibres by using a coaxial electrospinning method. The novel structure could improve the stress-transfer efficiency by significantly increasing the correlation between neighbouring piezoelectric particles. The output voltage and current were studied systematically for various types of PENGs based on BCZT, Tb-BCZT, and 3D core/shell structures. The open-circuit voltage (V_{oc}) and short-circuit current (I_{sc}) of the BCZT/PVDF composite fibres were only 15.6 V and 1.32 A, respectively. The V_{oc} and I_{sc} of the Tb-modified BCZT-doped PVDF composite fibres increased to 21.4 V and 1.71 A, respectively, and were approximately 1.37 and 1.29 times higher than that of BCZT/PVDF. Furthermore, the V_{oc} and I_{sc} of PENGs based on 3D-Tb-BCZT achieved a huge value of 48.5 V and 3.35 A, which were approximately 3.1- and 2.5-fold greater than for PENGs based on BCZT.

Similarly, Yan *et al* [33] used a photoreduction technique to produce silver (Ag)-decorated BCZT heterostructures, which were later introduced at 3 wt% into the matrix of PVDF-TrFE to develop the composite films. The surface of pure BCZT nanowires was relatively smooth, as seen in the SEM and TEM images in figures 2.5(B(i))(a)–(d). Figure 2.5(B(i))(f) shows spherical Ag nanoparticles on the surface of BCZT nanowires upon photoreduction. Furthermore, as the concentration of silver nitrate solution increased, the quantity of Ag nanoparticles is observed to increase considerably. The TEM images in figures 2.5(B(i))(g)–(h) showed Ag nanoparticles on the surface of BCZT nanowires. The TEM images also show an interplanar spacing distance of 0.291 nm, which is ascribed to the (110) planes of BCZT, and an interplanar spacing distance of 0.237 nm for the Ag decorated nanoparticles, which corresponded to the (111) plane of Ag. Furthermore, the energy-dispersive x-ray spectroscopy (EDS) detected elements such as Ba, Ca, Zr, Ti, and Ag, which were uniformly distributed in the Ag-BCZT nanowires. These findings demonstrated the successful deposition of Ag nanoparticles on the BCZT nanowires.

Also, the COMSOL 5.4 software was used to explore the impact of Ag nanoparticle modification on the electric field distribution and piezoelectric potential of PVDF-TrFE. The analysis was carried out in two two-dimensional (2D) models with applied electric fields of 300 kV cm^{-1} and mechanical loads of 10 N, respectively. The optimum electric field strength for the BCZT/PVDF-TrFE was 9.51×10^8 V m^{-1}, which was increased to 7.04×10^9 V m^{-1} in the Ag decorated composites (figure 2.5(B(ii))(a) and (b)), i.e. a nearly 6.4-fold enhancement over the bare BCZT NWs/PVDF-TrFE composite film was observed.

Furthermore, under the same mechanical stress, the piezoelectric potential produced by Ag-BCZT/PVDF-TrFE composite film was also greater than that of pure BCZT/PVDF-TrFE composite (figure 2.5(B(ii))(b)). Figure 2.5(B(iii)) depicts the output voltage of the composite film at various pressures and a constant frequency of 2 Hz. The open-circuit voltage showed a linear response with increasing pressure and attained a peak voltage of 3.5 V at 165 kPa, which was 2.3 higher than bare BCZT composite without Ag decoration. Furthermore, at 4 MΩ load resistance, the maximum output power of 4500 μW m^{-2} was achieved. The rapid development of the organic–inorganic lead halide perovskite materials (ABX_3, A = CH_3NH_3 (MA) or $CH(NH_2)_2$ (FA), B = Pb, and X = Cl, Br, or I) has acquired a significant amount of attention in optoelectronics. They can also be used in PENGs due to their cubic perovskite construction, which could lead to ferroelectric polarization. Chen *et al* [34] used a one-step electrospinning technique to develop *in situ* growth of $CsPbBr_3$ nanocrystals in the PVDF fibres with relatively uniform size and spatial distribution. They exhibited an outstanding open-circuit voltage and current density of 103 V and 170 A cm^{-2}, respectively. The obtained current density was 4.86-fold greater than the previously reported lead halide perovskites. They had an excellent open-circuit voltage and current density of 103 V and 170 A cm^{-2}, respectively. The obtained current density was 4.86 times greater than the previously reported lead halide perovskites equivalents. Furthermore, the PENG based on $CsPbBr_3$/PVDF composite fibres has significantly enhanced their durability markedly in thermal, water, humid, acid, and alkali environments. Their output ability was maintained by more than 79.16% at 220 °C, nearly 100% in water for approximately 6 months, and 63.53% and 82.35% in H_2SO_4 and NaOH solutions, respectively.

Kuntal Maity *et al* [35] investigated the coupling of photoexcitation and piezoelectric effects in $CsPbI_3$/PVDF composites to study the piezo-phototronics. The $CsPbI_3$ incorporated PVDF composite film was demonstrated to be a superior candidate for the nanogenerator and photodetector fabrication, as well as hybrid devices. The PVDF served as a $CsPbI_3$ stabilizing layer, protecting the $CsPbI_3$ from degradation in the surrounding atmosphere and boosting the total β-phase in the composite films to 96%. Under compressive pressure, the composite PENGs exhibited an output voltage and current values of 20 V and 6 A, respectively. Furthermore, the composite exhibited an outstanding photo responsivity 0.3 mA W^{-1} and detectivity of 0.122×10^{10} jones, served as highly sensitive visible light photodetector. Finally, the films were demonstrated as a hybrid piezo-phototronic device by demonstrating a 57% increase in the current values when compressive strain and light illumination were applied to the device at the same time.

2.3.2 Conductive fillers

Carbon nanotubes (CNTs), graphene oxide, carbon foam, Ag nanoparticles, and other conductive materials have been used widely in a variety of disciplines due to their exceptional features. These conductive materials aid in the processes of nucleation, dipole alignment, micro-capacitor formation, stress reinforcement, and other processes in the PVDF composites to improve the performance of piezoelectric nanogenerators. CNTs are well known for their surface-to-volume ratio, high electrical conductivity, and mechanical characteristics. Incorporating the CNTs into PVDF not only enhances the electrical conductivity greatly, but also acts as a nucleating agent, as previously discussed, to improve the PVDF β-phase, thereby improving the performance of the PENGs [36, 37]. Badatya *et al* [37] prepared a lightweight PVDF-CNT foam for the very first time to develop humidity resistant hydrophobic versatile PENGs. The solution method was used to create hydrophobic piezoelectric PVDF-CNT foam with a density of 0.15 g cm^{-3}. The PVDF-CNT foam had a crystalline and well-defined chain-like architecture with a 65% β-phase fraction. The PVDF-CNT foams' field emission scanning electron microscopy (FE-SEM) micrographs revealed a cellular architecture with a diameter of 3–4 μm (figure 2.6(A(i))(a)). Figure 2.6(A(i))(b) shows an enlarged view of the selected area, and confirmed the formation of porous structures in the foam and the distribution of the CNT within the PVDF matrix. The highly magnified SEM image clearly showed a spherulitic/chain-like structure, confirming the development of the β-phase in the

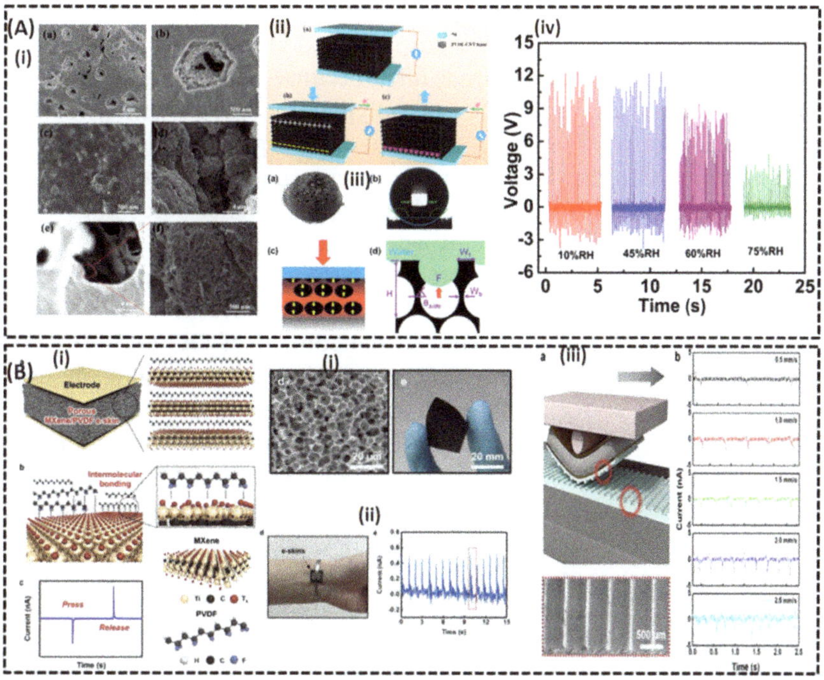

Figure 2.6. (A)(i) (a)–(f) FE-SEM images of PVDF/CNT foam. (A)(ii) Working mechanism of the PVDF-CNT foam device. (A)(iii) Surface contact angle measurement of the hydrophobic PVDF-CNT foam. (A)(iv) Output voltage of the PVDF-CNT foam nanogenerator under different humidity conditions. (Reproduced with permission from [37]. Copyright 2021 American Chemical Society.) (B)(i) (a) Schematic illustration of porous MXene/PVDF e-skin and molecular structures of MXene and PVDF. (b) Schematic diagram of the interior structure. (c) Output current of e-skin. (d)–(e) SEM micrograph and photograph of porous e-skin. (B)(ii) (d) Photograph of porous MXene/PVDF e-skins attached to the wrist for detecting radial artery pulse. (e) Real-time monitoring of radial artery pulse. (B) (iii) (a) Schematic illustration of texture perception, wherein porous MXene/PVDF e-skins with PDMS fingerprint patterns are attached to a microstage and fingerprint patterns are fabricated on PDMS. (b) Time dependent variation in piezoelectric currents when the patterned surface is scanned at different speeds. (Reproduced with permission from [40]. Copyright 2021 Elsevier.)

PVDF (figures 2.6(A(i))(c)–(d)). In the PVDF-CNT foam, the dispersion of the CNTs in the PVDF foam was clearly visible (figures 2.6(A(i))(c)–(f)).

The self-poled PVDF-CNT foam exhibited a high d_{33} value of 9.4 pC N^{-1}. The high d_{33} of the PVDF-CNT foam was caused by dipole alignment in the microcellular structure of the PVDF caused by a local electric field of CNTs. At 150 Hz the developed foam has an ultrahigh dielectric constant of 3048. At a low compressive pressure of 0.02 kgf, the flexible piezoelectric PVDF-CNT foam-based nanogenerator produced a high output voltage of 12 V and a current density of 30 nA cm^{-2}. The PENGs performance was determined under various humidity conditions, and an output voltage of up to 8 V was obtained even at 60% RH, as shown in figure 2.6(A(iv)). The obtained contact angle of the PVDF-CNT foam was approximately 139° (figure 2.6(A(ii))), with the large value due to the foam-like structure. CNT-induced self-alignment of electric dipoles in PVDF polymer produced a high output voltage even at a very low pressure, with no electrical poling, and in humid conditions.

Graphene oxide (GO) and reduced graphene oxide (rGO) have received a lot of attention as an alternative conductive filler because of their improving thermal, mechanical, and electrical properties of polymer matrices when these fillers are incorporated. Yang *et al* [38] studied the effect of incorporating the GO and rGO on PVDF. By adding GO to the PVDF phase, the current was increased to 200 nA and the voltage was increased to 1.5 V. By adding rGO a higher percentage of β-phase, short-circuit current, and open-circuit voltage of 87%, 700 nA,and 16 V were achieved. The high conductivity of the rGO enhanced the transfer of induced charge generated by the PVDF even further. The output performance of the PENGs increased significantly with the increase in the content of GO and rGO. rGO has four times more conductivity than GO, which had more significant impact on improving the PENG performance and rGO also provided heterogeneous polarization, and their interactions with the PVDF chain enhanced the PVDF's piezoelectric properties. Titanium carbide ($Ti_3C_2T_x$) (MXene) is another innovative form of 2D transition metal carbide, which has sparked widespread interest due to its distinctive electrical and mechanical properties, making it ideal for actuators, supercapacitors, batteries, and catalyst applications. MXene is also an interesting option for nanocomposites engineering due to its layered structure, large aspect ratio, and outstanding electric conductivity.

Zhao *et al* [39], for the first time, demonstrated the novel 2D MXene as an effective reinforcement to encourage the output and sensitivity of PENGs. The d_{33} value of MXene/PVDF hybrid film reached a maximum value of 43 pC N^{-1} at 0.4 wt% loading. In addition, the inclusion of MXene improved the mechanical performance of PVDF. The MXene/PVDF sensor was capable of high voltage sensitivity up to 0.0480 V N^{-1} due to its excellent piezoelectric response and mechanical properties.

Kim *et al* [40] demonstrated self-powered piezoelectric e-skins with good sensitivity and a wide detection range using three-dimensional (3D) porous MXene/PVDF structures. MXene was employed as a nucleation agent to improve the ferroelectric properties of PVDF. A piezoelectric e-skin made of MXene/PVDF with 3D porous microstructures and Ni/Cu electrodes on both sides is depicted schematically in figure 2.6(B(i))(a). MXene's large surface functional groups, such as oxide –O, –OH, and –F, could induce the intermolecular hydrogen bonding with PVDF's $-CH_2$ and $-CF_2$ groups, resulting in PVDF intercalation and confinement in MXene with the crystallization of highly polar β-phase PVDF. Furthermore, due to the large specific area of 2D geometry, MXene nanosheets could act as nucleation agents to effectively strengthen the electroactive properties of PVDF, as opposed to spherical or nanorod shapes (figure 2.6(B(i))(b)). Figure 2.6(B(i))(c) showed the piezoelectric output current. The porous MXene/PVDF e-skin was shown in figure 2.6(B(i))(d). The piezoelectric sensitivities of the porous piezoelectric e-skins were 11.9 and 1.4 nA kPa^{-1} in the low (2.5 kPa) and high (2.5–100 kPa) pressure ranges, respectively, and these were 31- and 3.7-fold greater compared to the planar MXene/PVDF e-skins (0.4 nA kPa^{-1}). The porous MXene/PVDF e-skin was attached to the wrist and measured the output values in real time to monitor the radial artery pulse (figure 2.6(B(ii))(d)). The measured pulse frequency was 68 beats per minute, which was consistent with a stable state of healthy adults (figure 2.6(B (ii))(e)). This improved the perception of the surface textures. The schematic design for the texture perception by e-skin is shown in figure 2.6(B(iii))(a). The line patterned polydimethylsiloxane (PDMS) (figure 2.6(B(iii))) was utilized to enhance the piezoelectric output stimulated by the surface texture, motivated by the amplification of tactile signals using fingerprint patterns. The piezoelectric current output values with respect to time when porous MXene/PVDF e-skin was scanned over the PDMS surface are shown in figure 2.6(B(iii))(b). The intensity and number of periodic spike peaks increased as the scanning speed increased.

2.3.3 Magnetic fillers

Incorporating magnetic ferrites into PVDF is also one of the best ways to improve the β-phase and energy harvesting performance. Tripathy *et al* [12] used a hydrothermal method to create BFO NPs, which are then incorporated into PVDF-TrFE and PVDF-HFP films to fabricate the magnetic composite films. The incorporation of BFO increased the nucleation and improved the electroactive β-phase to 81% and 79%, respectively. The composite nanogenerator produced a maximum instantaneous power density of 0.92 μW cm^{-2} at 100 MΩ and output voltage upon finger tapping of 18.5 V due to the influence of magnetic BFO particles on β-phase of polymer. Mu *et al* [41] proposed large self-biased 0–3 type P(VDF-TrFE)/$CoFe_2O_4$ flexible magneto-electric (ME) composite films for the development of AC magnetic sensors. The solution casting method was used to successfully fabricate the P(VDF-TrFE)/$CoFe_2O_4$ with a $CoFe_2O_4$ loading ranging from 5% to 30%, yielded the highest phase values of 97%. The composite films demonstrated exceptional ferroelectric and ferromagnetic properties. All of the P(VDF-TrFE)/$CoFe_2O_4$ exhibited a significant ME effect. At resonance frequency, the ME coefficient reached 47.1 mV cm^{-1} Oe^{-1} for a 20 wt% $CoFe_2O_4$ loaded sample. Furthermore, the composite films with a high linear correlation coefficient, sensitivity, and repeatability show the proposed P(VDF-TrFE)/$CoFe_2O_4$ as a potential candidate for flexible AC magnetic sensors.

Prasad *et al* [42, 43], using a two-step synthesis method, prepared a flexible PVDF/$ZnFe_2O_4$ multifunctional nanocomposite film. $ZnFe_2O_4$ nanofibers synthesized using an electrospinning process was incorporated into a PVDF to fabricate the PVDF/$ZnFe_2O_4$ through the solvent casting technique. With the addition of $ZnFe_2O_4$ the β-phase increased up to 88%, due to the role of $ZnFe_2O_4$ as the nucleating agent in the PVDF (figure 2.7(A(i))). The ferroelectric domain switching behaviours (figure 2.7(A(ii))) revealed high colour contrast in the composite films, which confirmed that the composite film exhibited a better ferroelectric nature when compared to the PVDF film. The fabricated flexible PVDF/$ZnFe_2O_4$ nanogenerator exhibited the highest open-circuit voltage of 7 V and an output power of 4 μW at 500 kΩ load resistance under a constant applied force of 1.5 N (figure 2.7(A(iii))).

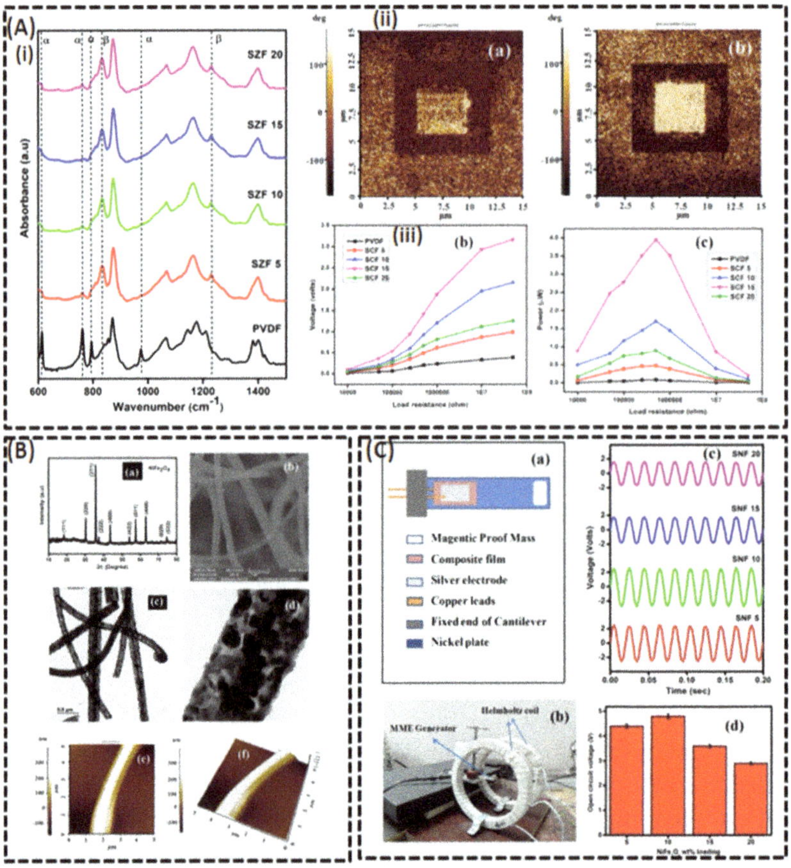

Figure 2.7. (A) (i) FTIR spectra, (ii) domain switching behaviour of PVDF/$NiFe_2O_4$, and (iii) output voltage and power of PVDF/$NiFe_2O_4$ nanogenerators. (Reproduced with permission from [42]. Copyright 2021 Elsevier.) (B) (a)–(f) XRD, SEM, TEM, and AFM images of $NiFe_2O_4$ fibres. (C) (a)–(d) Schematic, photograph of the experimental set-up, and output voltage responses of PVDF/$NiFe_2O_4$ composite MME generator. (Reproduced with permission from [43]. Copyright 2021 Elsevier.)

The group further studied PVDF/$NiFe_2O_4$ fibre composite films and the effect of magnetic material in enhancing the energy harvesting performance. The SEM and TEM images showed the formation of $NiFe_2O_4$ fibres with a diameter of 400 nm and figures 2.7(B)(e) and (f) showed 2D and 3D atomic force microscopy (AFM) images of the $NiFe_2O_4$ fibres. These fibres were incorporated into PVDF polymer and used to fabricate flexible magnetoelectric composite films.

A clear transformation of the α- to β-phase has been observed through SEM and AFM microstructures with the incorporation of magnetic $NiFe_2O_4$ fibres and obtained 68% β-phase. The composite film with 10 wt% $NiFe_2O_4$ displayed a high maximum power (P_{max}) value of 1.46 C cm^{-2} when the ferroelectric characteristics were examined in relation to $NiFe_2O_4$ loading. The PENGs were fabricated using composite films and generated an open-circuit voltage of 10 V. Further, these films were made into novel cantilever structured magneto-mechano-electric (MME) generators, as shown in figure 2.7(C), for harvesting ubiquitous magnetic energy via stray magnetic fields originating from electricity transmission systems, overhead electricity lines, industrial machines, and electronic parts. The device managed to generate a 4.8 V output voltage under a low AC magnetic field of 10 Oe at 50 Hz.

2.4 Conclusions and future scope

PVDF and its copolymers are flexible piezoelectric materials that are widely used in energy harvesting applications because of their outstanding versatility, low cost, high strength, strong mechanical properties, simple processing, and superior chemical stability. In this chapter we summarize the recent progress in flexible PVDF-based PENGs, focusing primarily on the fabrication process, structural design, and filler selection and incorporation in PVDFs to enhanced their crystalline phase, piezoelectric coefficient, and energy harvester applications. In addition, the application of these nanogenerators in various fields of study was discussed briefly. Despite significant efforts, the output power of PVDF-based PENGs is only sufficient to be applied for the low power small-scale electronic devices such as LEDs, as well as an electronic watches, calculators, and so on. However, more research is needed to address specific issues, and the following research aspects must be considered in the future:

- The incorporation of other functional materials could improve the output performance of PVDF-based nanogenerators significantly. As a result, a thorough understanding of the coupling mechanisms of various components would then aid in their coordination and provide a better understanding of the nanogenerators' mechanical energy harvesting ability.
- Along with the integration of functional materials, further consideration should be given to innovative device structural designs. The new structural designs could not only increase the sensitivity of energy harvesters, which sense a wide range of mechanical behaviours, but also fulfil the criteria of novel applications such as pacemakers and implantable biosensors and wearable electronics etc.
- Much more efficient PVDF-based nanogenerators could be developed using the combined effect of piezoelectricity and other electricity-generating

principles such as triboelectricity, magnetoelectricity, pyroelectricity, and electromagnetic effects.
- To establish self-powered systems, the PENGs should be constructed with other electronics. The combination of nanogenerators with energy storage devices such as batteries and supercapacitors, as well as an energy management system, would then enable wireless charging of PENGs to power low-power portable electronics continuously.
- While most nanogenerators produce AC voltage, DC voltage is required to run devices continuously in practical applications. As a result, developing DC nanogenerators is critical.
- Even though PVDF-based polymers have good biocompatibility and are thus considered good candidates for surgical implants, there is still a long way to go before actual human applications, especially when safety concerns are considered.

Acknowledgments

DPP acknowledges ANID-FONDECYT 2022 Project No: 3220360, Chile for financial support.

References

[1] Mitra R, Priyadarshini B S, Ramadoss A and Manju U 2022 Stretchable polymer-modulated PVDF-HFP/TiO_2 nanoparticles-based piezoelectric nanogenerators for energy harvesting and sensing applications *Mater. Sci. Eng.* B **286** 116029

[2] Hu D, Yao M, Fan Y, Ma C, Fan M and Liu M 2019 Strategies to achieve high performance piezoelectric nanogenerators *Nano Energy* **55** 288–304

[3] Arulraj A, Thachnatharen N, Rajkumar C and Mangalaraja R V 2022 Engineering 2D chalcogenides for energy and environmental remediation *Nanostructured Materials for Sustainable Energy and Environmental Remediation* (Bristol: IOP Publishing) pp 2-1–2-28

[4] Ghosh S K and Mandal D 2018 Synergistically enhanced piezoelectric output in highly aligned 1D polymer nanofibers integrated all-fiber nanogenerator for wearable nano-tactile sensor *Nano Energy* **53** 245–57

[5] Karan S K, Maiti S, Kwon O, Paria S, Maitra A, Si S K, Kim Y, Kim J K and Khatua B B 2018 Nature driven spider silk as high energy conversion efficient biopiezoelectric nanogenerator *Nano Energy* **49** 655–66

[6] Yan J, Liu M, Jeong Y G, Kang W, Li L, Zhao Y, Deng N, Cheng B and Yang G 2019 Performance enhancements in poly(vinylidene fluoride)-based piezoelectric nanogenerators for efficient energy harvesting *Nano Energy* **56** 662–92

[7] Khan A A, Mahmud A and Ban D 2019 Evolution from single to hybrid nanogenerator: a contemporary review on multimode energy harvesting for self-powered electronics *IEEE Trans. Nanotechnol.* **18** 21–36

[8] Huynh N D, Lin Z H and Choi D 2021 Dynamic balanced hybridization of TENG and EMG via Tesla turbine for effectively harvesting broadband mechanical pressure *Nano Energy* **85** 10598

[9] Jain A, J P K, Sharma A K, Jain A and Rashmi P N 2015 Dielectric and piezoelectric properties of PVDF/PZT composites: a review *Polym. Eng. Sci.* **55** 1589–616

[10] Liao Q, Zhang Z, Zhang X, Mohr M, Zhang Y and Fecht H J 2014 Flexible piezoelectric nanogenerators based on a fiber/ZnO nanowires/paper hybrid structure for energy harvesting *Nano Res.* **7** 917–28

[11] Zhang X, Xia W, Liu J, Zhao M, Li M and Xing J 2022 PVDF-based and its copolymer-based piezoelectric composites: preparation methods and applications *J. Electron. Mater.* **51** 5528–49

[12] Tripathy A *et al* 2023 Tuning of highly piezoelectric bismuth ferrite/PVDF-copolymer flexible films for efficient energy harvesting performance *J. Alloys Compd.* **932** 167569

[13] Lu L, Ding W, Liu J and Yang B 2020 Flexible PVDF based piezoelectric nanogenerators *Nano Energy* **78** 105251

[14] Mahale B, Bodas D and Gangal S A 2017 Study of beta-phase development in spin-coated PVDF thick films *Bull. Mater. Sci.* **40** 569–75

[15] Nishiyama T, Sumihara T, Sato E and Horibe H 2017 Effect of solvents on the crystal formation of poly (vinylidene fluoride) film prepared by a spin-coating process *Polym. J.* **49** 319–25

[16] Arulraj A, Bhuvaneshwari S, Senguttuvan G and Ramesh M 2018 Solution processed inverted organic bulk heterojunction solar cells under ambient air-atmosphere *J. Inorg. Organomet. Polym.* **28** 1029–36

[17] Nishiyama T, Sumihara T, Sato E and Horibe H 2016 Effect of solvents on the crystal formation of poly(vinylidene fluoride) film prepared by a spin-coating process *Polym. J.* **49** 319–25

[18] Ramasundaram S, Yoon S, Kim K J and Lee J S 2008 Direct preparation of nanoscale thin films of poly(vinylidene fluoride) containing β-crystalline phase by heat-controlled spin coating *Macromol. Chem. Phys.* **209** 2516–26

[19] Cardoso V F, Minas G, Costa C M, Tavares C J and Lanceros-Mendez S 2011 Micro and nanofilms of poly(vinylidene fluoride) with controlled thickness, morphology and electro-active crystalline phase for sensor and actuator applications *Smart Mater. Struct.* **20** 087002

[20] Baji A, Mai Y W, Li Q and Liu Y 2011 Electrospinning induced ferroelectricity in poly (vinylidene fluoride) fibers *Nanoscale* **3** 3068–71

[21] Persano L, Dagdeviren C, Su Y, Zhang Y, Girardo S, Pisignano D, Huang Y and Rogers J A 2013 High performance piezoelectric devices based on aligned arrays of nanofibers of poly (vinylidenefluoride-co-trifluoroethylene) *Nat. Commun.* **4** 1633

[22] Muduli S P, Parida S, Rout S K, Rajput S and Kar M 2019 Effect of hot press temperature on beta-phase, dielectric and ferroelectric properties of solvent casted poly(vinyledene fluoride) films *Mater. Res. Express* **6** 13

[23] Wu L, Jin Z, Liu Y, Ning H, Liu X, Alamusi and Ning H 2022 Recent advances in the preparation of PVDF-based piezoelectric materials *Nanotechnol. Rev.* **11** 1386–407

[24] Tusiime R, Zabihi F, Tebyetekerwa M, Yousry Y M, Wu Y, Eslamian M, Yang S, Ramakrishna S, Yu M and Zhang H 2020 High stress-driven voltages in net-like layer-supported organic–inorganic perovskites *J. Mater. Chem.* C **8** 2643–58

[25] Fu J, Hou Y, Zheng M and Zhu M 2020 Flexible piezoelectric energy harvester with extremely high power generation capability by sandwich structure design strategy *ACS Appl. Mater. Interfaces* **12** 9766–74

[26] Marandi M and Tarbutton J 2019 Additive manufacturing of single- and double-layer piezoelectric PVDF-TrFE copolymer sensors *Procedia Manuf.* **34** 666–71

[27] Pei H, Xie Y, Xiong Y, Qinniu L and Chen Y 2021 A novel polarization-free 3D printing strategy for fabrication of poly (vinylidene fluoride) based nanocomposite piezoelectric energy harvester *Composites* B **225** 109312

[28] Hwang G T, Park H, Lee J H, Oh S K, Park K I, Byun M, Park H, Ahn G, Chang K J and No K 2014 Self-powered cardiac pacemaker enabled by flexible single crystalline PMN-PT piezoelectric energy harvester *Adv. Mater.* **26** 4880–7

[29] Chamankar N *et al* 2020 A flexible piezoelectric pressure sensor based on PVDF nanocomposite fibers doped with PZT particles for energy harvesting applications *Ceram. Int.* **46** 19669–81

[30] Li C, Luo W, Liu X, Xu D and He K 2016 PMN-PT/PVDF nanocomposite for high output nanogenerator applications *Nanomaterials* **6** 67

[31] Shi K, Chai B, Zou H, Shen P, Sun B, Jiang P, Shi Z and Huang X 2021 Interface induced performance enhancement in flexible $BaTiO_3$/PVDF-TrFE based piezoelectric nanogenerators *Nano Energy* **80** 105515

[32] Lu H, Zhang J, Yang L, Zhang Y, Wu Y and Zheng H 2022 Enhanced output performance of piezoelectric nanogenerators by Tb-modified (BaCa)$(ZrTi)O_3$ and 3D core/shell structure design with PVDF composite spinning for microenergy harvesting *ACS Appl. Mater. Interfaces* **14** 12243–56

[33] Yan M *et al* 2022 Flexible PVDF–TrFE nanocomposites with Ag-decorated BCZT heterostructures for piezoelectric nanogenerator applications *ACS Appl. Mater. Interfaces* **14** 53261–73

[34] Chen H, Zhou L, Fang Z, Wang S, Yang T, Zhu L, Hou X, Wang H and Wang Z L 2021 Piezoelectric nanogenerator based on *in situ* growth all-inorganic $CsPbBr_3$ perovskite nanocrystals in PVDF fibers with long-term stability *Adv. Funct. Mater.* **31** 2011073

[35] Maity K, Pal U, Mishra H K, Maji P, Sadhukhan P, Mallick Z, Das S, Mondal B and Mandal D 2022 Piezo-phototronic effect in highly stable $CsPbI_3$-PVDF composite for self-powered nanogenerator and photodetector *Nano Energy* **92** 106743

[36] De Volder M F, Tawfick S H, Baughman R H and Hart A J 2013 Carbon nanotubes: present and future commercial applications *Science* **339** 535–9

[37] Badatya S, Bharti D K, Sathish N, Srivastava A K and Gupta M K 2021 Humidity sustainable hydrophobic poly(vinylidene fluoride)-carbon nanotubes foam based piezoelectric nanogenerator *ACS Appl. Mater. Interfaces* **13** 27245–54

[38] Yang J, Zhang Y, Li Y, Wang Z, Wang W, An Q and Tong W 2021 Piezoelectric nanogenerators based on graphene oxide/PVDF electrospun nanofiber with enhanced performances by *in-situ* reduction *Mater. Today Commun.* **26** 101629

[39] Zhao Q, Yang L, Ma Y, Huang H, He H, Ji H, Wang Z and Qiu J 2021 Highly sensitive, reliable and flexible pressure sensor based on piezoelectric PVDF hybrid film using MXene nanosheet reinforcement *J. Alloys Compd.* **886** 161069

[40] Kim J *et al* 2021 MXene-enhanced β-phase crystallization in ferroelectric porous composites for highly-sensitive dynamic force sensors *Nano Energy* **89** 106409

[41] Xuejian Mu H, Zhang C, Zhang S, Yang J, Xu Y, Huang J, Xu Y, Zhang Q, Li X and Wang D 2021 Poly(vinylidene fluoride-trifluoroethylene)/cobalt ferrite composite films with a self-biased magnetoelectric effect for flexible AC magnetic sensors *J. Mater. Sci.* **56** 9728–40

[42] Durga Prasad P and Hemalatha J 2021 Multifunctional films of poly(vinylidene fluoride)/ $ZnFe_2O_4$ nanofibers for nanogenerator applications *J. Alloys Compd.* **854** 157189

[43] Durga Prasad P and Hemalatha J 2021 Energy harvesting performance of magnetoelectric poly(vinylidene fluoride)/$NiFe_2O_4$ nanofiber films *J. Magn. Magn. Mater.* **532** 167986

IOP Publishing

Materials Technology for the Energy and Environmental Nexus, Volume 2

R V Mangalaraja, Arunachalam Arulraj, Radhamanohar Aepuru and Chan Siew Hwa

Chapter 3

Copper based metal oxide nanocomposites toward solar water oxidation

V Balasubramanian and B Shunmugapriya

Potential materials for energy conversion and storage applications include nanostructured metal oxides, which are now the focus of extensive research. In particular, copper-based oxides have been discovered to be significant candidates for solar-driven hydrogen generation since they provide a practical pathway for hydrogen production from solar energy. Cu-based metal oxides, which show enhanced photoelectrochemical performance in the quest for possible TiO_2 photoanode substitutes, have their best band gaps in the visible spectrum. This chapter addressed the photoelectrochemical and synthesis properties of various copper (Cu) containing oxides for applications in water splitting. Cu-based ternary oxide nanocomposites showed better resistance to photocorrosion than Cu binary oxides. It was discovered that the band gap of Cu ternary oxides is within the ideal range for solar water oxidation. It also has a similar conduction band edge and a more negative valence band edge when compared to the analogous binary oxides. As a result, it is simple to oxidize the water for photoelectrochemical water splitting. The chapter also re-emphasizes the critical part that solid state chemistry can play in the ongoing hunt for an inorganic semiconductor for solar water oxidation. Finally, the future prospects for creating functional solar water splitting devices are provided.

3.1 Introduction

The interaction of solid state chemistry (SSC) and photoelectrochemistry (PEC) is examined in this chapter using copper-based complex (ternary and quaternary) oxides as a discussion framework. Since their inception, the two disciplines of SSC and PEC have essentially advanced simultaneously. SSC and PEC are both interested in chemicals (mostly oxides and chalcogenides), but for quite different

doi:10.1088/978-0-7503-5729-6ch3

reasons. While PEC has focused on inorganic semiconductors, SSC has largely focused on magnetic or electrically and ionically conductive solid materials. However, SSC has eventually advanced to superconducting (also known as high T_c) materials. However, in the area of energy conversion, the two disciplines have come together. Significant advancements have been made over the past few decades in the production of solid state materials for Li-ion batteries, thermoelectric conversion, photovoltaic solar cells, and solid-oxide fuel cells. Even high T_c superconductivity research has produced specialized uses in practical technologies. Solar fuels (such as those that split water or reduce CO_2 or N_2) have been studied for only a short time and they have not yet reached the market stage.

The most well-known wet chemical process for turning sunlight into electrical energy or chemical fuels is PEC systems. PEC systems were created in the middle of the 1970s and the 1980s to gather solar energy and transform it into fuels and electricity using semiconductor electrodes in contact with electrolyte solutions. The following four benefits are provided by the solid–liquid arrangement:

- By submerging the semiconductor in a suitable electrolyte solution, the junction needed for effective charge separation of photogenerated electrons and holes can be created readily.
- A readily conformable and strain-free junction is possible with liquid electrolyte.
- PEC cells can have a third electrode added to them to enable *in situ* chemical storage for 24 h power.
- The absence of separate wires and an electrolyzer due to the direct conversion of light energy into fuel.

PEC has successfully converted solar energy into high electrical power (>15%) and hydrogen production (>10%) efficiently. However, the lifetime of the photo-electrode and cost issues restrict commercialization efforts [1, 2]. As can be seen from figure 3.1, which lists the number of publications on H_2 generation [3], it has been the focus of researchers' attention over the past two decades on a global scale.

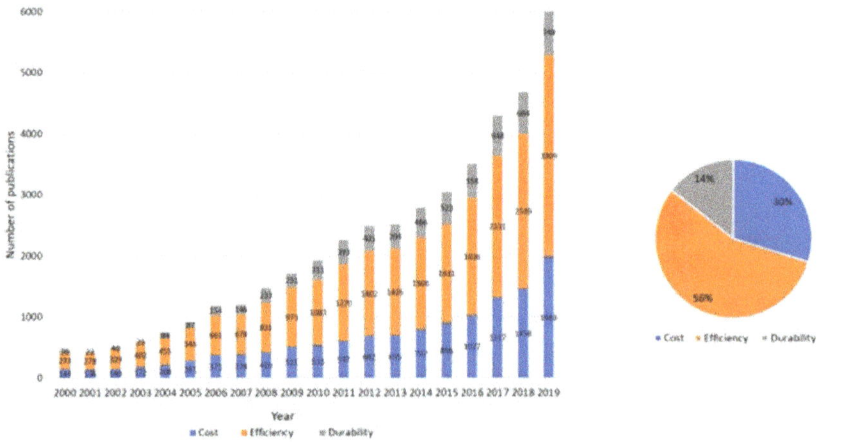

Figure 3.1. Yearly publications on hydrogen generation for the last two decades.

Physicists and material scientists conduct extensive studies in the field of solid state sciences, but chemists' contributions to this field have a unique identity. There has been a worldwide resurgence in solids chemistry and physics in recent years. The availability of new inorganic and molecular solids, which advance our understanding of structure and chemical bonding and have significant practical implications in fields such as electronics, optics, catalysis, and superconductivity, is what propels this field's development. Modern solid state synthesis techniques and our growing understanding of chemical interactions involving solids have enabled the development of these novel materials. The exponential growth of this discipline over the past 25 years is proof that solid state sciences have had a greater and broader influence on more recent branches of research and industry. Solid state physics, solid state chemistry, materials science, ceramic engineering, mineralogy, and metallurgy are just a few of the areas of interest and overlap in this field. Of all the solid state sciences, solid state chemistry is the most important since it focuses on the synthesis, structures, characteristics, and uses of solid materials. The idea of solid–solid interaction has undergone a great deal of fundamental research over the years, particularly in the 1960s, as solid devices began to appear. Interest in solids studies has been sparked by applications [4, 5]. Their importance has expanded due to their expanding uses in fields such as polymer chemistry, ceramic technology, laser chemistry, artificial gem manufacturing, geochemical processes, and propellants. One of the main factors influencing solid state chemistry research has been the use of solid fast ions conductors, which have gained attention due to their potential applications as solid electrolytes in a range of electrochemical devices [6–9]. Compared to reactions in fluids, where intimate contact between the reactant molecules is a natural result of the kinetic character of the reactants, the access that the reactants have to one another in solid state processes is far more limited. In solid state reactions, the diffusion of at least one reactant into the other is necessary for the initiation and growth of the reaction. The impacts of the lattice structure and diffusion mechanism are thus the primary factors that set chemical reactions in solids apart from those that take place in liquid or gaseous phases [10]. In order to create new bonds and create products, whether it be a change in the crystal structure or the production of a solid with a different chemical composition [11], previous connections must be broken. Two phases are involved in solid state reactions. The production of the product's nuclei occurs during the first stage of the reaction, while the expansion of the product layer occurs during the second. Although nucleation is a challenging process, the next phase, which involves the growth of the product layer, is more challenging. Counter diffusion of ions must take place directly through the current product layer for further reaction to take place and for the product layer to thicken.

In this review, we will primarily focus on the application of copper-based semiconductors as photoelectrodes. We will first provide a thorough explanation of the water splitting process before highlighting the importance of Cu oxides. We review the synthesis processes used to create these materials, taking into account all photoelectrochemical modifications that might be made to improve the performance of the finished photoelectrode.

3.1.1 Detailed process of PEC water splitting

When PEC is exposed to radiation, significant processes such photon absorption, charge separation, and the extraction of electrons and holes to the photoelectrodes for the H_2 and O_2 evolution reactions take place. The system's overall performance is significantly affected by each stage. It is useful to look at each phase separately in order to comprehend the system.

3.1.1.1 Photon absorption of the photoanode

The morphology, dimensionality, and crystallinity of the semiconductor with a suitable band gap should be designed to promote light absorption and boost the photocurrent in order to maximize energy conversion efficiency. The equation for the absorption coefficient can be used to calculate a crucial parameter, the semiconductor's band gap. The absorption coefficient (units: m^{-1}) can be expressed as follows:

$$\alpha = A\left(h\nu - E_g\right)^m / h\nu, \tag{3.1}$$

where A is a constant and m depends on the nature of the optical transition: $m = 1/2$ for a direct band gap, and $m = 2$ for an indirect band gap. With reference to equation (3.1), the extrapolation of the linear region of $(\alpha h\nu)^{1/2}$ versus $h\nu$ gives the material's indirect band gap, whereas the extrapolation of the linear region of $(\alpha h\nu)^2$ versus $h\nu$ gives the direct band gap of the material. Such a plot is known as a 'Tauc's plot'. Indirect band gap semiconductors often have lower absorption coefficients than direct band gap semiconductors; the difference between the two can be up to two orders of magnitude or more. Consider the case of indirect semiconductors, such as Si and haematite. Due to a momentum mismatch, it is not possible to move from the top of the valence band, E_v, to the bottom energy of the conduction band, E_c. As a result, it is inherently wasteful to absorb photons with energy near to the band gap since doing so necessitates interactions between electrons and phonons, which are the vibrations of the crystal lattice [6]. Three geometric parameters must be taken into account when tailoring the photoanode materials to promote hole transfer to the electrode/electrolyte interface. First, the photoelectrode's thickness must be sufficient to absorb photons with energies greater than the material's band gap. Second, the width of the space charge area (W_{sc}), and third, the minority carrier diffusion length (L_p), which is influenced by the lifetime and mobility of the hole, are significant. If the holes are produced too far away from the surface, they are more likely to recombine with electrons than to be utilized for oxidizing water. The performance of the majority of water-stable photoanode materials is severely constrained by the mismatch between the absorption depth (long) and charge diffusion distance (long). This issue can be effectively solved by separating the two channels using anisotropic devices with nanoscale-specific diameters. The usual examples that have been examined thoroughly in the literature are high aspect ratio nanowires or nanorods. The nanoscale spacing across the diameters (smaller than L_p) for these nanostructures allows for effective charge separation while the longitudinal length scale of micrometres ensures adequate light absorption.

3.1.1.2 Charge separation

To understand the charge separation in the PEC system, we must become familiar with the energetics involved in the semiconductor/electrolyte interface.

Energetics of the semiconductor/electrolyte interface

In a perfect crystal there are two different sorts of energy bands: permitted bands, which have filled energy levels, and band gaps, which have empty energy levels. As seen in figure 3.2, the lower almost filled band for a semiconductor is known as the valence band, while the higher empty band is known as the conduction band. The valence band edge, conduction band edge, and Fermi level all serve to identify the energy levels of semiconductors. The Fermi level, which also denotes the chemical potential of electrons in the semiconductor, reflects the equilibrium distribution of carriers in the bands. The strength of the chemical bonds determines the width of the band gap. Silicon possess a band gap of about 1.12 eV at 300 K. A band gap reduction of more than 100 mV is linked to a dopant concentration of 10^{19} cm^3 [7]. Heavy doping, larger than silicon, reduces the effective band gap.

According to the band structure, the electrons of the basic substance in the crystal can produce intrinsic conductivity, whereas electrons of impurity atoms or the existence of defects can produce extrinsic conductivity. In intrinsic semiconductors, the production of current carriers results from the thermal excitation of some electrons from the valence band to the conduction band and the accompanying thermal rupture of some chemical bonds. At the same time, the valence band develops an equal number of positively charged holes. These holes act like positive charge particles with an electric charge that is equivalent to the electron's charge in absolute terms when they are in an electric field. Impurities and defects in extrinsic semiconductors are categorized as either acceptors or donors based on their energy levels, which are situated in the band gap, as depicted in figure 3.2. Electron conductivity (n-type semiconductors) is produced by donors, which are often found at energy levels just below the conduction band. Donors do this by donating extra electrons to the conduction band. The basic substance's valence electrons are

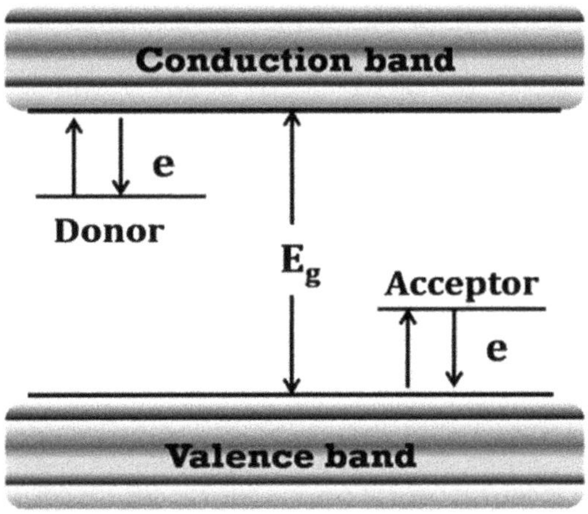

Figure 3.2. Schematic energy band diagram for a semiconductor.

captured by acceptors, which are positioned at energy levels just above the valence band, resulting in hole conductivity (p-type semiconductors).

Equilibrium between an electrolyte and a semiconductor is reached when the Fermi levels in the two phases are equal, or when $E_f = E_{redox}$. Electrons will flow from an n-type semiconductor to the solution phase if the semiconductor's E_f is higher than that of the solution. The surplus charge that results in the solid semiconductor is spread in an area close to the surface known as the space charge zone rather than remaining at the surface as it would in a metal. A bending of the bands demonstrates the ensuing electric field in the region of the space charge. The bands are bent upward (with respect to the level in the bulk semiconductor) when the semiconductor is positively charged relative to the solution. A double electric layer is instead created between the solid surface and solution by an ionic layer on the solution side. To understand better, let us examine the following key concepts involved in the process of photoelectrochemical water splitting.

Semiconductor band bending and the built-in electric field

In a PEC system, the semiconductor/electrolyte interface forms a junction similar to a Schottky diode. The built-in electric field due to charge distribution within the semiconductor and at the semiconductor/electrolyte interface is the driving force that separates photoexcited carriers. The phenomenon can be understood using the model developed by Gerischer [8] that describes the relationship between the solution electrochemical potential and the solid state Fermi level. These electric fields will be further elaborated after introducing the concept of band bending.

The band bending was first proposed by Schottky and Mott in their study of metal–semiconductor contacts, which is an important concept for understanding the photo-excitation process of semiconductor materials and the development of highly efficient photoactive materials [9]. The energy band diagrams of an n-type semiconductor and the electrolyte is shown in figure 3.3 [10]. For an n-type semiconductor, the Fermi level is close to its conduction band minimum (CBM) and higher than the electro-chemical potential of the electrolyte. Since the semiconductor's Fermi level and the

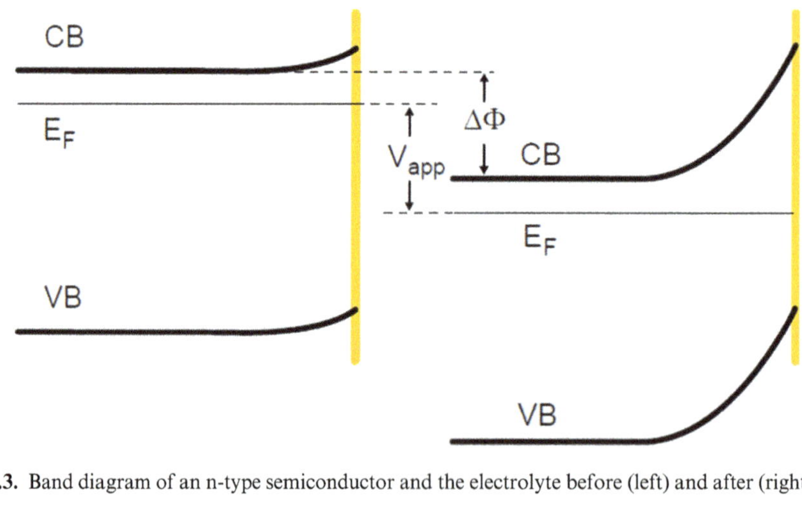

Figure 3.3. Band diagram of an n-type semiconductor and the electrolyte before (left) and after (right) contact.

electrochemical potential of the electrolyte are aligned after contact, the majority of electron carriers will go from the semiconductor to the electrolyte. The shift in carrier concentration also causes the semiconductor's energy band edges to move. Positive charges are created by the ionized donors remaining inside the semiconductor, whereas negative charges are concentrated at the contact with the electrolyte. Consequently, at the interaction, an electric field develops. This field maintains a dynamic equilibrium by counterbalancing the ongoing passage of charges.

The space charge area is also frequently referred to as the depletion region due to the depletion of electrons within it. The depletion width, also known as the space charge region width, or W_{sc}, is a crucial statistic used to describe the typical length of this region. W_{sc} depends on the doping density N, the relative permittivity, ε, and the potential drop across the space charge region, $\Delta\Phi_{SC}$, which corresponds to a band bending of $q\Delta\Phi_{sc}$, as follows:

$$W_{sc} = (2\Delta\varphi_{sc}\varepsilon\varepsilon_0/qN)^{1/2}. \tag{3.2}$$

W_{sc} decreases with increasing N doping density. A larger W_{sc} value is obtained by increasing the band bending degree. The Helmholtz layer typically has a characteristic length scale of few nanometres within the electrolyte, but the W_{sc} might range from several microns to several nanometres.

Upon irradiation, electrons are excited from the valence band to the conduction band. As the majority carrier, the change of electron concentration is relatively small, and the quasi-Fermi level of electrons remains unchanged in comparison with that in the dark. As the minority carrier, the concentration of holes increases dramatically, leading to a significant separation of its quasi-Fermi level from that in the dark. Note that the same discussion is also valid for p-type semiconductors (photocathodes), although the polarity should be reversed (i.e. the direction of band bending as shown in figure 3.3 would be opposite for the photocathode).

3.1.1.3 Surface state induced band bending

An electron has a 0.5% chance of occupying a given energy level, which is known as the Fermi level. The Fermi level is the energy that separates the occupied states from the unoccupied ones for a metal at $T = 0$ K (whose density of states is continuous). The band gap divides the occupied (valence band) and unoccupied levels (conduction band) in semiconductors. At $T = 0$ K, the Fermi level for an intrinsic semiconductor is located in the middle of the band gap. The Fermi level goes towards the conduction band maximum (CBM) as relative electron concentration increases, and towards the valence band maximum (VBM) as relative hole concentration increases. The Fermi level at the surface is desired to be free floating in relation to the band edge positions for a semiconductor/electrolyte system so that it can track the electrochemical potential of the electrolyte for maximum band bending upon contact. However, non-ideal factors such as surface states or chemisorption often result in states within the gap that pin the Fermi level.

Due to the termination of the lattice periodicity, surface states may be electronic states on the surface. They might also be brought on by the adsorption of chemical species such as water molecules (and their derivatives, such OH and H^+). The Fermi level (E_f), which is where there is a 50% probability of occupation by electrons, is situated in the middle of the band gap for intrinsic semiconductors (undoped), since the VB is filled with electrons (100%), while the CB is empty (0%), and as a result, the semiconductor's Fermi level is there. In order to keep things simple, the mid-energy gap is used as the Fermi level for the surface state (E_s). There is no charge transfer between the bulk and surface of undoped semiconductors since their Fermi level is identical to that of their surface state. In semiconductors that have surface impurities doped, the Fermi level is moved near to either the CB for n-type semiconductors or the VB for p-type semiconductors. The formation of an upward band bending (depletion layer formation) or a downward band bending (accumulation layer formation) for n-type and p-type semiconductors, respectively, is caused by the flow of electrons from the bulk to the surface or from the surface to the bulk due to the energy difference. This process continues until the Fermi levels of the semiconductor and the surface state become aligned (at the equilibrium state).

3.1.1.4 Absorption induced band bending

An acceptor molecule (A) will be absorbed by an n-type semiconductor when it is submerged in aqueous electrolyte. The electron flow from the semiconductor to the molecule causes upward bending as the LUMO of the acceptor shifts downward as the molecule approaches the semiconductor surface until equilibrium is attained. At the same time, a Helmholtz layer forms on the semiconductor surface with a space distance of d (about 2–5 angstroms), which is the area between the particularly adsorbed ions and the closest ions in the solution. At 25 °C, V_H varies by −59 mV per unit pH, indicating the potential drop across this layer.

3.1.1.5 Applied bias induced band bending

Applying a voltage between the working electrode and the reference electrode in a PEC cell can cause band bending. The space charge layer and the Helmholtz layer will receive the potential difference. Any adjustment to the applied bias will be distributed across the semiconductor's depletion layer since V_H will remain constant. When a positive bias is put on an n-type semiconductor, it will result in an increase in the depletion layer and an increase in the upward band bending. However, by using a negative bias, the amount of band bending can be diminished or even completely avoided when a strong enough negative bias is used. Similar to this, a p-type semiconductor and a reference electrode require a negative bias to increase the depletion layer.

3.2 Copper oxide based materials

Photoelectrochemical water splitting is typically one of the most promising processes for generating renewable hydrogen, which converts sunlight into hydrogen in one step under benign conditions [12–15]. The oxygen-evolution reaction (OER) serves as the anodic process in photoelectrochemical water splitting (PEC-WS)

$$H_2O \rightarrow 1/2O_2 + 2H^+ + 2e^-, \quad (3.3)$$

while the hydrogen evolution reaction (HER) is the cathodic one:

$$2H_2O + 2e^- \rightarrow H_2 + 2OH^-. \quad (3.4)$$

The following characteristics should be present in an effective semiconductor:

- A high yield of excited states inside the semiconductor can be generated with a significant amount of sunlight absorption.
- A suitable band gap energy (E_g) to enable sunlight absorption.
- Effective charge separation that guarantees high quantum efficiency and prevents recombination.
- Proper band positioning in relation to the appropriate half-reactions' equilibrium potentials.
- High stability and photostability.

The solar spectrum that can be absorbed by semiconductors that can carry out operations (3.1) and (3.2) without photodegradation is frequently constrained by their wide band gaps (for instance, TiO_2 with a 3.2 eV band gap can only absorb in the UV region) [16]. In the research for suitable photocathodes, Cu_2O is one of the most studied materials since:

- The band gap is found to be 2.17 eV [17]. The energy required to drive water electrolysis by visible light absorption is low enough at this value.
- It offers appropriate band positions that permit both the HER and the OER [18].
- It is made of abundant and low-cost elements.
- Since it is non-toxic, industrialization can be done more easily. Compared to PEC-WS, which contains As, Cd, and other hazardous metals, and to other semiconductors, this is advantageous.
- It is easily and consistently synthesized using a variety of techniques, including electrodeposition.

With a conduction band (CB) edge potential, E_c, of about 1.16 V, which is significantly higher than the energy corresponding to the H^+/H_2 couple (0.65 V) and E_v (about +1.0 V), and only slightly lower than the water oxidation potential (+0.82 V) at pH = 7, the band position in Cu_2O satisfies the aforementioned requirements. CuO, the equivalent Cu(II) oxide, frequently co-exists with Cu_2O when being co-synthesized for the creation of Cu-based photoelectrodes, despite Cu_2O being the most promising Cu-based semiconductor [21]. CuO is a p-type semiconductor like Cu_2O (because of the presence of Cu vacancies), has a monoclinic structure with the space group C2/c [22], and has a band gap of roughly 1.7 eV [23]. CuO is more conductive than Cu_2O [24, 25]. CuO is a promising photoactive substance, especially for the oxidation of organic contaminants. When combined with Cu_2O, CuO has synergistic effects on the CO_2 reduction reaction and water splitting. CuO has specifically been recommended frequently as an overlayer for Cu_2O to promote electron transit towards the electrolyte, reducing the risk of charge recombination, and extending the lifetime. This combination produced remarkable photocurrents up to 19.12 mA cm^2 at 1 V versus RHE when employed as a photocathode [26]. Due to the low band gap energy of CuO, Cu_2O/CuO complexes have a longer absorption spectrum than pure Cu_2O. While the latter has an absorption edge at around 600 nm, the former has a larger absorption in the visible range, up to near-infrared (NIR), whose edge is at about 900 nm [27]. This system can be partly hydrogenated to provide a thin layer of $Cu(OH)_2$ and increase activity [28], or a carbon-based coating can be deposited to reduce charge recombination and increase charge transfer [29]. It is interesting to note that given operating conditions, a CuO/Cu_2O composite can be created by first partly reducing pure CuO. This material exhibits a good photocurrent for 6 h at 0.35 V (RHE) [30]. First, a similar technique conducted without the fluorine doped tin oxide (FTO) coated glass support produced a CuI/CuO core/shell powder with good faradaic efficiency and photocathodic and photoanodic characteristics [31]. It is worth noting that this material's electrical properties are pH-tunable because the band edges of the two components have different pH dependences [32].

3.3 Preparation methods

In the literature many application-specific preparation techniques for photoelectrochemical water splitting are reported. Below are brief descriptions of the most reliable methods.

Stability, quick charge carrier transfer, and separation are crucial for improving photocatalyst efficiency. Researchers have focused on semiconductor-based heterostructures in order to support the expansion of photoabsorption into the infrared region of the solar spectrum, with charge carrier separation and quick transportation. A single material called a heterostructured nanocomposite is created by combining two or more semiconductor nanoparticles [33, 34]. To create a heterostructured nanocomposite, two or more procedures are needed, such as the integration of a separate semiconductor as a co-catalyst on the surface of the bare photocatalyst or shell formation to cover the bare photocatalyst [35–38]. The connection between the photocatalyst and co-catalyst considerably aids in the decrease of electron–hole pair recombination in the heterostructure. Recent research has shown some novel photocatalysts with increased photocatalytic activity, including those built on heterostructured nanocomposites of $CuWO_4$, $CoWO_4$, and Ag_2WO_4 [39–41]. Noble metals (Ag, Au, Pt, Pd, Ir, Rh, and Ru), non-metals, and a few binary oxides have also been utilized as co-catalysts to alter photocatalyst performance and lengthen the life of the charge carrier. In an effort to extend the life of charge carriers and lessen the demand for noble metals, researchers have recently worked to create heterostructure nanocomposite materials without noble metals [42–46]. Co-catalysts based on copper (Cu) (Cu, $Cu(OH)_2$, Cu_2O, and CuO) have drawn particular attention for their role in the oxidation of organic contaminants and the creation of photocatalytic hydrogen [47–50].

Cu-based ternary oxides ($CuWO_4$, $CuRhO_2$, $CuRh_2O_4$, $CuNbO_3$, $CuNb_2O_6$, and Cu_3VO_4) are stable photocatalysts in aqueous electrolytes with neutral to basic conditions [37, 51–56]. Inevitably, copper oxide (CuO), which exists as a type-II heterostructure compound, was generated as an island-like nanoscale layer or as a thin layer on the surface of the ternary oxides during the particle formation. The type-II heterostructured compounds are crucial for PEC applications because they contain CuO on the surface of Cu-based ternary oxides. Obviously, the potential of the CuO was considered to be an impurity when preparing Cu-based oxides in the past. Thus it was removed by washing with soluble agents (HCl or NH_4OH) to obtain the bare compound. However, a greater population of CuO decreased the photoconversion efficiency due to the suppression of the light absorption of the core. To allow more light absorption, the sensitizer CuO should be very thin or exist as island-like particles on the surface of the core. It can be achieved only by the solid state reaction, in comparison to the solution process, since the loss of copper is at the trace level. Similarly, various copper (II) based ternary metal oxides such as $CuWO_4$ [57–61], CuV_2O_6 [62–64], $CuNb_2O_6$ [65], $CuFe_2O_4$ [66–68], and $CuBi_2O_4$ [69–74], and copper (I) based ternary metal oxides such as $CuRhO_2$ [75], $CuNbO_3$ [55, 76], and $CuFeO_2$ [77–80] were synthesized using the solid state reaction method in a stoichiometric ratio using their respective binary oxide phases.

Since the energy input for the synthesis itself must be reduced to a minimum, it is evident that this will shorten the time it takes for the energy to be paid back. When creating a substance that can harvest energy, this is true. Solid state (i.e. ceramic) synthesis techniques are limited in this regard since they need to operate at relatively high temperatures. However, techniques such as solution combustion synthesis are interesting because the thermal energy required for the synthesis is already present in the heat of the combustion of the fuel:oxidizer precursor combination.

3.4 Synthesis and structural aspects

Ternary oxides, such as delafossites, have a quasi-two-dimensional layered superlattice structure. Linear O–A–O dumbbell layers and oblique BO6 octahedra layers are placed alternately in this arrangement [81]. Because of their tiny band gap energy, high onset potential, and strong stability in aqueous electrolytes, delafossite materials based on copper have received a lot of attention as prospective photoanode and photocathode materials for the production of solar fuel. Table 3.1

Table 3.1. Ternary oxides of copper synthesized through a solid state reaction method.

S.No.	Ternary oxide	A	B	A:B	Crystal system	References
1	$CuBi_2O_4$	CuO	Bi_2O_3	1: 1	Tetragonal	[45–56]
2	$CuFe_2O_4$		Fe_2O_3	1: 1	Tetragonal	[44, 67–69]
3	$CuNb_2O_6$		Nb_2O_5	1: 1	Monoclinic	[74]
4	$CuCr_2O_4$		Cr_2O_3	1: 1	Tetragonal	[44, 60–66]
5	$CuGa_2O_4$		Ga_2O_3	1: 1	Cubic	[70]
6	CuV_2O_6		V_2O_5	1: 1	Triclinic	[76, 77]
7	$CuWO_4$		WO_3	1: 1	Triclinic	[85–93]

summarizes the various copper-based ternary oxides synthesized using the solid state reaction method in a stoichiometric ratio.

$CuWO_4$ belongs to the triclinic crystal structure that resembles the distorted wolframite structure with space group P-1. It exhibits more stable photocurrent (remaining at 85% for up to 12 h) in neutral and basic electrolyte solution. Visible light-active Cu_2O/WO_3 nanocomposite was prepared by Ali *et al* in 2021 via a direct Z-scheme splitting method [82]. In 2016 Ma *et al* a synthesized $CuWO_4/Zn_{1-x}Cu_xWO_4/ZnWO_4$ hybrid photocatalyst using a one-port precipitation-calcination method with doping of Zn ions in various concentrations (0, 0.1, 0.5, 1, and 5 wt%) [83]. Likewise, $CuWO_4$ based photoanodes were synthesized by various methods which can facilitate charge transfer faster than its respective binary oxides [57–59]. Currently, n-type $CuWO_4$ with a suitable band gap of 2.2–2.4 eV is a desirable and promising ternary metal oxide for photoanodes. $CuWO_4$ has an excellent band gap property for water splitting in a neutral pH electrolyte because the hybridized Cu-3d and O-2p orbitals contribute to the valence band maxima. However, no discernible increase in $CuWO_4$ photocatalyst for the oxidation of water is seen. The carrier mobility may suffer as a result of the midgap states that the empty Cu (3d x2-y2) introduces. Therefore, it is important to enhance the transport characteristics and electron transfer kinetics of $CuWO_4$, either by changing the structure, doping, or fabricating a heterojunction. It has been demonstrated recently that hydrogen treatment can significantly enhance the PEC performance of TiO_2, WO_3, and $BiVO_4$ photoanodes due to the enhanced donor density and electrical conductivity brought about by the introduction of a minor amount of oxygen vacancies. Therefore, it is anticipated that $CuWO_4$ will respond similarly to hydrogen treatment [60, 61].

Metal niobates are desirable candidates in the search for photocathode materials because of their exceptional electrochemical properties. $CuNb_2O_6$ has a monoclinic crystal structure and exhibits p-type semiconductor behaviour. Six oxygen atoms circle each Cu:Nb atom centre in the $CuNb_2O_6$ structure, forming zigzag chains of CuO_6. Edge sharing connects the Jahn–Teller-distorted octahedral NbO_6 structure [84]. According to the reaction temperatures, $CuNb_2O_6$ often exhibits two polymorphs. Temperatures below 740 °C yield the monoclinic polymorph, whereas temperatures above 740 °C yield the orthorhombic form. It is interesting to note that p-type $CuNb_2O_6$ demonstrated promising CO_2 PEC reduction activity [85, 86]. Columbite $CuNb_2O_6$ anode materials were created by Li *et al* in 2021 using the solid state reaction technique [87]. For use in Li-ion batteries, $CuNb_2O_6$, $ZnNb_2O_6$, and $MnNb_2O_6$ nanoparticles were synthesized by Lee *et al* in 2020 [88]. Using the solid state reaction approach, in 2018 Kamimura *et al* examined the impact of two $CuNb_2O_6$ polymorphs on the photocatalytic activity and its solar H_2 evolution process [91]. In 2016 Kormanyos *et al* [90] reported on the solution combustion synthesis, characterization, and photoelectrochemistry of $CuNb_2O_6$ and $ZnNb_2O_6$ nanoparticles. Interestingly, the Cu and O vacancies in $CuNb_2O_6$ photocathodes encourage the holes to move freely in the valence band, leading to high transparency, high carrier mobility, and a more favourable flat-band potential. In 2013 the Maggard research group proposed that the band gap might be reduced from

3.9 eV to 1.3 eV due to filled $3d^{10}$ copper orbitals, which results in a higher energy valence band through the solid solution synthesis of $Li_{1-x}Cu_xNb_3O_8$. The band gap energy of $NaCu(Ta_{1-x}Nb_x)_4O_{11}$ is regulated by increasing the substitution of lower energy Nb 4d orbitals [65].

Researchers discovered that metal ferrite (MFe_2O_4) nanoparticles are visible light-active photocatalysts that have been studied extensively for water purification and the removal of various organic contaminants from industrial effluent. In addition, MFe_2O_4 are typical spinel-type nanoparticles and possess a lower band gap. $CuFe_2O_4$ nanoparticles were prepared using a solid state reaction, co-precipitation, and the sol–gel method [67, 68]. The indirect band gap of $CuFe_2O_4$ is between 1.54 and 1.95 eV. In principle, it has a maximum photocurrent density of around 27 mA cm^{-2} and a maximum solar to hydrogen conversion efficiency of about 33%. It is used as a suitable p-type photocathode in PEC water splitting. The high energy CB and indirect band gap of $CuFe_2O_4$ offer the photo-electrons a potent reducing ability and are located close to the optimal wavelength of the Sun. The constituent elements Cu and Fe are the most ideal when considering toxicity, and are also inexpensive.

$CuBi_2O_4$ is a p-type semiconductor with a number of advantageous characteristics as a photocathode for solar water splitting. First, a band gap of 1.5–1.8 eV is said to exist. It may therefore utilize a sizable portion of the visible spectrum. Its conduction band minimum (CBM) is projected to be at a greater negative potential than the thermodynamic potential for water reduction, allowing solar H_2 to be produced. Third, its valence band maximum (VBM) is located at a much higher positive potential than the p-type Si and p-type Cu_2O, which are currently considered to be the most promising photocathodes. As a result, it has the ability to create H_2 with a photovoltage (i.e. the difference between the thermodynamic reduction potential of water and the photocurrent onset potential) more than 1 V. It also has the potential to have a more positive flat-band potential (>1.0 V versus RHE). $CuBi_2O_4$ has previously been made as a powder-type photocatalyst via metal–organic decomposition, solid state synthesis, the sol–gel method, hydrothermal synthesis, and solid state synthesis. $CuBi_2O_4$ was identified as a promising photocathode material in 2007 through the Sayama group's combinatorial chemistry research [91]. Since then, there have not been many investigations on $CuBi_2O_4$ as a PEC material, particularly when compared to metal oxides such as TiO_2, Fe_2O_3, $BiVO_4$, and Cu_2O, that have been studied more frequently. Table 3.1 summarizes the crystal systems of these various copper-based ternary metal oxides, and figure 3.4 shows a schematic representation of them.

3.5 Photoelectrochemical aspects

In-depth research is being conducted on PEC cells, which provide a means of converting solar energy into a chemical form that can be stored (molecular bonds). For instance, using PEC cells to power the water splitting reaction that yields H_2 could be a clean and sustainable approach to convert solar energy into chemical fuels. As a result of Fujishima and Honda's and others' discoveries of catalytic water

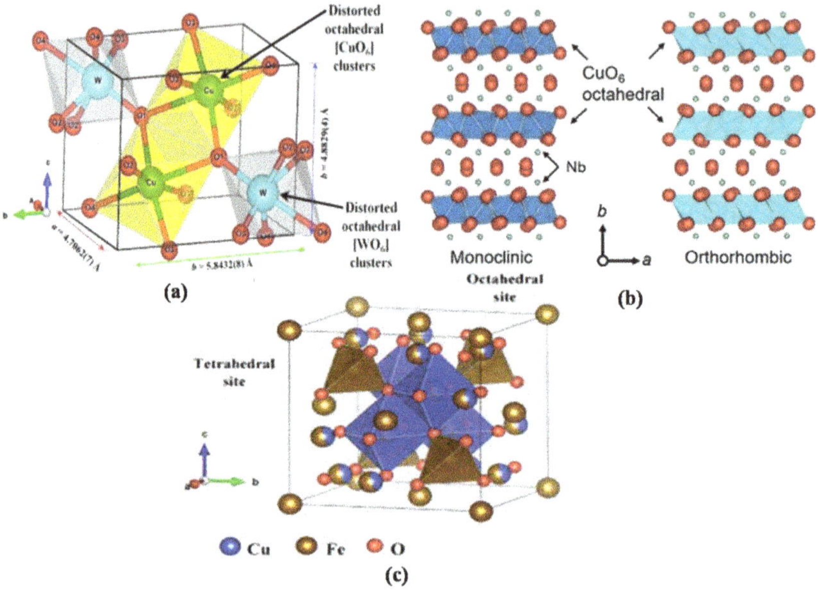

Figure 3.4. Schematic representations of (a) $CuWO_4$, (b) $CuNb_2O_6$, and (c) $CuFe_2O_4$. ((a) Reproduced with permission from [109]. Copyright 2017 Elsevier.)

splitting on TiO_2 photoelectrodes, much research has been put into creating high-efficiency metal oxides that can be used as photocatalysts for this reaction. The development of highly catalytic, chemically stable anodes for the OER is necessary to increase the performance of PEC cells. The water oxidation half reaction process is limited by the requirement for four protons and four electrons to couple in two steps to generate O_2.

Copper-based delafossite materials ($CuMO_2$; M = Fe, Rh, Cr, Al, and Ga) have drawn a lot of interest as prospective cathodic photomaterials for the production of solar fuel due to their tiny band gap energy, high onset potential, and exceptional stability in aqueous electrolytes [81]. However, due to weak charge transport properties, photoelectrochemical H_2 evolution efficiency is limited [79]. Copper-iron oxide ($CuFeO_2$) is one of the few oxides with naturally occurring p-type conductivity, and unlike copper oxide (Cu_2O), it is stable in reductive environments [78]. Under light, water can be reduced to hydrogen in $CuFeO_2$ conduction band, which is at about 0.4 V versus RHE. The most used synthesis method is the solid state reaction of iron and copper oxides at 800 °C in an oxygen-free atmosphere. The photoactivity of delafossite material has been suggested to be increased using a unique production technique. As a post-treatment, hybrid microwave annealing (HMA) was used to prepare the $CuFeO_2$ [79]. The citrate-nitrate sol–gel method was proposed by Prévot *et al* [78], in which metal nitrates were combined with citric acid, ethylene glycol, and ethanol. Due to the use of sacrificial electron acceptors and a beneficial photocurrent onset at +0.9 V versus RHE, a thin layer was successfully produced directly on FTO glass, revealing significant photocurrent densities (up to 1.4 mA cm^{-2}). The incident photon to current efficiency (IPCE) onset for this material is at λ = 830 nm, and it absorbs light over a wide wavelength range. Additional research revealed a significant surface state density, which may result in Fermi level pinning at the semiconductor liquid-junction (SCLJ) [99]. Both in the dark and in the light, these surface states function as electron traps, causing an inversion of the depletion layer upon filling and encouraging charge recombination at the surface. Additional interesting p-type copper-based delafossite photoelectro-catalysts are $CuCrO_2$, $CuAlO_4$, $CuGaO_2$, and $CuRhO_2$. The delafossite $CuCrO_2$ photocatalyst can be produced using both hydrothermal and solid state reaction methods, yielding materials that are extremely different from one another. Hydrothermal $CuCrO_2$ samples display greater purity, lower crystallinity, smaller particle size, and stronger photocatalytic activity when compared to samples created using a solid state method [100]. The electrodeposition of Cu^{2+} and Al^{3+} onto FTO glass, followed by thermal treatment, produced the $CuAlO_2$ photocathodes [101]. Under solar light, $CuAlO_2$ possesses the required electrical properties to create H_2. Additionally, it possesses a faradaic efficiency of roughly 70% at +0.3 V and a photocurrent onset potential of +0.9 V against RHE, both of which can be significantly improved by the addition of sacrificial hole scavengers. A hydrothermal method has also been used to create $CuGaO_2$ [81]. The $CuGaO_2$ electrode demonstrated high photocurrent densities, long-term stability, and hydrogen production potential in PEC tests. The $CuGaO_2$ electrode exhibits hydrogen production with a small overpotential (0.1 V) and long-term stability. Bocarsly's group described a $CuRhO_2$ material that is a more active photomaterial in oxygenated conditions [75] because of O_2-driven self-reparation. $CuRhO_2$ can undergo reductive deactivation when exposed to radiation, however, when oxygen is present, this regenerates the active photocatalyst once more. $CuRhO_2$ has a faradaic efficiency of about 80%, allowing H_2 to evolve. The conduction band edge of Cu-based ternary oxides areas follows, $CuFeO_2$: 1.5 eV–0.4 V versus RHE [78]; $CuCrO_2$: 2.4 or 3.7 eV (depending on defects) [100]; $CuAlO_2$: 1.4 eV–0.24 V versus RHE [101]; $CuGaO_2$: 1.4 eV [81]; $CuRhO_2$: 1.9 eV [75]; and $CuBi_2O_4$: 1.8 eV–0.3 V versus RHE [102]. Numerous other ternary oxides are also described in the literature as having rather high photocatalytic or photoelectrochemical activity. A new review summarizes several of these instances [103]. $CuNbO_3$, $Cu_2Ta_4O_{11}$, and Cu_3VO_4 are examples of copper (I)-based compounds, whereas $CuAl_2O_4$, $CuCo_2O_4$, $CuFe_2O_4$, $CuMn_2O_4$, CuV_2O_6, and $CuWO_4$ are examples of copper (II)-based oxides. It is also important to mention a few other materials. $CuBi_2O_4$ photocathodes made using a simple drop-casting technique were described by Berglund *et al* [102]. $CuBi_2O_4$ was said to have a modest optical absorption because, for light energies above the band gap value, there is a steady rather than abrupt increase in absorption, which lowers the photocurrent density. Poor charge carrier (mostly holes) conveyance is another constraint that drastically lowers the achievable photocurrent density. Copper ferrite was employed as a photocathode in a water splitting reaction because it is a p-type semiconductor. $CuFe_2O_4$ was directly created on FTO glass using the sol–gel technique and thermal treatment, yielding a photocurrent density of 1.82 mA cm^{-2} at 0.4 V versus RHE [104]. This discovery is substantially more encouraging when compared to several other copper oxide compounds. The reduced oxygen

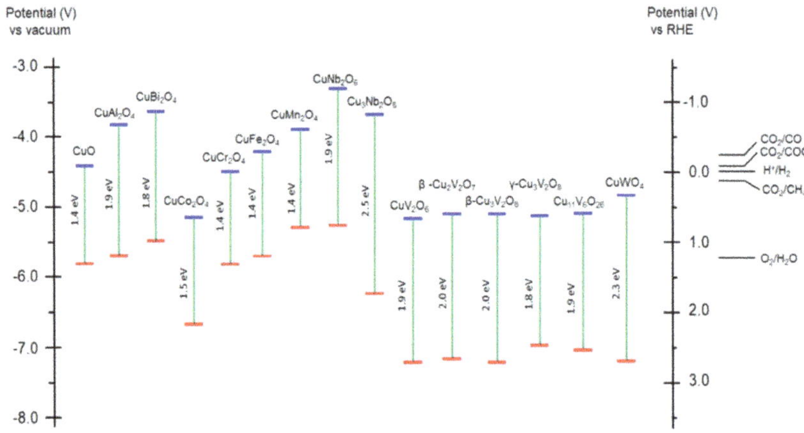

Figure 3.5. Band edge positions of Cu(II) based metal ternary oxides. (Reproduced from [103]. CC BY 4.0.)

vacancy concentration promoted charge transfer and enhanced interfacial charge transfer efficiency. Figure 3.5 depicts the band edge position of a copper-based ternary metal oxide nanocomposite. The n- and p-type semiconductors' conduction band and valence band boundaries are both close to the voltage that is beneficial for photoelectrochemical water splitting.

It is important to note that the lack of a framework for reporting PEC data or a figure-of-merit for photocatalysis makes it difficult to compare the efficacy of one photoelectrode (or photocatalyst powder) material to other candidates in a given application. In contrast, there are such standards and certified cell efficiencies for solar power devices. Reports on the effectiveness of ternary oxides in CO_2 photoreduction, water splitting, or pollutant degradation are still possible despite this limitation. One can rapidly ascertain the oxide material's photoelectrochemical quality using a photovoltammetry scan or by measuring the photocurrent at a specific bias potential in a suitable aqueous solution. It is envisaged that the OER, or (anodic) oxidation of solution species such as water, will produce the photocurrent for an n-type material. The proper water splitting reaction for the p-type material in a deoxygenated aqueous medium would be the hydrogen evolution reaction (HER). The photocurrent production of several copper-based metal ternary oxides is shown in table 3.2.

The concept that the source of the photocurrent, rooted in the OER or HER, depends on this property is supported by the idea that the extent of photoelectrochemical corrosion of the oxide semiconductor itself is assumed to be small in the specific medium. This needs to be thoroughly verified by collecting, measuring, and comparing the product to the amount expected from the charge conveyed (i.e. the current efficiency) for a product such as O_2 or H_2. This has not been done in the vast majority of the studies listed in table 3.2. The other important variables in these data are the medium pH and photon flux, both of which are mentioned in the compilation. When compared to the other compounds as a whole, $CuBi_2O_4$ is unquestionably superior and appears to have a greater photocurrent level. Cu-based

Table 3.2. Photoelectrochemistry of various copper-based metal ternary oxides toward water splitting. (Reproduced from [107]. CC BY 4.0.)

S. No.	Oxide	Electrolytes	pH	Light source and intensity	Photocurrent, $\|j\|/\mu A\ cm^{-2}$	Potential, V versus RHE	n- type/ p- type	Comments
1	$CuBi_2O_4$	0.3 M K_2SO_4 with 0.2 M phosphate buffer	6.65	Solar simulator	2500	0.6	p	—
2	$CuNb_2O_6$	0.1 M $NaHCO_3$ solution, bubbled	7.0	150 W tungsten–	~150	0.2	p	—
3	$Cu_3Nb_2O_8$	CO_2 0.5 M $NaHCO_3$ solution, bubbled	7.3	Solar simulator	180	0.3	p	—
4	CuV_2O_6	CO_2 0.1 M borate buffer with 0.1 M	9.2	150 W xenon lamp	~220	1.58	n	—
5	α-$Cu_2V_2O_7$	Na_2SO_3 M borate buffer with 0.1 M Na_2SO_3	9.2	385 nm light emitting diode, illumination power was 3.0 mW	600	1.0	n	PEC was performed in scanning droplet cell
6	β-$Cu_2V_2O_7$	0.1 M borate buffer with 0.1 M	9.2	150 W xenon lamp	~120	1.58	n	—
		Na_2SO_3 0.1 M borate buffer with 0.1 M	9.2	150 W xenon lamp	~40	1.0	n	—
7	β-$Cu_3V_2O_8$	Na_2SO_3 0.1 M borate buffer with 0.1 M	9.2	300 W Xe arc lamp	~100	1.0	n	—
		Na_2SO_3 0.1 M borate buffer with 0.1 M	9.2	300 W Xe arc lamp	~150	1.0	n	Mo-doped sample
		Na_2SO_3 0.1 M borate buffer with 0.1 M	9.2	300 W Xe arc lamp	~100	1.0	n	Cr-doped sample

(*Continued*)

Table 3.2. (*Continued*)

S. No.	Oxide	Electrolytes	pH	Light source and intensity	Photocurrent, $\lvert j \rvert/\mu A\ cm^{-2}$	Potential, V versus RHE	n- type/ p- type	Comments
8	γ-$Cu_3V_2O_8$	Na_2SO_3 0.1 M borate buffer with 0.1 M	9.2	150 W Xe arc	91	1.23	n	—
		Na_2SO_3 0.1 M borate buffer with 0.1 M	9.2	150 W Xe arc lamp	~53	1.0	n	—
9	$Cu_{11}V_6O_{26}$	Na_2SO_3 0.1 M borate buffer with 0.1 M	9.2	300 W Xe arc lamp	55	1.0	n	—
		Na_2SO_3 0.1 M borate buffer with 0.1 M	9.2	300 W Xe arc lamp	85	1.0	n	Mo-doped sample
		Na_2SO_3 0.1 M borate buffer with 0.1 M	9.2	300 W Xe arc lamp	145	1.0	n	W-doped sample
10	$CuWO_4$	Na_2SO_3 0.1 M potassium phosphate	7.0	150 W Xe arc lamp	200	1.23	n	—

oxide semiconductor stability problems have often been resolved by swiftly moving the photogenerated electrons to a second phase, such as carbon. Carbon has the benefit that the injected electrons delocalize in the π-electron manifold very quickly. There are also numerous nanocarbons that can enhance the performance of the photoelectrodes.

3.6 Application of copper oxide based materials in photoelectrochemistry

One of the most promising methods for producing and storing renewable hydrogen in the hydrogen economy of the future is photoelectrochemical water splitting. The only components needed for the process are water, sunlight, and a semiconducting material. It goes without saying that, in relation to the HER and OER, respectively, the photoanode and photocathode must have the correct band alignment and band gap. In the working mode, the semiconductor is immersed in a water-based electrolyte, with sunlight acting as the main energy source for the generation of charges inside the valence and conduction bands. These charges serve as the source of energy for the water splitting reaction. When exposed to light, photons with energies greater than E_g are absorbed by the substance, causing electrons in the VB to be excited to the CB and creating electron–hole pairs. Electrons and holes produced by photosynthesis can be forced to separate by an applied electric field. In doped semiconductors, band banding is produced during immersion in the electrolyte, which results in an internal potential gradient. This draws holes to the bulk semiconductor and then to the counter electrode in a p-type material while drawing electrons to the solution. By using an external bias, potential overpotentials can be avoided, band bending can be increased for efficient charge separation, or band edges that are not in the right place in regard to the semi-reactions of interest can be compensated for [105, 106]. There are some requirements for a PEC-WS system to work properly:

- The effective production of excited states in the semiconductor by the absorption of solar radiation.
- A high light-to-chemicals conversion rate and strong charge separation to prevent the electron–hole couple from recombining.
- Correct band positions from a thermodynamic and kinetic perspective relative to the HER and OER.
- High stability against photodegradation processes.

It is difficult to meet all of these needs. Numerous semiconductors that can carry out the necessary process without photodegradation have a large band gap, which affects the range of solar spectra that can be absorbed (for instance, TiO_2 with a 3 eV band gap can only absorb in the UV). Despite this, PEC-WS offers the benefit of having a reasonably straightforward and affordable procedure compared to other H_2 manufacturing approaches. It does not need expensive set-ups, is entirely powered by renewable resources, and produces no by products. If the temperature is above 0 °C, it can operate in a wide range of temperatures and without the need for warming up. The PEC reaction begins once all conditions at the photoanode and pathocathode are met. The hydrogen created in this way can be collected and stored for later use, such as in fuel cells or as a chemical in the oil industry. In addition to the planned electron transfer, there are other choices for holes and electrons. When the semiconductor is immersed in an electrolyte that contains a donor and an acceptor, light-excited electrons and holes are transported at the appropriate edges of energy bands. As a result, they could undergo radiative and nonradiative recombination as well as (reactive and non-reactive) trapping on surface locations. The efficiency of sunlight conversion to chemical compounds will be decreased by these alternative paths for the electron–hole couple.

The following issues with PEC-WS must be resolved in this system:

- Longer lifetime for photoelectrode components, resulting in reduced (photo) corrosion processes.
- From a thermodynamic perspective, it should possess a proper band edge.
- Fast kinetics for the HER/OER.
- A sufficient optical band gap for high photon efficiency and ideal light absorption.
- Fast charge transport in the bulk semiconductor.
- Low internal electrical resistance.
- Limited phenomena such as trapping and recombination.
- Low-cost materials with wide availability and reduced plant capital costs.

It is a practical choice for near-term energy storage given that PEC-WS technology is based on today's almost mature semiconductors. Although the cost of creating H_2 in PEC-WS systems is significant, current developments in the sector give promise for the ability of this technology to scale up in the future.

3.7 Conclusion

This review looked at and discussed the use of Cu-based compounds as photoactive materials in water splitting. Cu is a fairly common element, making it easy to make and work with its compounds. Utilizing Cu's oxidation state, the ability to synthesize ternary and quaternary oxides, and the ability to change the electrical characteristics of these materials through doping, it is possible to produce a semiconductor with the necessary attributes. This review can be concluded by stating that these findings enable the design of a Cu-based photocathode to be optimized through doping, the addition of under/overlayers, co-catalysts, and plasmonic effect-inducing particles, as well as synthesis (with a preference for low-cost, easily scalable methods). This will also be a difficult task because there are so many different factors to take into account, including the type of precursor, the semiconductor synthesis method and its parameters, the underlayer synthesis method and its parameters, the overlayer synthesis method and its parameters, the co-catalyst synthesis method and its parameters, and, last but not least, the possible ways to build this complex architecture.

Acknowledgments

The author acknowledges financial support provided to the Department of Physics at Manonmaniam Sundaranar University in Tirunelveli by DST-FIST and UGC-SAP:DRS-II.

References

[1] Peh C K N, Gao M and Ho G W 2015 Harvesting broadband absorption of the solar spectrum for enhanced photocatalytic H_2 generation *J. Mater. Chem.* A **3** 19360–7

[2] Barham P J and Burkhard K 2019 Synthetic Photoelectrochemistry Angew *Chemie Int. Ed.* **59** 11732–47

[3] Yue M, Lambert H, Pahon E, Roche R, Jemei S and Hissel D 2021 Hydrogen energy systems: a critical review of technologies, applications, trends and challenges *Renew. Sustain. Energy Rev.* **146** 111180

[4] Hedvall J A 1953 The reactivity of solids *J. Chem. Educ.* **30** 638

[5] Hedvall J A 1963 *Reactions Fahigkeit. Fester Stoffe, Photolitho Print Reproduction* (Ann Arbor, MI: Edwards Brothers) pp 67

[6] Liu Q, Sun X and Wu W 1990 A new SO_2/SO_3 sensor with Ag_2SO_4-based electrolyte *Solid State Ion.* **40/41** 456

[7] Neelakanta P S, Abello J and Gu C 1992 Microwave reflection at an active surface imbedded with fast-ion conductors *IEEE Trans. Micro. J. Theory. Tech.* **40** 1028

[8] Neelakant P S 1993 *Electrical Engineering Handbook* (Boca Raton, FL: CRC Press) pp 173

[9] Moseley P T and Crocker A J 1996 *Sensor Materials* (Bristol: Institute of Physics)

[10] Wagner C 1938 The mechanism of the movement of ions and electrons in solids and the interpretation of reactions between solids *Trans. Faraday Soc.* **34** 851

[11] Lewis N S, Crabtree G, Nozik A J, Wasielewski M R, Alivisatos P, Kung H and Nault R M 2005 Basic research needs for solar energy utilization *Report of the Basic Energy Sciences Workshop on Solar Energy Utilization, 18-21. DOESC* (USDOE Office of Science)

[12] Arulraj A, Praveen Kumar M, Alejandra Tello Z and Mangalaraja R V 2023 Nanoarchitectonics of layered metal chalcogenides-based ternary electrocatalyst for water splitting *Energies.* **16** 1669

[13] Praveen Kumar M, Sasikumar M, Arulraj A, Rajasudha V, Murugadoss G, Rajesh Kumar M, Gouse Peera S and Mangalaraja R V 2022 NiFe layered double hydroxide electrocatalyst prepared via an electrochemical deposition method for the oxygen evolution reaction *Energies.* **12** 1470

[14] Arulraj A, Mehana Usmaniya U, Senguttuvan G, Sivakumar V and Khalid M 2021 Chalcogenides nanocrystals and its applications ed N M Mubarak, M Khalid, R Walvekar and A Numan *Contemporary Nanomaterials in Material Engineering Applications. Engineering Materials* (Cham: Springer) pp 201–27

[15] Arulraj A, Thachnatharen N, Rajkumar C and Mangalaraja R V 2022 Engineering 2D chalcogenides for energy and environmental remediation *Nanostructured Materials for Sustainable Energy and Environmental Remediation* (Bristol: IOP Publishing) pp 2-1–2-28

[16] Jiménez Reinosa J, Leret P, Álvarez-Docio C M, Del Campo A and Fernández J F 2016 Enhancement of UV absorption behavior in $ZnO–TiO_2$ composites *Bol. Soc. Esp. Ceram. Vidrio.* **55** 55–62

[17] Tang S J, Moniz S J A, Shevlin S A, Martin D J, Guo Z X and Tang J 2015 Unparalleled lithium and sodium superionic conduction in solid electrolytes with large monovalent cage-like anions *Energy Environ. Sci.* **8** 731–59
[18] Chen S and Wang L W 2012 Thermodynamic oxidation and reduction potentials of photocatalytic semiconductors in aqueous solution *Chem. Mater.* **24** 3659–66
[19] Jiang Y, Yuan H and Chen H 2014 Enhanced visible light photocatalytic activity of Cu_2O via cationic-anionic passivated codoping *Phys. Chem. Chem. Phys.* **17** 630–7
[20] De Jongh P E, Vanmaekelbergh D and Kelly J J 1999 Cu_2O: a catalyst for the photochemical decomposition of water? *Chem. Commun.* **12** 1069–70
[21] Siavash Moakhar R *et al* 2021 Photoelectrochemical water-splitting using CuO-based electrodes for hydrogen production: a review *Adv. Mater.* **33** 2007285
[22] Forsyth J B and Hull S 1991 The effect of hydrostatic pressure on the ambient temperature structure of CuO *J. Phys. Condens. Matter* **3** 5257–61
[23] Koffyberg F P and Benko F A 1982 A photoelectrochemical determination of the position of the conduction and valence band edges of *p*-type CuO *J. Appl. Phys.* **53** 1173–7
[24] Rakhshani A E 1986 Preparation, characteristics and photovoltaic properties of cuprous oxide—a review *Solid. State. Electron.* **29** 7–17
[25] Chaudhary Y S, Agrawal A, Shrivastav R, Satsangi V R and Dass S 2004 A study on the photoelectrochemical properties of copper oxide thin films *Int. J. Hydrog. Energy* **29** 131–4
[26] Aktar A, Ahmmed S, Hossain J and Ismail A B M 2020 Solution-processed synthesis of copper oxide (CuxO) thin films for efficient photocatalytic solar water splitting *ACS Omega* **5** 25125–34
[27] Yang Y, Xu D, Wu Q and Diao P 2016 Cu_2O/CuO bilayered composite as a high-efficiency photocathode for photoelectrochemical hydrogen evolution reaction *Sci. Rep.* **6** 1–13
[28] Jun Seo Y, Arunachalam M, Ahn K S and Hyung K S 2021 Integrating heteromixtured Cu_2O/CuO photocathode interface through a hydrogen treatment for photoelectrochemical hydrogen evolution reaction *Appl. Surf. Sci.* **551** 1–12
[29] Xiao C, Lei B, Jin W and Xu L 2021 Candle flame fabrication of carbon coated CuO-Cu_2O composite photocathode for photoelectrochemical water reduction *Mater. Lett.* **298** 130006
[30] Baran T, Wojtyla S, Lenardi C, Vertova A, Ghigna P, Achilli E, Fracchia M, Rondinini S and Minguzzi A 2016 An efficient Cu_xO photocathode for hydrogen production at neutral pH: new insights from combined spectroscopy and electrochemistry *ACS Appl. Mater. Interfaces* **8** 21250–60
[31] Visibile A, Baran T, Rondinini S, Minguzzi A and Vertova A 2020 Determining the efficiency of photoelectrode materials by coupling cavity-microelectrode tips and scanning electrochemical microscopy *Chem. Electro. Chem.* **7** 2440–7
[32] Baran T *et al* 2018 Reverse type I core–CuI /shell–CuO: a versatile heterostructure for photoelectrochemical applications *Electrochim. Acta* **266** 441–51
[33] Uddin M T, Nicolas Y, Olivier C, Toupance T, Servant L, Müller M M, Kleebe H-J, Ziegler J and Jaegermann W 2012 Nanostructured SnO_2–ZnO heterojunction photocatalysts showing enhanced photocatalytic activity for the degradation of organic dyes *Inorg. Chem.* **51** 7764–73
[34] Shi J 2013 On the synergetic catalytic effect in heterogeneous nanocomposite catalysts *Chem. Rev.* **113** 2139–81
[35] Pelletier F and Thiébaut B 2016 Improvement of noble metal based photocatalysts by spray pyrolysis processes *Johnson Matthey Technol. Rev.* **60** 39–54

[36] Wei Z, Kowalska E, Endo M, Wang K, Szczupak A M and Ohtani B 2017 Noble metal-modified octahedral anatase titania particles with enhanced activity for decomposition of chemical and microbiological pollutants *Chem. Eng. J.* **318** 121–34
[37] Zhang H, Yilmaz P, Ansari J O, Khan F F, Binions R, Krause S and Dunn S 2015 Incorporation of Ag nanowires in $CuWO_4$ for improved visible light-induced photoanode performance *J. Mater. Chem.* A **3** 9638–44
[38] Su R *et al* 2012 Promotion of phenol photodecomposition over TiO_2 using Au, Pd, and Au–Pd nanoparticles *ACS Nano* **6** 6284–92
[39] Gohari M S and Yangjeh A H 2016 Fabrication of novel magnetically separable visible-light-driven photocatalysts through photosensitization of Fe_3O_4/ZnO with $CuWO_4$ *J. Ind. Eng. Chem.* **44** 174–84
[40] Gohari M S and Yangjeh A H 2017 Fe_3O_4/ZnO/$CoWO_4$ nanocomposites: novel magnetically separable visible-light-driven photocatalysts with enhanced activity in degradation of different dye pollutants *Ceram. Int.* **43** 3063–71
[41] Pirhashemi M and Yangjeh A H 2017 Ultrasonic-assisted preparation of plasmonic ZnO/Ag/Ag_2WO_4 nanocomposites with high visible-light photocatalytic performance for degradation of organic pollutants *J. Colloid Interface Sci.* **491** 216–29
[42] Yi S S, Yan J M, Wulan B R, Li S J, Liu K H and Jiang Q 2017 Noble-metal-free cobalt phosphide modified carbon nitride: an efficient photocatalyst for hydrogen generation *Appl. Catal.* B **200** 477–83
[43] Chang K, Hai X and Ye J 2016 Transition metal disulfides as noble-metal-alternative co-catalysts for solar hydrogen production *Adv. Energy Mater.* **6** 1502555
[44] Mejía E, Luo S P, Karnahl M, Friedrich A, Chierlei S, Surkus A E, Junge H, Gladiali S, Lochbrunner S and Beller M 2013 A noble-metal-free system for photocatalytic hydrogen production from water *Chem. Eur. J.* **19** 15972–8
[45] Luo S P *et al* 2013 Photocatalytic water reduction with copper-based photosensitizers: a noble-metal-free system *Chem. Int. Ed.* **52** 419–23
[46] Li B, Hao Y, Zhang B, Shao X and Hu L 2016 A multifunctional noble-metal-free catalyst of CuO/TiO_2 hybrid nanofibers *Appl. Catal.* A **531** 1–12
[47] Xiao S, Liu P, Zhu W, Li G, Zhang D and Li H 2015 Copper nanowires: a substitute for noble metals to enhance photocatalytic H_2 generation *Nano Lett.* **15** 4853–8
[48] Ran J, Zhang J, Yu J, Jaroniec M and Qiao S Z 2014 Earth-abundant cocatalysts for semiconductor-based photocatalytic water splitting *Chem. Soc. Rev.* **43** 7787–812
[49] Yu J and Ran J 2011 Facile preparation and enhanced photocatalytic H_2-production activity of $Cu(OH)_2$ cluster modified TiO_2 *Energy Environ. Sci.* **4** 1364–71
[50] Moon G D, Joo J B, Lee I and Yin Y 2014 Decoration of size-tunable CuO nanodots on TiO_2 nanocrystals for noble metal-free photocatalytic H_2 production *Nanoscale* **6** 12002–8
[51] Chen H and Xu Y 2015 Photocatalytic organic degradation over W-rich and Cu-rich $CuWO_4$ under UV and visible light *RSC Adv.* **5** 8108–13
[52] Skorupska K, Maggard P A, Eichberger R, Schwarzburg K, Shahbazi P, Zoellner B and Parkinson B A 2015 Combinatorial investigations of high temperature CuNb oxide phases for photoelectrochemical water splitting *ACS Comb. Sci.* **17** 742–51
[53] Ismunandar, Kennedy B J and Hunter B A 1999 Phase transformation in $CuRh_2O_4$: a powder neutron diffraction study *Mater. Res. Bull.* **34** 135–43

[54] Guczi L, Lazar K, Boskovic G, Kis E and Putanov P 1991 Promoter Effects of Ca, Mo, Al, ON Fe/MgO catalyst on CO hydrogenation *Studies in Surface Science and Catalysis* **61** 251–56

[55] Zoellner B, Stuart S, Chung C C, Dougherty D B, Jones J L and Maggard P A 2016 $CuNb_{1-x}Ta_xO_3$ ($x \leq 0.25$) solid solutions: impact of Ta(v) substitution and Cu(i) deficiency on their structure, photocatalytic, and photoelectrochemical properties *J. Mater. Chem.* A **4** 3115–26

[56] Choi J, King N and Maggard P A 2013 Metastable Cu(I)-niobate semiconductor with a low-temperature, nanoparticle-mediated synthesis *ACS Nano* **7** 1699–708

[57] Benko F A, Maclaurin C L and Koffyberg F P 1982 $CuWO_4$ and Cu_3WO_6 as anodes for the photoelectrolysis of water *Mater. Res. Bull.* **17** 133–6

[58] Doumerc J P, Hejtmanek J, Chaminade J P, Pouchard M and Krussanova M 1984 A photoelectrochemical study of $CuWO_4$ single crystals *Phys. Stat. Sol.* **82** 285–94

[59] Yourey J E, Pyper K J, Kurtz J B and Bartlett B M 2013 Chemical stability of $CuWO_4$ for photoelectrochemical water oxidation *Phys. Chem.* C **117** 8708–18

[60] Yourey J E and Bartlett B M 2011 Electrochemical deposition and photoelectrochemistry of $CuWO_4$, a promising photoanode for wateroxidation *J. Mater. Chem.* **21** 7651–60

[61] Chang Y, Braun A, Deangelis A, Kaneshiro J and Gaillard N 2011 Effect of thermal treatment on the crystallographic, surface energetics, and photoelectrochemical properties of reactively cosputtered copper tungstate for water splitting *J. Phys. Chem.* C **115** 25490–5

[62] Guo W 2015 Synthesis and characterization of CuV_2O_6 and $Cu_2V_2O_7$: two photoanode candidates for photoelectrochemical water oxidation *J. Phys. Chem.* C **119** 27220

[63] Newhouse P F, Boyd D A, Shinde A, Guevarra D, Zhou L, Soedarmadji E, Li G, Neaton J B and Gregoire J M 2016 Solar fuel photoanodes prepared by inkjet printing of copper vanadates *J. Mater. Chem.* A **4** 7483

[64] Kim M W, Joshi B, Yoon H, Ohm T Y, Kim K, Al-Deyab S S and Yoon S S 2017 Electrosprayed copper hexaoxodivanadate (CuV_2O_6) and pyrovanadate ($Cu_2V_2O_7$) photoanodes for efficient solar water splitting *J. Alloys Compd.* **708** 444

[65] Kormányos A, Thomas A, Huda M N, Sarker P, Liu J P, Poudyal N, Janáky C and Rajeshwar K 2016 Solution combustion synthesis, characterization, and photoelectrochemistry of $CuNb_2O_6$ and $ZnNb_2O_6$ nanoparticles *J. Phys. Chem.* C **120** 16024

[66] Yang H, Yan J, Lu Z, Cheng X and Tang Y 2009 Photocatalytic activity evaluation of tetragonal $CuFe_2O_4$ nanoparticles for the H_2 evolution under visible light irradiation *J. Alloys Compd.* **476** 715

[67] Kezzim N, Nasrallah A, Abdi F and Trari M 2011 Visible light induced hydrogen on the novel hetero-system $CuFe_2O_4/TiO_2$ *Energy Convers. Manag* **52** 2800

[68] Uddin M R, Khan M R, Rahman M W, Yousuf A and Cheng C K 2015 Photocatalytic reduction of CO_2 into methanol over $CuFe_2O_4/TiO_2$ under visible light irradiation *React. Kinet. Mech. Cat.* **116** 589

[69] Hahn N T, Holmberg V C, Korgel B A and Mullins C B 2012 Electrochemical synthesis and characterization of p-$CuBi_2O_4$ thin film photocathodes *J. Phys. Chem.* C **116** 6459

[70] Xie Y, Zhang Y, Yang G, Liu C and Wang J 2013 Hydrothermal synthesis of $CuBi_2O_4$ nanosheets and their photocatalytic behavior under visible light irradiation *Mater. Lett.* **107** 291

[71] Deng Y, Chen Y, Chen B and Ma J 2013 Preparation, characterization and photocatalytic activity of $CuBi_2O_4/NaTaO_3$ coupled photocatalysts *Alloys Compd.* **559** 116

[72] Patil R, Kelkar S, Naphade R and Ogale S 2014 Low temperature grown $CuBi_2O_4$ with flower morphology and its composite with CuO nanosheets for photoelectrochemical water splitting *J. Mater. Chem.* A **2** 3661

[73] Nakabayashi Y, Nishikawa M and Nosaka Y 2014 Fabrication of $CuBi_2O_4$ photocathode through novel anodic electrodeposition for solar hydrogen production *Electrochim. Acta* **125** 191

[74] Wang M, Zai J, Wei X, Chen W, Liang N, Xu M, Qi R and Qian X 2015 N-type hedgehog-like $CuBi_2O_4$ hierarchical microspheres: room temperature synthesis and their photoelectrochemical properties *Cryst. Eng. Comm.* **17** 4019

[75] Gu J, Yan Y, Krizan J W, Gibson Q D, Detweiler Z M, Cava R J and Bocarsly A B 2014 *p*-type $CuRhO_2$ as a self-healing photoelectrode for water reduction under visible light *J. Am. Chem. Soc.* **136** 830–3

[76] Joshi U A, Palasyuk A and Maggard P A 2011 Photoelectrochemical investigation and electronic structure of a p-type $CuNbO_3$ photocathode *J. Phys. Chem.* C **115** 13534

[77] Read C G, Park Y and Choi K S 2012 Electrochemical synthesis of p-type $CuFeO_2$ electrodes for use in a photoelectrochemical cell *J. Phys. Chem. Lett.* **3** 1872

[78] Prevot M S, Guijarro N and Sivula K 2015 Enhancing the performance of a robust sol–gel-processed p-type delafossite $CuFeO_2$ photocathode for solar water reduction *Chem. Sus. Chem.* **8** 1359

[79] Jang Y J, Park Y B, Kim H E, Choi Y H, Choi S H and Lee J S 2016 Oxygen-intercalated $CuFeO_2$ photocathode fabricated by hybrid microwave annealing for efficient solar hydrogen production *Chem. Mater.* **28** 6054–64

[80] Prevot M S, Jeanbourquin X A, Bourée W S, Abdi F, Friedrich D, van de Krol R, Guijarro N, Le Formal F and Sivula K 2017 Evaluating charge carrier transport and surface states in $CuFeO_2$ photocathodes *Chem. Mater.* **29** 4952

[81] Zhao Q M, Zhao Z Y, Liu Q L, Yao G Y and Dong X D 2020 Delafossite $CuGaO_2$ as promising visible-light-driven photocatalyst: synthesize, properties, and performances *J. Phys. D: Appl. Phys.* **53** 135102

[82] Ali H, Guler A C, Masar M, Urbanek P, Urbanek M, Skoda D, Suly P, Machovsky M, Galusek D and Kuritka I 2021 Solid-state synthesis of direct Z-scheme Cu_2O/WO_3 nanocomposites with enhanced visible-light photocatalytic performance *Catalysts* **11** 293

[83] Ma L, Su J, Liu M, Zhang L, Li Y and GuO L 2016 Enhanced photocatalytic activity over a novel $CuWO_4/Cu_{1-x}Zn_xWO_4/ZnWO_4$ hybrid material with sandwiched heterojunction *J. Mater. Res.* **31** 1616–21

[84] Pullar R C 2009 The synthesis, properties, and applications of columbite niobates (M_2+Nb_2O_6): a critical review *J. Am. Ceram. Soc.* **92** 563–77

[85] Wahlstrom E and Marinder B O 1977 Phase analysis studies in the copper-niobium-oxygen system *Inorg. Nucl. Chem. Lett.* **13** 559–64

[86] Kratzheller B and Gruehn R 1992 Beiträge zum thermischen Verhalten von Übergangsmetallniobaten III. Darstellung und Struktur von orthorhombischem $CuNb_2O_6$ *J. Alloys Compd.* **183** 75–84

[87] Li M, Su A, Qin Q, Qin Y, Dou A, Zhou Y, Su M and Liu Y 2021 High-rate capability of columbite $CuNb_2O_6$ anode materials for lithium-ion batteries *Mater. Lett.* **284** 128915

[88] Lee S Y, Lim A S, Kwon Y M, Cho K Y and Yoon S 2020 Copper, zinc, and manganese niobates ($CuNb_2O_6$, $ZnNb_2O_6$, and $MnNb_2O_6$): structural characteristics, Li^+ storage properties, and working mechanisms *Inorg. Chem. Front.* **7** 3176–83

[89] Kamimura S, Abe S, Tsubota T and Ohno T 2018 Solar-driven H_2 evolution over $CuNb_2O_6$: effect of two polymorphs (monoclinic and orthorhombic) on optical property and photocatalytic activity *J. Photochem. Photobiol.* A **356** 263–71

[90] Kormanyos A, Thomas A, Huda M N, Sarker P, Liu J P, Poudyal N, Janaky C and Rajeshwar K 2016 Solution combustion synthesis, characterization, and photoelectrochemistry of $CuNb_2O_6$ and $ZnNb_2O_6$ nanoparticles *J. Phys. Chem.* C **120** 16024–34
[91] Arai T, Konishi Y, Iwasaki Y, Sugihara H and Sayama K 2007 High-throughput screening using porous photoelectrode for the development of visible-light-responsive semiconductors *J. Comb. Chem.* **9** 574–81
[92] Wei L, Shifu C, Sujuan Z, Wei Z, Huaye Z and Xiaoling Y 2010 Preparation and characterization of p–n heterojunction photocatalyst p-$CuBi_2O_4$/n-TiO_2 with high photocatalytic activity under visible and UV light irradiation *J. Nanopart. Res.* **12** 1355–66
[93] Elaziouti A, Laouedj N and Bekka A 2015 Synergetic effects of Sr-doped CuBi 2O4 catalyst with enhanced photoactivity under UVA–light irradiation *Environ. Sci. Pollut. Res.* 1–15
[94] Abdelkader E, Nadjia L and Ahmed B 2012 Synthesis, characterization and UV-A light photocatalytic activity of 20 wt%SrO–$CuBi_2O_4$ composite *Appl. Surf. Sci.* **258** 5010–24
[95] Zhang J, Jiang Y, Gao W and Hao H 2015 Synthesis and visible photocatalytic activity of new photocatalyst MBi_2O_4 (M = Cu, Zn) *J. Mater. Sci., Mater. Electron.* **26** 1866–73
[96] Deng Y, Chen Y, Chen B and Ma J J 2013 Preparation, characterization and photocatalytic activity of $CuBi_2O_4$/$NaTaO_3$ coupled photocatalysts *Alloys Compd.* **559** 116–22
[97] Oh W D, Lua S K, Dong Z and Lim T T 2016 Rational design of hierarchically-structured $CuBi_2O_4$ composites by deliberate manipulation of the nucleation and growth kinetics of $CuBi_2O_4$ for environmental applications *Nanoscale* **8** 2046–54
[98] Yuvaraj S, Karthikeyan K, Kalpana D, Lee Y S and Selvan R K 2016 Surfactant-free hydrothermal synthesis of hierarchically structured spherical $CuBi_2O_4$ as negative electrodes for Li-ion hybrid capacitors *J. Colloid Interface Sci.* **469** 47–56
[99] Prévot M S, Jeanbourquin X A, Bourée W S, Abdi F, Friedrich D, Van De Krol R, Guijarro N, Le Formal F and Sivula K 2017 Evaluating charge carrier transport and surface states in $CuFeO_2$ photocathodes *Chem. Mater.* **29** 4952–62
[100] Zhao R D, Zhang Y M, Liu Q L and Zhao Z Y 2020 Effects of the preparation process on the photocatalytic performance of delafossite $CuCrO_2$ *Inorg. Chem.* **59** 16679–89
[101] Choi S Y, Kim C D, Han D S and Park H 2017 Facilitating hole transfer on electrochemically synthesized p-type $CuAlO_2$ films for efficient solar hydrogen production from water *J. Mater. Chem.* A **5** 10165–72
[102] Berglund S P, Abdi F F, Bogdanoff P, Chemseddine A, Friedrich D and Van De Krol R 2016 Comprehensive evaluation of $CuBi_2O_4$ as a photocathode material for photoelectrochemical water splitting *Chem. Mater.* **28** 4231–42
[103] Rajeshwar K, Hossain M K, Macaluso R T, Janáky C, Varga A and Kulesza P J 2018 Review—copper oxide-based ternary and quaternary oxides: where solid-state chemistry meets photoelectrochemistry *J. Electrochem. Soc.* **165** H3192–206
[104] Park S, Baek J H, Zhang L, Lee J M, Stone K H, Cho I S, Guo J, Jung H S and Zheng X 2019 Rapid flame-annealed $CuFe_2O_4$ as efficient photocathode for photoelectrochemical hydrogen production *ACS Sustain. Chem. Eng.* **7** 5867–74
[105] O'Regan B and Grätzel M 1991 A low-cost, high-efficiency solar cell based on dye-sensitized colloidal TiO_2 films *Nature* **353** 737–40
[106] Lewis N S 2001 Light work with water *Nature* **414** 589–90
[107] Souza E L S 2017 Structural evolution, growth mechanism and photoluminescence properties of CuWO4 nanocrystals *Ultrason. Sonochem.* **38** 256–70

Chapter 4

Aerogels and carbon-based materials for photoelectrocatalytic CO_2 reduction into solar fuels

Prabhu Saravanan, Kavery Elangovan, Cristian H Campos, Felipe Sanhueza-Gómez, Padmanaban Annamalai, Kiruthiga Ramakrishnan and R V Mangalaraja

The conversion of carbon dioxide (CO_2) into useful value-added chemicals or fuels is one of the best ways to address the global energy and environmental issues of recent years. The solar light utilizing phtotoelectrocatalytic (PEC) process that combines the photocatalytic and electrocatalytic principles is one of the efficient and sustainable methods to achieve this. This process increases the reaction kinetics and reduces the energy requirement. The photocatalyst is the key component of a PEC CO_2 reduction process. Titanium dioxide (TiO_2) has been a widely investigated photocatalyst for the photoconversion of CO_2 due to its inexpensive, more stable, and non-toxic nature. However, pure TiO_2 can only be active under ultraviolet (UV) light irradiation. In addition, several metal oxide photocatalysts have also been employed for the PEC CO_2 reduction. However, the quantum efficiency of the above photocatalysts is still low. Therefore, the modification of photocatalysts with non-metal and metal doping and coupling with other semiconductors has been attempted. The carbon-based materials as a component in photocatalysts enhance the performance due to their excellent characteristics, such as high surface area and excellent electrical properties. The carbon-based and aerogel (AG) catalysts have greatly improved the PEC CO_2 reduction efficiency in several studies owing to their excellent physicochemical characteristics. This chapter initially discusses an overview of the PEC CO_2 reduction technology, then its advancement with some significant recent catalysts, carbon-based and AG catalysts. Lastly, the challenges

doi:10.1088/978-0-7503-5729-6ch4

and future prospects of the carbon-based and AG catalysts are presented for practical applications of PEC CO_2 reduction.

4.1 Introduction

The carbon dioxide (CO_2) imbalance is a significant concern for the energy and environmental problems that the world is facing today. The industrial revolution, declining CO_2 consumption by plants, an exponential rise in fossil fuel consumption, and more deforestation are the main reasons for the drastically increasing emission of CO_2. The increasing emission of CO_2 causes environmental air pollution and global climate change and also affects human health [1]. Therefore, the capture and reuse or conversion of CO_2 and/or CO_2 emission minimization are in great demand to resolve the above energy and environmental problems. There are several methods to transform CO_2 into beneficial products, such as thermochemical conversion [2, 3], electrocatalytic (EC) reduction [4], photocatalytic (PC) reduction [5], photoelectrocatalytic (PEC) reduction [6, 7], and biotransformation [8]. The processes that use light to convert CO_2 into valuable products can be sustainable methods. For example, photosynthesis in leaves involves the production of chemical energy using water, CO_2, and solar energy through a complex chemical reaction. This photosynthesis process is essential for living organisms on Earth [9]. Inspired by the above, an artificial photosynthesis method could be a potential suitable strategy for the conversion of CO_2 into useful products. Since the demonstration of water splitting under UV light on titanium dioxide (TiO_2) by Fujishima and Honda in 1972, the artificial photosynthesis methodology has been investigated widely in the area of catalysis [10]. The solar light utilizing PEC process combining the PC and EC principles is one of the sustainable and efficient methods, because it reduces the energy requirement and increases the reaction kinetics [6, 7]. In the PEC CO_2 reduction process, the semiconductor catalyst produces electron–hole (e^-–h^+) pairs by the absorption of photons. The potential applied along with the produced e^- will reduce the CO_2 into beneficial organic compounds. The band structure of the semiconductor photocatalysts used in the PEC CO_2 reduction applications should enable the harvesting of most of the solar/visible light. The conduction band (CB) edge of the semiconductor photocatalysts should be more negative than the CO_2 reduction potential in order to overcome the overpotentials for the CO_2 reduction reactions. The PEC CO_2 reduction process has several advantages, such as being economical, mimicking the natural photosynthesis processes, and producing pure high-yield compounds. Thus, researchers have focused significantly more on the PEC CO_2 reduction process [5, 6, 11–13].

This chapter discusses the following aspects: (i) the fundamentals of CO_2 conversion to beneficial chemical compounds or fuels by the PEC process; (ii) some recent significant advances on the different catalysts for the photocathodes in the PEC CO_2 reduction system; (iii) the significant carbon and aerogel (AG) based materials for the photocathode and electrode or catalytic materials for the CO_2 reduction reactions; and (iv) the challenges and prospects for the development of efficient photocathode materials for the CO_2 reduction reaction.

4.2 Fundamentals of PEC CO_2 reduction

4.2.1 Basic principles of PEC CO_2 reduction

The PEC CO_2 reduction route combines the principles of photocatalysis and electrocatalysis. Photocatalysis involves the generation of charge carriers on a semiconductor catalyst by using light irradiation whose energy is equal to or greater than the semiconductor catalyst, as shown in figure 4.1(a). When the light is irradiated, the electrons (e^-) in the valance band (VB) of the semiconductor catalyst will transition to its conduction band (CB) and leave holes (h^+) in the VB. Thus, the generated e^-–h^+ pairs are then separated and transported to the surface of the semiconductor catalyst. At the active sites present in the semiconductor surface, the redox reactions occur, such as the oxidation of water (H_2O) into oxygen (O_2) by h^+ and the reduction of CO_2 by e^- into several products such as CH_3OH, CO, CH_4, etc. In addition, the surface and volume recombination of the photogenerated e^-–h^+ pairs will occur [14]. In the EC reduction of CO_2, the redox reactions will occur on the system's electrodes (anode and cathode) using external electric energy. The oxidation of H_2O into O_2 and protons (H^+) occurs on the anode, then the H^+ migrates through the proton exchange membrane separator to become involved in the reaction of CO_2 reduction on the cathode, as shown in figure 4.1(b) [15].

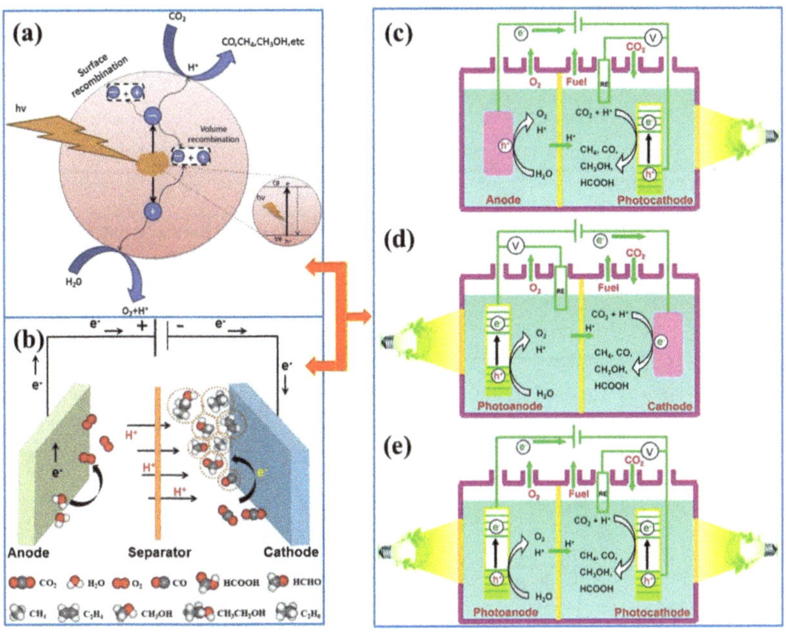

Figure 4.1. Schematic illustration of (a) the mechanism of photocatalytic CO_2 reduction, (b) electrocatalytic CO_2 reduction with different possible products, and the most common three different phtotoelectrocatalytic (PEC) CO_2 reduction set-ups consisting of (c) a photocathode and dark anode, (d) photoanode and dark cathode, and (e) photocathode and photoanode. ((a) Reproduced with permission from [14]. Copyright 2019 Elsevier. (b) Reproduced from [15]. CC BY 4.0. (c)–(e) Reproduced with permission from [16]. Copyright 2016 Royal Society of Chemistry.)

Owing to the several limitations of PC and EC, the CO_2 reduction in practical applications is also limited, so researchers have developed the PEC technique by combining the above two (PC and EC) technologies for the effective conversion of CO_2. In the PEC CO_2 reduction process, the electrons generated by the irradiation of the semiconductor photoelectrode are transported with the aid of external voltage to the electrode (photocathode) surface where the CO_2 reduction reaction occurs [17]. The applied external potential can stimulate the directional transfer of charge carriers, and efficiently separates the photogenerated e^-–h^+ pairs to enhance the redox reaction. The redox potential can be adjusted by applying a suitable bias potential when the band positions of the semiconductor photocatalyst are insufficient to derive the CO_2 reduction and the water oxidation reactions [18]. The PEC CO_2 reduction process has the advantage of reducing the overpotential at the photocathode by utilizing the external bias compared to the PC process and reducing the energy consumption by utilizing a clean and sustainable solar energy source when compared to the EC process. The most common PEC CO_2 reduction set-up (three-electrode system) is two-compartment cells separated by a proton exchange membrane and consisting of the working, counter, and reference electrodes. The system can be categorized into three types, namely (i) photocathode and dark anode, (ii) photoanode and dark cathode, and (iii) photocathode and photoanode, as shown in figures 4.1(c)–(e) [16]. Generally, the p-type semiconductors connected with a metallic electrode are used as the photocathode in a PEC CO_2 reduction electrochemical cell due to the downward band bending characteristic in the interface of the p-type semiconductor/electrolyte interface, as presented in figure 4.2 [13].

4.2.2 Thermodynamics

The CO_2 is a most stable molecule and is difficult to break because of the enormous 750 kJ mol^{-1} bonding energy of the C=O double bond [19]. The carbon atom in the

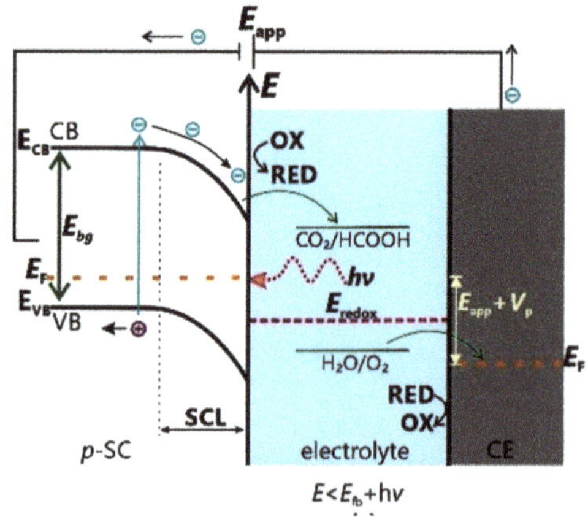

Figure 4.2. A schematic illustration of downward band bending in the interface of a p-type semiconductor (p-SC)/electrolyte. (Reproduced with permission from [13]. Copyright 2022 Elsevier.)

Table 4.1. Different CO_2 reduction reactions with standard Gibbs free energy changes (ΔG^0) and the corresponding half-reactions with required electrochemical redox potentials (E^0) with reference to the reversible hydrogen electrode (RHE). (Reproduced with permission from [20]. Copyright 2022 Wiley.)

CO_2 reduction reaction	ΔG^0 (kJ mol^{-1})	Corresponding reductive electrochemical half-reaction	E^0 [V_{RHE}]
$H_2O(l) \rightarrow H_2(g) + 1/2O_2(g)$	237.2	$2H^+ + 2e^- \rightarrow H_2(g)$	0.000
—	—	$CO_2(g) + e^- \rightarrow CO_2^{\bullet -}$	−1.436
$CO_2(g) \rightarrow CO(g) + 1/2O_2(g)$	257.4	$CO_2(g) + 2H^+ + 2e^- \rightarrow CO(g) + H_2O(l)$	−0.107
$CO_2(g) + H_2O(l) \rightarrow HCOOH(l) + 1/2O_2(g)$	270.4	$CO_2(g) + 2H^+ + 2e^- \rightarrow HCOOH(l)$	−0.171
$CO_2(g) + H_2O(l) \rightarrow HCHO(l) + O_2(g)$	521.9	$CO_2(g) + 4H^+ + 4e^- \rightarrow HCHO(l) + H_2O(l)$	−0.071
$CO_2(g) + 2H_2O(l) \rightarrow CH_3OH(l) + 3/2O_2(g)$	702.8	$CO_2(g) + 6H^+ + 6e^- \rightarrow CH_3OH(l) + H_2O(l)$	+0.016
$CO_2(g) + 2H_2O(l) \rightarrow CH_4(g) + 2O_2(g)$	919.6	$CO_2(g) + 8H^+ + 8e^- \rightarrow CH_4(g) + 2H_2O(l)$	+0.169
$2CO_2(g) + 2H_2O(l) \rightarrow CH_3COOH(l) + 2O_2(g)$	963.8	$2CO_2(g) + 8H^+ + 8e^- \rightarrow CH_3COOH(l) + 2H_2O(l)$	+0.098
$2CO_2(g) + 2H_2O(l) \rightarrow C_2H_4(g) + 3O_2(g)$	1332	$2CO_2(g) + 12H^+ + 12e^- \rightarrow C_2H_4(g) + 4H_2O(l)$	+0.085
$2CO_2(g) + 3H_2O(l) \rightarrow C_2H_5OH(l) + 3O_2(g)$	1327	$2CO_2(g) + 12H^+ + 12e^- \rightarrow C_2H_5OH(l) + 3H_2O(l)$	+0.084
$3CO_2(g) + 4H_2O(l) \rightarrow C_3H_7OH(l) + 9/2O_2(g)$	1962	$3CO_2(g) + 18H^+ + 18e^- \rightarrow C_3H_7OH(l) + 5H_2O(l)$	+0.095

CO_2 is in its highest oxidation state. According to the number of electrons injected and the number of H^+ in the CO_2 reduction reaction, several products would be produced, including carbon monoxide (CO), formate ($HCOO^-$), methane (CH_4), methanol (CH_3OH), ethanol (C_2H_5OH), ethylene (C_2H_4), and acetic acid (CH_3COOH). The different CO_2 reduction reactions with standard Gibbs free energy changes (ΔG^0) and the half-reactions with required electrochemical redox potentials (E^0) with reference to the reversible hydrogen electrode (RHE) are given in table 4.1 [20]. The ΔG^0 for all the CO_2 reduction reactions listed in table 4.1 are greater than the ΔG^0 of the water-splitting reaction of 237.3 kJ.mol^{-1}. Also, the E^0 for the different products from the CO_2 reduction reactions are close to those for the H_2 evolution from the water-splitting reaction. Therefore, the H_2 evolution is a severe competition reaction for the CO_2 reduction reactions.

4.2.3 General mechanism pathways

The possible mechanistic pathways for the reduction of CO_2 into various products are illustrated in figure 4.3 [21]. For any particular product, several mechanistic pathways are possible. Generally, different steps such as proton (H^+) transfer, electron (e^-) transfer, both H^+ and e^- transfer, insertion, dimerization, and water

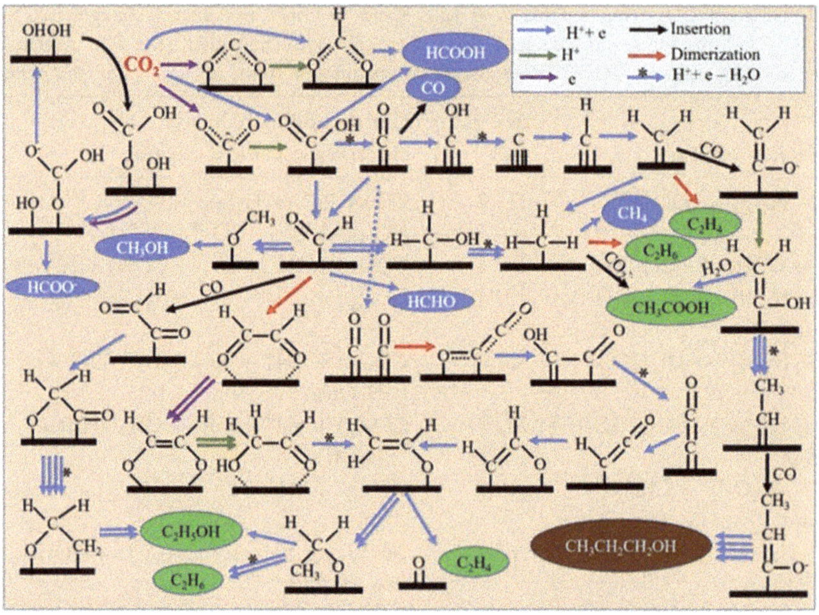

Figure 4.3. Illustration of the possible mechanistic pathways for the reduction of CO_2 into different products. (Reproduced with permission from [21]. Copyright 2022 the Royal Society of Chemistry.)

elimination occur based on the surface of the catalyst used [13, 21]. The possible mechanism, in brief, is that the $CO_2^{\bullet-}$ radicals can initially be formed by reducing CO_2. Meanwhile, the hydrogen radicals ($H^\bullet$) can result by reacting electrons with H^+, which can be formed by reacting H^+ with water. Then, by reacting $CO_2^{\bullet-}$ with $H^\bullet$, the CO molecules can be generated, which can be used as a starting material for further higher carbon compounds [22].

4.2.4 Evaluation of PEC CO_2 conversion efficiency

The efficiency of PEC CO_2 conversion reactions can be evaluated using different parameters such as the rate of product conversion (R) (μmol $h^{-1}g^{-1}$), turnover number (TON), turnover frequency (TOF), Faradaic efficiency (FE), and quantum yield (QY) [18]:

$$R = \frac{n_{\text{product}}}{t \times m_{\text{catalyst}}} \tag{4.1}$$

$$\text{TON} = \frac{n_{\text{product}}}{n_{\text{catalyst}}} \tag{4.2}$$

$$\text{TOF} = \frac{n_{\text{product}}}{n_{\text{catalyst} \times t}} \tag{4.3}$$

$$\text{FE} = \frac{a \times n_{\text{product}} \times F}{Q} \quad (4.4)$$

$$\text{Internal QY} = \frac{a \times n_{\text{product}}}{n_{\text{photon}}} \quad (4.5)$$

$$\text{Apparent QY} = \frac{a \times n_{\text{product}}}{n'_{\text{photon}}}, \quad (4.6)$$

where $n_{product}$, $m_{catalyst}$, $n_{catalyst}$, t, α, F, Q, n_{photon}, and n'_{photon} are the molar numbers of products (in μmol), the weight of the catalyst (in g), molar numbers of catalysts (in mol), reaction time (in h), number of reacted electrons for the products (e.g. 2 for CO, 8 for CH_4, etc), Faraday's constant, total passed charge (in coulombs), number of photons absorbed, and number of incident photons, respectively.

The PEC CO_2 reduction efficiency can also be expressed in terms of solar-to-fuel (STF) conversion efficiency and applied bias photon-to-current efficiency (ABPE) when an external voltage (V_{bias}) is applied as given in the following equations, respectively [23]:

$$\text{STF} = \frac{r_{\text{fuel}} \times \Delta G^0}{P_{\text{solar}} \times A} \quad (4.7)$$

$$\text{STF} = \frac{J_{\text{SC}} \times \Delta E^0 \times \text{FE}}{P_{\text{solar}}} \quad (4.8)$$

$$\text{ABPF} = \frac{J_{\text{ph}} \times [\Delta E^0 - V_{\text{bias}}] \times \text{FE}}{P_{\text{solar}}}, \quad (4.9)$$

where r_{fuel}, ΔG^0, P_{solar}, A, J_{SC}, ΔE^0, FE, J_{ph}, and V_{bias} are the amount of the chemical product generated (in mmol s^{-1}), Gibbs free energy (in kJ mol^{-1}), irradiation light power density (AM 1.5 G) (in mW cm^{-2}), area of the photoelectrode irradiated (in cm^2), short circuit photocurrent density observed (in mA cm^{-2}), the energy of thermodynamics stored in the PEC cell (in V), Faradaic efficiency (FE) calculated using equation (4.4), and photocurrent observed (in mA cm^{-2}) under applied external voltage (in V), respectively.

4.3 Role of photocatalysts in PEC CO_2 conversion

The photocatalyst is a crucial component of a PEC CO_2 reduction process. Figure 4.4 illustrates the commonly used semiconductor photocatalysts with the band positions for the CO_2 reduction reactions [20]. A photocatalyst should hold the four essential properties to achieve better PEC CO_2 reduction, such as (i) high solar light absorption, (ii) reduced recombination rate of e^-–h^+ pairs, (iii) effective collection and transportation of photogenerated electrons to the reaction site, and (iv) high reactant adsorption capacity [24]. Several photocatalysts based on metal oxides, metal complexes, metal organic frameworks (MOF), carbon, and aerogels

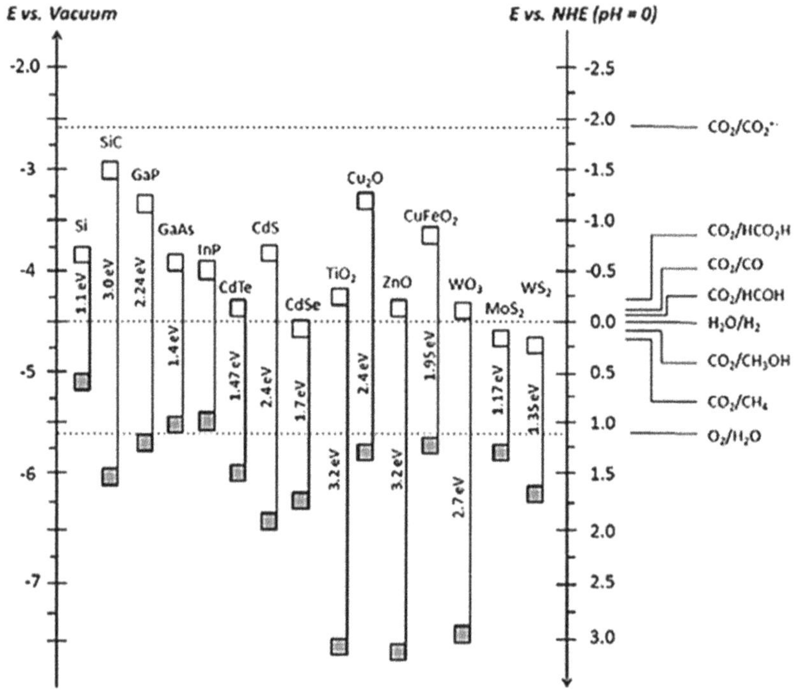

Figure 4.4. Different commonly used semiconductor photocatalysts with their band positions for the CO_2 reduction reactions. (Reproduced with permission from [20]. Copyright 2022 Wiley.)

(AGs) have demonstrated good CO_2 reduction performances. Among the above, the AGs and carbon-based catalysts have attracted tremendous interest due to their outstanding characteristics. The presence of carbon as a component in the catalysts improved the CO_2 reduction activities due to their high surface area, high electrical conductivity, high carrier transportation ability, and efficient e^- collection and utilization for the reaction [25–28]. The AGs are high-potential materials in several applications such as environmental clean-up [29], energy storage [30], chemical sensors [31], thermal insulation [32], and pharmaceuticals [33]. They possess low density, outstanding electrical conductivity, a high number of pores, better stability, small pore size (<10 nm), environmental compatibility, large surface area (>1000 $m^2\ g^{-1}$), elastic structural and surface properties, and ~99z of optical transmission in the visible region [34]. Moreover, the AGs show outstanding adsorption of reactants [35] and catalytic activities [36]. In addition, the AGs displayed an improved capacity for the adsorption of CO_2, mainly due to their hierarchical pore structures [37–40].

4.4 Overview of recent significant advances in photocathodes for PEC CO_2 reduction

The metal-based photocatalysts were modified using different strategies to improve the efficiency of CO_2 conversion by the PEC process. Among others, the strategy of combining with other semiconductor materials has gained more benefit for enhanced efficiency owing to the formation of the heterojunction and other potential characteristics [41–44]. The copper (Cu) based materials have been explored as photocatalysts for the PEC CO_2 reduction due to their Earth-abundance, low cost, and high efficiency. The nanorod arrays of CuO–Cu_2O hybrid semiconductors showed higher efficiency with a selectivity of ~95% for the generation of CH_3OH [45]. The demonstrated higher efficiency under solar illumination of CuO–Cu_2O nanorod arrays was attributed to their high surface area [46]. Cu_2O/Fe_2O_3 nanotubes prepared using the electrodeposition method showed 93% of FE with high selectivity for the production of CH_3OH with a yield of 4.94 mmol l^{-1} cm^{-2} after 6 h due to the low overpotential (0.18 V) and band gap (1.96 eV) [47]. The layered CuO/Cu_2O films decorated with several transition metals such as silver (Ag), gold (Au), cadmium (Cd), Cu, lead (Pb), and tin (Sn) showed good performance for PEC CO_2 reduction under visible light. Among all the above photocatalysts, Pb/CuO/Cu_2O showed better performance for the production of HCOOH and CH_3OH with a yield of 0.524 μmol h^{-1} cm^{-2} and 0.102 μmol h^{-1} cm^{-2}, respectively, and 40.45% of total FE. However, the stability of the photocatalysts decreased as the compositional destruction occurred [48]. The n-type and p-type Cu_2O thin films were electrodeposited on the copper metal electrode and produced CH_4 and C_2H_4 in the PEC CO_2 reduction reaction upon visible light irradiation. The higher efficiency over the p-type Cu/Cu_2O photoelectrode than over the n-type Cu/Cu_2O was observed with higher selectivity for C_2H_4 [49]. The PEC CO_2 reduction reaction under different parameters such as reaction time, pH, and applied bias over Cu/Cu_2O photocatalysts showed acetaldehyde (C_2H_4O) and acetone (CH_3COCH_3) as their main products [50]. The Cu_2O coated with Cu^+-incorporated TiO_2 (TiO_2-Cu^+) showed 56.5% of FE for the reduction of CO_2 which was 23.6% higher than that of bare Cu_2O photocathodes. The higher efficiency was attributed to their lower charge transfer resistance and enhanced CO_2 adsorption on the TiO_2 surface [51]. Similarly, Cu_2O coated with Au-TiO_2 enhanced the PEC CO_2 reduction efficiency through the *Z*-scheme pathway mechanism under solar light [44]. The monophase Cu_2O nanofiber photocathode was prepared and demonstrated an excellent performance for the reduction of CO_2 into CH_3OH with 90% selectivity in the water [52]. The Cu_2O nanowires modified with titanium carbide (Ti_3C_2) quantum dots showed 8.25-fold higher PEC CO_2 reduction for the production of CH_3OH than that of bare Cu_2O nanowires, which was due to their higher photon absorption and charge transfer characteristics [53]. The amine-functionalized polysiloxane as a protective layer for the Cu_2O photocathode demonstrated enhanced stable PEC CO_2 reduction efficiency for the selective generation of $HOOC^-$ with 61% of FE [54]. The Cu_2O-SnO_x hybrid nanowires exhibited enhanced PEC CO_2 reduction into CO and H_2 (syngas) with 90.32% of total FE. The electrodeposition of SnO_x on the Cu_2O nanowires improved the CO_2 adsorption capacity and charge transfer efficiency and resulted in enhanced PEC CO_2 reduction performance [55].

The $CuFe_2O_4$ photocathode exhibited 68% CH_3OH selectivity with 14.4% quantum efficiency for the PEC CO_2 reduction under visible light irradiation [56]. Similarly, the $CuFeO_2$ layer on the Cu_2O nanorod enhanced the stability of the photocathode for the PEC reduction of CO_2 into $HOOC^-$ and acetate with 21.6% and 68.6% FE, respectively [57]. The Mg-doped $CuFeO_2$ electrode exhibited higher PEC CO_2 reduction into $HOOC^-$ with slight H_2 generation [58]. The $CuFeO_2$/CuO photocathode showed higher $HOOC^-$ generation at 0.15 V_{RHE} with 90% of selectivity. The photocathode demonstrated ~1% $HOOC^-$ generation efficiency even without any external potential for over one week [59]. In addition, the $CuFeO_2$/CuO exhibited 80% FE for the production of acetate with high selectivity. However, the selectivity was shifted to $HOOC^-$ generation due to the rapid leaching of Fe^{3+} from the electrode surface [60]. The efficiency of the $CuFeO_2$ (CFO) photocathode was improved by decorating it on the surface of the $CuInS_2$ (CIS) thin film surface. The CFO/CIS photocathode showed three times higher CH_3OH generation with a rate of 15.40 mol h^{-1} m^{-2} than that of the CIS photocathode, which was due to the lower mass-transfer resistance [43]. Nevertheless, the CIS photocathode fabricated by electrodeposition of Cu-In alloy followed by sulfurization exhibited 97% FE for the PEC CO_2 reduction to CH_3OH [61].

The PEC CO_2 transformation performance using zinc (Zn) based photocatalysts was improved by adopting several approaches. Bismuth (Bi) and α-Fe_2O_3 co-modified zinc oxide (ZnO) nanorod array photocathodes exhibited an impressive performance of the PEC reduction of CO_2 to HCOOH with 61.2% FE. The Bi@α-Fe_2O_3/ZnO heterojunction enhanced the CO_2 adsorption and activation and separation of photogenerated charge carriers which resulted in higher PEC CO_2 reduction efficiency [62]. Zinc telluride (ZnTe) on Zn/ZnO nanowire as photocathode demonstrated efficient and stable PEC reduction of CO_2 to CO [63]. Au as a co-catalyst for ZnTe/ZnO nanowire highly enhanced the PEC reduction of CO_2 to CO as a significant product. The higher efficiency was due to the improved charge separation and active sites for the PEC CO_2 reduction reaction [42]. Likewise, the Au decorated ZnO@ZnTe@CdTe triple layer core–shell nanorod array photocathode showed highly enhanced STF of 0.43% for the selective production of CO with a tandem $CH_3NH_3PbI_3$ perovskite solar cell [64]. An interesting Ag_3Cu/TiO_2/ZnTe metal–insulator–semiconductor photocathode was fabricated for the PEC reduction of CO_2 to syngas. This photocathode exhibited excellent stability due to its unique structure and fast transfer of photogenerated charge carriers through the interface of the photocatalyst [65]. The ZnO@ZnSe nanostructure photocathode demonstrated drastic improvement in the production of CO in the PEC CO_2 reduction reaction with 52.2 and 96.6% FE and selectivity, respectively [41].

Mixed metal-based catalysts were employed to achieve outstanding PEC activity for the reduction of CO_2. CdSeTe nanosheet coupled TiO_2 nanotube photocatalysts demonstrated enhanced CH_3OH generation of 1166.77 μmol L^{-1} after 5 h from the PEC CO_2 reduction reaction with good stability [66]. However, CdSeTe nanoparticle coupled TiO_2 nanotube photocatalysts achieved 2.72 times higher performance compared to the previous photocatalyst and reached 18.57 mmol L^{-1} CH_3OH generation after 300 min with 88% FE [67]. The Cu_2ZnGeS_4 thin film on a Mo coated glass photocathode demonstrated CO generation from the PEC reduction of CO_2, and the performance was further improved by covering the Cu_2ZnGeS_4 surface with a ZnS layer [68]. The Cu_2ZnSnS_4/CdS photocathode demonstrated significantly increased PEC CO_2 performance for producing the CO and alcohols with tunable selectivity based on the S vacancies and O-doping [69]. The $In_{0.4}Bi_{0.6}$ eutectic alloy integrated over a lead halide perovskite PV cell as a photocathode demonstrated almost 100% FE for the PEC reduction of CO_2 to HCOOH [70].

Gallium nitride (GaN) based catalytic materials were proven to be potential candidates for PEC solar fuel generation from CO_2. A photocathode of GaN nanowire on silicon (Si) solar cells demonstrated excellent and more selective solar PEC CO_2 conversion to CH_4 with 19% FE. The monolithically integrated photocathode system showed more absorption of the solar spectrum and high charge carrier extraction efficiency [71]. The unique type of nanoarchitecture of GaN:Sn on Si substrate demonstrated a better benchmarking HCOOH generation efficiency of 76.9% FE in the PEC CO_2 reduction reaction due to their excellent solar light absorption, charge separation, and catalytic activity [72]. The photocathodes consisting of silver halide (AgX, X = Cl, Br, I) decorated heterojunctions of GaN nanowires/ n^+–p Si exhibited superior syngas generation performance from the PEC reduction of CO_2 [73].

Silver (Ag) or silver ions (Ag^+) significantly enhance the PEC CO_2 reduction efficiency in several photocathode systems. Ag^+ ions on a Si wafer achieved 90% PEC CO_2 reduction efficiency for the generation of CO [74]. By controlling the Ag nanoparticle size on the p-Si nanowire, the PEC CO_2 reduction to syngas varied in the molar ratio range between 1 and 4 [75]. The adsorption of a thin layer of 2-aminobenzenthiol onto a Ag nanoparticle decorated Si substrate photocathode displayed more selectivity toward PEC CO_2 reduction to CO [76]. Dendritic Cu catalysts supported with Ag on a Si substrate as the photocathode integrated with $CH_3NH_3PbI_3$ perovskite semitransparent solar cells as a self-powered PEC CO_2 reduction device demonstrated excellent performance for as long as 10 days. The device generated C_2 and C_3 products with 3.5% sunlight to hydrocarbon conversion efficiency [77].

Si materials with different metals or metal ions displayed superior activity. Bi^{3+} incorporation by the etching of Si wafers exhibited highly enhanced and stable $HCOO^-$ generation from the PEC reduction of CO_2 with about 90% FE [78]. In contrast, the Bi nanotube with a Si substrate reached a stable 90% FE for the production of $HCOO^-$ from the PEC CO_2 reduction reaction [79]. The incorporation of Cu into arrays of n^+p-Si microwires or nanowires showed the production of several selective products over different reaction times from the simulated sunlight-driven PEC CO_2 reduction reaction [80, 81]. The Sn porous nanowires on Si substrate as the photocathode showed selective production of HCOOH from the PEC reduction of CO_2 with about 60% of FE [82].

Further, graphitic carbon nitride (g-C_3N_4) based catalysts displayed better PEC performance for the conversion of CO_2. The CuO/g-C_3N_4 performed 4.1 times higher for the production of CH_3OH from the PEC reduction of CO_2 compared to that of bare g-C_3N_4 photocathode with 75% FE [83]. A g-C_3N_4/ZnTe photocathode exhibited impressive PEC CO_2 reduction to C_2H_5OH with a 17.1 μmol cm^{-2} h^{-1} generation rate [84]. A Ti_3C_2/g-C_3N_4 heterojunction photocatalyst showed ten times higher PEC CO_2 reduction performance than that of neat g-C_3N_4, which was due to the high solar light absorption and efficient separation of photogenerated charge carriers through the photocatalyst heterojunction [24].

Several metal complex-based photocathodes exhibited higher performance of PEC reduction of CO_2. Immobilized Ru(II)–Re(I) supramolecular metal complexes on a NiO electrode as a NiO–RuRe photocathode demonstrated high selectivity for the production of CO with a maximum TON of 32 [85]. A metal complex of Ru(II)–Re(I) electrodeposited using a methyl phosphoric acid anchoring group on a NiO electrode with electro-polymerized ligand vinyl groups as the photocathode showed an improved FE of 85% and 2.5 times more production of CO [86]. The photocathode consisting of ruthenium complex and TiO_2/N, Zn-Fe_2O_3/Cr_2O_3 multi-heterojunction structures exhibited higher PEC CO_2 reduction efficiencies for the generation of HCOOH and CO with 0.15% solar to chemical energy conversion efficiency [87]. In addition, the photobiocathode consisting of formate dehydrogenase from *Clostridium ljungdahlii* (ClFDH) and the mixture of $CuFeO_2$ and CuO coated with TiO_2 exhibited 2.45 times higher PEC reduction of CO_2 to $HOOC^-$ compared to that of photocathode without ClFDH with 33.5% FE. This PEC system was the first wireless solar driven semi-biological system [88].

4.5 Carbon-based materials for PEC CO_2 reduction

Several graphene (G) or carbon-based photocatalysts, such as graphene-$Zn_{0.5}Cd_{0.5}S$ [89], reduced titania wrapped with N-doped graphene oxide (GO) [27], Ni nanoparticles@graphitic carbon (GC) [28], C_{60}/$CsPbBr_3$ composite [90], carbon nanotube (CNT)-TiO_2 [91], GO-TiO_2 [92], B-doped G-TiO_2 [93], N-doped G-CdS [94], α-Fe_2O_3-ZnO/reduced graphene oxide (rGO) [95], CoP/CNT or rGO [96], Co quaterpyridine-G [97], ordered mesoporous carbon-CNT-g-C_3N_4 composites [98], and functionalized G-quantum dots [99, 100] demonstrated good PC CO_2 reduction performances. The PEC CO_2 reduction performance was also enhanced by incorporating carbon or graphene (G) based materials in the photocathode catalysts. The G-TiO_2 composite catalyst exhibited higher PEC CO_2 reaction performance at high pH due to the reduced resistance of charge transfer [101]. The Cu-rGO–TiO_2 catalyst in a PEC CO_2 reaction cell demonstrated stable and selective performance for the generation of HCOOH and CH_3OH with a current efficiency of 32.47%. The catalyst showed a production rate of 255 μmol h^{-1} cm^{-2} and 189.06 μmol h^{-1} cm^{-2} for HCOOH and CH_3OH, respectively. The higher efficiency was due to its efficient separation of e^-–h^+ pairs [102]. rGO covered TiO_2 nanotube arrays exhibited enhanced photocurrent generation efficiency as well as higher photochemical conversion of CO_2 to CH_4. The higher PEC efficiency of the rGO covered TiO_2 nanotube arrays was due to the enhanced capture of the incident photons and efficient separation of the photogenerated charge carriers [103]. Pt-rGO as the cathode in a PEC cell displayed improved CO_2 reduction performance [104]. Highly promising photocatalysts consisting of different proportions of Cu, palladium (Pd), and platinum (Pt) in small amounts, graphene nanoribbons, and TiO_2 nanotube arrays demonstrated excellent PEC CO_2 conversion to CH_3OH and C_2H_5OH. The superior PEC CO_2 reduction activity was due to the better utilization of photogenerated e^-, avoiding H_2 generation and efficient inhibition of the recombination of charge carriers [105]. CNT/Cu_2O films demonstrated higher and more stable PEC performance for CO_2 reduction applications compared to that of Cu_2O due to the excellent electrical conductivity of the CNT/Cu_2O films [106]. Similarly, different types of graphene supported Cu_2O photoelectrodes exhibited better PEC activity for CO_2 reduction applications, which was achieved by harvesting photogenerated e^- efficiently through the higher surface area of the catalysts [107]. The photostability of the Cu_2O photocatalyst was improved by incorporating a mixture of carbide derived carbons (CDC) with tungsten carbide (WC-CDC) or CDC. WC-CDC modified Cu_2O showed higher PEC CO_2 reduction to CH_3OH than the CDC modified one due to the efficient collection and transportation of photogenerated e^- and a lower e^-–h^+ pair recombination rate [108].

A CIS/graphene thin film photocathode exhibited about 1.4 times higher PEC reduction of CO_2 to CH_3OH compared to the neat CIS thin film photocathode [109]. GO incorporated $CuFe_2O_4$ photocathodes significantly improved the PEC reduction of CO_2 to CH_3OH. The GO in the catalyst reduced the e^-–h^+ pair recombination and trapped the photogenerated e^- from the $CuFe_2O_4$ to derive the PEC CO_2 reaction. Therefore, a high CH_3OH generation of 28.8 μmol l^{-1} cm^{-2} was achieved with 20.5 and 87% quantum efficiency and FE, respectively [26]. rGO modified Pt as the cathode in the PEC CO_2 reduction cell demonstrated higher performance to produce different valuable chemicals compared to that of Pt modified with CNTs and carbon catalysts. The total production rate was 600 nmol h^{-1} cm^{-2}, and the enhanced activity was attributed to the high adsorption of reactants and efficient transportation of charge carriers [25].

A photocathode consisting of multiple-pixel BiOI with graphite paste as the conductive part in PEC tandem devices displayed enhanced and stable production of CO along with H_2 [110]. Cobalt phthalocyanine (CoPc)/CNT-C-CsFAPb$(IBr)_3$ as a photocathode of sandwich type demonstrated superior CO generation efficiency with high stability in the PEC CO_2 reduction reaction. An unassisted PEC cell with tandem perovskite and amorphous Si as the photocathode and photoanode, respectively, achieved 3.34% CO conversion efficiency and 3.85% total STF efficiency [111]. Similarly, a GO/$CH_3NH_3PbBr_3$ hybrid displayed excellent PEC CO_2 reduction to CO of 1.05 μmol cm^{-2} h^{-1} yield in organic media [112]. The remarkable performance of the previous two photocathodes was due to the incorporation of carbon materials such as CNTs and GO, which enhanced the durability of the catalyst and improved the transportation of the photogenerated electrons [111, 112].

Metal-free carbon-based catalytic materials showed excellent PEC CO_2 reduction performance. Carbon dots with nitrogen groups exhibited superior PEC CO_2 reduction to more selective CH_4 generation of 74.8% (electronic selectivity). The π-electron cloud in the carbon dots was responsible for stabilizing the photogenerated electrons and thus higher e^-–h^+ pair separation efficiency, resulting in better CO_2 reduction activity [113]. The cobalt(II) chlorin modified multiwalled CNT as a cathode exhibited higher and selective production of CO with an FE of 83% from the PEC reduction of CO_2 [114].

4.6 Aerogels for CO_2 conversion

The AG materials such as g-C_3N_4/GO AG [115], polypyrrole/GO AG [116], porous carbon AG composites [117], perylene imide, and GO modified g-C_3N_4 AG [118], $CeVO_4$/rGO-AG/$BiVO_4$ [119], and MOF(2Fe/Co)/carbon AG [120], exhibited PC and EC applications including CO_2 reduction activities. AG-based catalysts demonstrated superior CO_2 reduction performance. Surprisingly, only a few research articles are available on AG-based materials for the PEC reduction of CO_2. Therefore, only some significant research achievements on EC, PC, and PEC CO_2 reduction for the production of valuable chemicals are presented in this section. Pt/G AGs incorporated in Cu formed as the cathode in a PEC cell demonstrated excellent CO_2 reduction into several liquid products such as CH_3OH, C_2H_5OH, CH_3COOH, HCOOH, and propionic acid. A higher conversion rate of 5040 nmol h^{-1} cm^{-2} was achieved. The better performance of the Pt/G AGs catalysts was due to the higher adsorption of reactant and more efficient transportation of charges than those of Pt/rGO catalyst [121]. Interestingly, the Au AG achieved EC reduction of CO_2 to CO with 95.6% FE. The superior efficiency was attributed to the enormous surface area for electrochemical activity, porous structure, and higher active sites of the Au AG [122]. The Sn/CNT-AGs displayed highly enhanced and stable EC CO_2 reduction to $HCOO^-$ compared to Sn/CNT with a maximum of 82.7% FE. The enhanced performance was due to the high surface area, unique 3D structure, and better conductivity [36].

Pd-g-C_3N_4/rGO-AG synthesized using a simple route exhibited better PEC activity. The PC CO_2 reduction to CH_4 was achieved with a 6.4 μmol g^{-1} h^{-1} production rate. The Pd-g-C_3N_4/rGO-AG showed 12.8-fold more CH_4 generation performance than neat g-C_3N_4. The Pd performed as an e^- sink, and the rGO-AG enhanced the adsorption of the reactant and effective light energy utilization and photogenerated charge carrier separation, thus achieving higher PEC and PC activities [123]. The three-dimensional (3D) porous g-C_3N_4/GO AG demonstrated excellent photoreduction of CO_2 into CO. A CO production yield of 23 mmol g^{-1} in 6 h was achieved, which was 2.3 times higher compared to that of neat g-C_3N_4. The higher efficiency of 3D porous g-C_3N_4/GO AG was attributed to superior characteristics such as multireflection of the incident photon, higher light absorption, mass transportation, increased active sites, and a higher rate of e^- transfer [124]. NiAl layered double hydroxide (LDH)/g-C_3N_4/G-AG demonstrated PC reduction of CO_2 to CO with a special 28.83 μmol g^{-1} h^{-1} production rate upon irradiation of visible light. The catalysts showed 24- and 16-fold higher performance than bare NiAl LDH and g-C_3N_4, respectively. The enhanced efficiency was attributed to the higher absorption of light, excellent catalytic active sites, efficient separation and transportation of photogenerated e^-–h^+ pairs, and energetic mass-transfer properties of the catalysts [125].

Cadmium selenide (CdSe) quantum dot AGs achieved higher and selective PC CO_2 reduction to CO with a 15 μmol g^{-1} h^{-1} generation rate. The catalyst demonstrated 12 times higher CO production efficiency due to the unique porous structure than bare aggregated CdSe quantum dots [126]. The CuO loaded Fe^{3+}

doped TiO_2 nanofiber shell combined SiO_2 AG hierarchical photocatalyst exhibited the conversion of CO_2 into CH_3OH yield of ~2400 μmol g_{cat}^{-1}. The hierarchical structure of the photocatalyst provided efficient movement of charge carriers and, thus, higher photocurrent generation, which was beneficial for the higher CO_2 conversion activity [127].

4.7 Challenges and prospects

The conversion of CO_2 to solar fuels or value-added chemical compounds is still highly challenging. The semiconductor catalysts used in the PEC cells should possess outstanding characteristics to achieve better performance. Several solar active photocatalysts or photocathodes have been developed for the PEC CO_2 reduction reaction but still show low efficiency, selectivity, and stability. Also, a greater number of active sites and high CO_2 adsorption ability were rarely demonstrated by the photocatalysts. The use of metal-based semiconductors for the fabrication of photocathodes involves challenges such as high cost and lower robustness. These problems were addressed by coupling the metal-based catalysts with other materials, such as other semiconductors, organic compounds, and metal complexes. In addition, doping with metals and introducing nanoarchitectures and multi-heterojunctions were also adopted to improve the PEC CO_2 reduction performance. Nevertheless, the benchmarking or significant outstanding performance by metal-based photocatalysts for the PEC reduction of CO_2 is yet to be achieved. The above significant problems were also attempted to be resolved by modifying the photocatalysts with carbon-based materials due to their excellent surface structural and electrical characteristics. However, the mechanisms of transportation of photogenerated charge carriers to the active sites and the reduction reaction must be studied further to gain a better understanding. Surprisingly, the AG-based materials displayed excellent CO_2 conversion activity, but AGs for PEC CO_2 reduction applications are rarely studied. Therefore, the development of alternative highly promising AG-based photocathodes should be demonstrated further. On the other hand, the mechanism of the product formation, including intermediates analysis, should be explored more clearly. Above all, the PEC CO_2 reduction technology is still in laboratory scale research. Therefore, real-time CO_2 capture and energy storage should be explored more for practical applications.

4.8 Conclusions

In conclusion, CO_2 conversion to value-added chemicals or fuels using the PEC technique under solar light is suitable for addressing current energy and environmental problems. The PEC CO_2 conversion method has several advantages, such as being highly efficient by combining the PC and EC techniques, sustainable, generating pure compounds with high yield, and mimicking natural photosynthesis. The highly stable CO_2 molecule can be converted into different value-added chemicals by reacting with solar energy generated electrons and protons using a semiconductor catalyst with an applied external potential in the PEC systems. The semiconductor catalysts play a crucial role in achieving improved CO_2 conversion efficiency. Several attempts were demonstrated to develop efficient catalysts. The formation of heterojunctions by combining with other semiconductor materials was shown to enhance performance owing to the efficient separation of the charge carriers. In addition, the outstanding properties such as higher surface area, excellent electrical conductivity, superior porosity, environmental compatibility, and efficient utilization of the charge carriers of AGs and carbon-based materials improved the PEC CO_2 reduction performance greatly. However, using AG-based materials for PEC CO_2 reduction applications was reported only rarely. Thus, more attention should be paid to the development of efficient AG-based photocatalysts for PEC CO_2 reduction applications.

Acknowledgments

The authors PS, PA, and KR are grateful to the Government of Chile, Santiago, for the financial assistance through FONDECYT/ANID Postdoctoral Project Nos. 3220357, 3220176, and 3200232, respectively.

References

[1] Olah G A, Prakash G K S and Goeppert A 2011 Anthropogenic chemical carbon cycle for a sustainable future *J. Am. Chem. Soc.* **133** 12881–98

[2] Aresta M, Dibenedetto A and Quaranta E 2015 Reaction mechanisms in carbon dioxide conversion *React Mech Carbon Dioxide Convers* (Berlin: Springer) pp 237–310

[3] Yang H, Zhang C, Gao P, Wang H, Li X, Zhong L, Wei W and Sun Y 2017 A review of the catalytic hydrogenation of carbon dioxide into value-added hydrocarbons *Catal. Sci. Technol.* **7** 4580–98

[4] Noda H, Ikeda S, Oda Y and Ito K 1989 Potential dependencies of the products on electrochemical reduction of carbon dioxide at a copper electrode *Chem. Lett.* **18** 289–92

[5] Albero J, Peng Y and García H 2020 Photocatalytic CO_2 reduction to C_{2+} products *ACS Catal.* **10** 5734–49

[6] Inoue T, Fujishima A, Konishi S and Honda K 1979 Photoelectrocatalytic reduction of carbon dioxide in aqueous suspensions of semiconductor powders *Nature* **277** 637–8

[7] Li X, Wang L, Xing Y and Su W 2022 Research progress in photo/electrocatalytic reduction of CO_2 by functionalized metal-organic frameworks *Sustain. Energy Fuels* **6** 3516–26

[8] Gao K and McKinley K R 1994 Use of macroalgae for marine biomass production and CO_2 remediation: a review *J. Appl. Phycol.* **6** 45–60

[9] Barber J 2009 Photosynthetic energy conversion: natural and artificial *Chem. Soc. Rev.* **38** 185–96

[10] Fujishima A and Honda K 1972 Electrochemical photolysis of water at a semiconductor electrode *Nature* **238** 37–8

[11] Sohn Y, Huang W and Taghipour F 2017 Recent progress and perspectives in the photocatalytic CO_2 reduction of Ti-oxide-based nanomaterials *Appl. Surf. Sci.* **396** 1696–711

[12] Wang P, Wang S, Wang H, Wu Z and Wang L 2018 Recent progress on photo-electrocatalytic reduction of carbon dioxide *Part. Part. Syst. Charact.* **35** 1700371

[13] Brito J F, de, Bessegato G G, Perini J A L, Torquato L D, de M and Zanoni M V B 2022 Advances in photoelectroreduction of CO_2 to hydrocarbons fuels: contributions of functional materials *J. CO_2 Util.* **55** 101810
[14] Fu Z *et al* 2019 Photocatalytic conversion of carbon dioxide: from products to design the catalysts. *J. CO_2 Util.* **34** 63–73
[15] Zhang W, Hu Y, Ma L, Zhu G, Wang Y, Xue X, Chen R, Yang S and Jin Z 2018 Progress and perspective of electrocatalytic CO_2 reduction for renewable carbonaceous fuels and chemicals *Adv. Sci.* **5** 1700275
[16] Xie S, Zhang Q, Liu G and Wang Y 2016 Photocatalytic and photoelectrocatalytic reduction of CO_2 using heterogeneous catalysts with controlled nanostructures *Chem. Commun.* **52** 35–59
[17] Castro S, Albo J and Irabien A 2018 Photoelectrochemical reactors for CO_2 utilization *ACS Sustain. Chem. Eng.* **6** 15877–94
[18] Chen P, Zhang Y, Zhou Y and Dong F 2021 Photoelectrocatalytic carbon dioxide reduction: fundamental, advances and challenges *Nano Mater. Sci.* **3** 344–67
[19] White J L *et al* 2015 Light-driven heterogeneous reduction of carbon dioxide: photocatalysts and photoelectrodes *Chem. Rev.* **115** 12888–935
[20] Li D, Yang K, Lian J, Yan J and Liu S F 2022 Powering the world with solar fuels from photoelectrochemical CO_2 reduction: basic principles and recent advances *Adv. Energy Mater.* **12** 2201070
[21] Mandal S, Ghosh D and Kumar P 2022 Recent advancement in heterogeneous CO_2 reduction processes in aqueous electrolyte *J. Mater. Chem.* A **10** 20667–706
[22] Dhabarde N, Selvaraj J, Yuda A, Kumar A and Subramanian V R 2022 Review of photocatalytic and photo-electrocatalytic reduction of CO_2 on carbon supported films *Int. J. Hydrogen Energy* **47** 30908–36
[23] Kumaravel V, Bartlett J and Pillai S C 2020 Photoelectrochemical conversion of carbon dioxide (CO_2) into fuels and value-added products *ACS Energy Lett.* **5** 486–519
[24] Xu Y, Wang S, Yang J, Han B, Nie R, Wang J J, Dong Y, Yu X, Wang J J and Jing H 2018 Highly efficient photoelectrocatalytic reduction of CO_2 on the Ti_3C_2/g-C_3N_4 heterojunction with rich Ti^{3+} and pyri-N species *J. Mater. Chem.* A **6** 15213–20
[25] Cheng J, Zhang M, Wu G, Wang X, Zhou J and Cen K 2014 Photoelectrocatalytic reduction of CO_2 into chemicals using Pt-modified reduced graphene oxide combined with Pt-modified TiO_2 nanotubes *Environ. Sci. Technol.* **48** 7076–84
[26] Rezaul Karim K M, Tarek M, Ong H R, Abdullah H, Yousuf A, Cheng C K and Khan M M R 2019 Photoelectrocatalytic reduction of carbon dioxide to methanol using $CuFe_2O_4$ modified with graphene oxide under visible light irradiation *Ind. Eng. Chem. Res.* **58** 563–72
[27] Hiragond C B, Lee J, Kim H, Jung J W, Cho C H and In S I 2021 A novel N-doped graphene oxide enfolded reduced titania for highly stable and selective gas-phase photocatalytic CO_2 reduction into CH_4: an in-depth study on the interfacial charge transfer mechanism *Chem. Eng. J.* **416** 127978
[28] Lin X, Wang S, Tu W, Wang H, Hou Y, Dai W and Xu R 2019 Magnetic hollow spheres assembled from graphene-encapsulated nickel nanoparticles for efficient photocatalytic CO_2 reduction *ACS Appl. Energy Mater.* **2** 7670–8
[29] Gan G, Li X, Fan S, Wang L, Qin M, Yin Z and Chen G 2019 Carbon aerogels for environmental clean-up *Eur. J. Inorg. Chem.* **2019** 3126–41

[30] Pan Z, Liu M, Yang J, Qiu Y, Li W, Xu Y, Zhang X and Zhang Y 2017 High electroactive material loading on a carbon nanotube@3D graphene aerogel for high-performance flexible all-solid-state asymmetric supercapacitors *Adv. Funct. Mater.* **27** 1701122

[31] Wang R, Lu K Q, Zhang F, Tang Z R and Xu Y J 2018 3D carbon quantum dots/graphene aerogel as a metal-free catalyst for enhanced photosensitization efficiency *Appl. Catal.* B **233** 11–8

[32] Jia X, Dai B, Zhu Z, Wang J, Qiao W, Long D and Ling L 2016 Strong and machinable carbon aerogel monoliths with low thermal conductivity prepared via ambient pressure drying *Carbon* **108** 551–60

[33] García-González C A, Alnaief M and Smirnova I 2011 Polysaccharide-based aerogels—promising biodegradable carriers for drug delivery systems *Carbohydr. Polym.* **86** 1425–38

[34] Mao J, Iocozzia J, Huang J, Meng K, Lai Y and Lin Z 2018 Graphene aerogels for efficient energy storage and conversion *Energy Environ. Sci.* **11** 772–99

[35] Liu X, Liu H, Mi M, Kong W, Ge Y and Hu J 2019 Nitrogen-doped hierarchical porous carbon aerogel for high-performance capacitive deionization *Sep. Purif. Technol.* **224** 44–50

[36] Chen Z, Yao S and Liu L 2017 3D hierarchical porous structured carbon nanotube aerogel-supported Sn spheroidal particles: an efficient and selective catalyst for electrochemical reduction of CO_2 to formate *J. Mater. Chem.* A **5** 24651–6

[37] Kong Y, Shen X, Cui S and Fan M 2015 Facile synthesis of an amine hybrid aerogel with high adsorption efficiency and regenerability for air capture via a solvothermal-assisted sol–gel process and supercritical drying *Green Chem.* **17** 3436–45

[38] Martín-Jimeno F J, Suárez-García F, Paredes J I, Martínez-Alonso A and Tascón J M D 2015 Activated carbon xerogels with a cellular morphology derived from hydrothermally carbonized glucose-graphene oxide hybrids and their performance towards CO_2 and dye adsorption *Carbon* **81** 137–47

[39] Oh Y, Le V D, Maiti U N, Hwang J O, Park W J, Lim J, Lee K E, Bae Y S, Kim Y H and Kim S O 2015 Selective and regenerative carbon dioxide capture by highly polarizing porous carbon nitride *ACS Nano* **9** 9148–57

[40] Hu Y, Tong X, Zhuo H, Zhong L, Peng X, Wang S and Sun R 2016 3D hierarchical porous N-doped carbon aerogel from renewable cellulose: an attractive carbon for high-performance supercapacitor electrodes and CO_2 adsorption *RSC Adv.* **6** 15788–95

[41] Cai C, Xu Y F, Chen H Y, Wang X D and Kuang D B 2018 Porous ZnO@ZnSe nanosheet array for photoelectrochemical reduction of CO_2 *Electrochim. Acta* **274** 298–305

[42] Jang Y J, Jang J W, Lee J, Kim J H, Kumagai H, Lee J, Minegishi T, Kubota J, Domen K and Lee J S 2015 Selective CO production by Au coupled ZnTe/ZnO in the photoelectrochemical CO_2 reduction system *Energy Environ. Sci.* **8** 3597–604

[43] Yuan J, Gu C, Ding W and Hao C 2020 Photo-electrochemical reduction of carbon dioxide into methanol at $CuFeO_2$ nanoparticle-decorated $CuInS_2$ thin-film photocathodes *Energy Fuels* **34** 9914–22

[44] Li J M, Tsao C W, Fang M J, Chen C C, Liu C W and Hsu Y J 2018 TiO_2–Au–Cu_2O photocathodes: Au-mediated Z-scheme charge transfer for efficient solar-driven photoelectrochemical reduction *ACS Appl. Nano Mater.* **1** 6843–53

[45] Ghadimkhani G, de Tacconi N R, Chanmanee W, Janakyab C and Rajeshwar K 2013 Efficient solar photoelectrosynthesis of methanol from carbon dioxide using hybrid CuO–Cu_2O semiconductor nanorod arrays *Chem. Commun.* **49** 1297–9

[46] Rajeshwar K, de Tacconi N R, Ghadimkhani G, Chanmanee W and Janáky C 2013 Tailoring copper oxide semiconductor nanorod arrays for photoelectrochemical reduction of carbon dioxide to methanol *Chem. Phys. Chem.* **14** 2251–9

[47] Li P, Jing H, Xu J, Wu C, Peng H, Lu J and Lu F 2014 High-efficiency synergistic conversion of CO_2 to methanol using Fe_2O_3 nanotubes modified with double-layer Cu_2O spheres *Nanoscale* **6** 11380–6

[48] Won D H, Choi C H, Chung J and Woo S I 2014 Photoelectrochemical production of formic acid and methanol from carbon dioxide on metal-decorated CuO/Cu_2O-layered thin films under visible light irradiation *Appl. Catal.* B **158–159** 217–23

[49] Ba X, Yan L L, Huang S, Yu J, Xia X J and Yu Y 2014 New way for CO_2 reduction under visible light by a combination of a Cu electrode and semiconductor thin film: Cu_2O conduction type and morphology effect *J. Phys. Chem.* C **118** 24467–78

[50] de Brito J F, Araujo A R, Rajeshwar K and Zanoni M V B 2015 Photoelectrochemical reduction of CO_2 on Cu/Cu_2O films: product distribution and pH effects *Chem. Eng. J.* **264** 302–9

[51] Lee K, Lee S, Cho H, Jeong S, Kim W D, Lee S and Lee D C 2018 Cu^+-incorporated TiO_2 overlayer on Cu_2O nanowire photocathodes for enhanced photoelectrochemical conversion of CO_2 to methanol *J. Energy Chem.* **27** 264–70

[52] Kang H Y, Nam D H, Yang K D, Joo W, Kwak H, Kim H H, Hong S H, Nam K T and Joo Y C 2018 Synthetic mechanism discovery of monophase cuprous oxide for record high photoelectrochemical conversion of CO_2 to methanol in water *ACS Nano* **12** 8187–96

[53] Zeng Z, Yan Y, Chen J, Zan P, Tian Q and Chen P 2019 Boosting the photocatalytic ability of Cu_2O nanowires for CO_2 conversion by MXene quantum dots *Adv. Funct. Mater.* **29** 1806500

[54] Galante M T, Santiago P V B, Yukuhiro V Y, Silva L A, Dos Reis N A, Pires C T G V M T, Macedo N G, Costa L S, Fernandez P S and Longo C 2021 Aminopolysiloxane as Cu_2O photocathode overlayer: photocorrosion inhibitor and low overpotential CO_2-to-formate selectivity promoter *Chem. Cat. Chem.* **13** 859–63

[55] Zhang Y, Pan D, Tao Y, Shang H, Zhang D, Li G and Li H 2022 Photoelectrocatalytic reduction of CO_2 to syngas via SnO_x-enhanced Cu_2O nanowires photocathodes *Adv. Funct. Mater.* **32** 2109600

[56] Rezaul Karim K M, Ong H R, Abdullah H, Yousuf A, Cheng C K and Rahman Khan M M 2018 Photoelectrochemical reduction of carbon dioxide to methanol on p-type $CuFe_2O_4$ under visible light irradiation *Int. J. Hydrogen Energy* **43** 18185–93

[57] Oh S H, Kang H Y, Joo W H and Joo Y C 2020 Photoelectrochemical CO_2 reduction via $Cu_2O/CuFeO_2$ hierarchical nanorods photocatalyst *Chem. Cat. Chem.* **12** 5185–91

[58] Gu J, Wuttig A, Krizan J W, Hu Y, Detweiler Z M, Cava R J and Bocarsly A B 2013 Mg-doped $CuFeO_2$ photocathodes for photoelectrochemical reduction of carbon dioxide *J. Phys. Chem.* C **117** 12415–22

[59] Kang U, Choi S K, Ham D J, Ji S M, Choi W, Han D S, Abdel-Wahab A and Park H 2015 Photosynthesis of formate from CO_2 and water at 1% energy efficiency via copper iron oxide catalysis *Energy Environ. Sci.* **8** 2638–43

[60] Yang X, Fugate E A, Mueanngern Y and Baker L R 2017 Photoelectrochemical CO_2 reduction to acetate on iron-copper oxide catalysts *ACS Catal.* **7** 177–80

[61] Yuan J and Hao C 2013 Solar-driven photoelectrochemical reduction of carbon dioxide to methanol at $CuInS_2$ thin film photocathode *Sol. Energy Mater. Sol. Cells* **108** 170–4

[62] Ouyang T, Ye Y Q, Tan C, Guo S T, Huang S, Zhao R, Zhao S and Liu Z Q 2022 1D α-Fe_2O_2/ZnO junction arrays modified by Bi as photocathode: high efficiency in photo-electrochemical reduction of CO_2 to HCOOH *J. Phys. Chem. Lett.* **13** 6867–74

[63] Jang J W, Cho S, Magesh G, Jang Y J, Kim J Y, Kim W Y, Seo J K, Kim S, Lee K H and Lee J S 2014 Aqueous-solution route to zinc telluride films for application to CO_2 reduction *Angew. Chem. Int. Ed.* **53** 5852–7

[64] Jang Y J, Jeong I, Lee J, Lee J, Ko M J and Lee J S 2016 Unbiased sunlight-driven artificial photosynthesis of carbon monoxide from CO_2 using a ZnTe-based photocathode and a perovskite solar cell in tandem *ACS Nano* **10** 6980–7

[65] Wen P, Li H, Ma X, Lei R, Wang X, Geyer S M and Qiu Y 2021 A colloidal ZnTe quantum dot-based photocathode with a metal–insulator–semiconductor structure towards solar-driven CO_2 reduction to tunable syngas *J. Mater. Chem.* A **9** 3589–96

[66] Li P, Zhang J, Wang H, Jing H, Xu J, Sui X, Hu H and Yin H 2014 The photoelectric catalytic reduction of CO_2 to methanol on CdSeTe NSs/TiO_2 NTs *Catal. Sci. Technol.* **4** 1070–7

[67] Wei W, Yang Z, Song W, Hu F, Luan B, Li P and Yin H 2017 Different CdSeTe structure determined photoelectrocatalytic reduction performance for carbon dioxide *J. Colloid Interface Sci.* **496** 327–33

[68] Ikeda S, Fujikawa S, Harada T, Nguyen T H, Nakanishi S, Takayama T, Iwase A and Kudo A 2019 Photocathode characteristics of a spray-deposited Cu_2ZnGeS_4 thin film for CO_2 reduction in a CO_2-saturated aqueous solution *ACS Appl. Energy Mater.* **2** 6911–8

[69] Zhou S, Sun K, Huang J, Lu X, Xie B, Zhang D, Hart J N, Toe C Y, Hao X and Amal R 2021 Accelerating electron-transfer and tuning product selectivity through surficial vacancy engineering on CZTS/CdS for photoelectrochemical CO_2 reduction *Small* **17** 2100496

[70] Chen J, Yin J, Zheng X, Ait Ahsaine H, Zhou Y, Dong C, Mohammed O F, Takanabe K and Bakr O M 2019 Compositionally screened eutectic catalytic coatings on halide perovskite photocathodes for photoassisted selective CO_2 reduction *ACS Energy Lett.* **4** 1279–86

[71] Wang Y, Fan S, AlOtaibi B, Wang Y, Li L and Mi Z 2016 A monolithically integrated gallium nitride nanowire/silicon solar cell photocathode for selective carbon dioxide reduction to methane *Chem. Eur.* J. **22** 8809–13

[72] Zhou B *et al* 2019 A GaN:Sn nanoarchitecture integrated on a silicon platform for converting CO_2 to HCOOH by photoelectrocatalysis *Energy Environ. Sci.* **12** 2842–8

[73] Dong W J, Zhou P, Xiao Y, Navid I A, Lee J L and Mi Z 2022 Silver halide catalysts on GaN nanowires/Si heterojunction photocathodes for CO_2 reduction to syngas at high current density *ACS Catal.* **12** 2671–80

[74] Hu Y, Chen F, Ding P, Yang H, Chen J, Zha C and Li Y 2018 Designing effective Si/Ag interface via controlled chemical etching for photoelectrochemical CO_2 reduction *J. Mater. Chem.* A **6** 21906–12

[75] Wei L, Lin J, Xie S, Ma W, Zhang Q, Shen Z and Wang Y 2019 Photoelectrocatalytic reduction of CO_2 to syngas over Ag nanoparticle modified p-Si nanowire arrays *Nanoscale* **11** 12530–6

[76] Kan M, Yan Z W, Wang X, Hitt J L, Xiao L, McNeill J M, Wang Y, Zhao Y and Mallouk T E 2020 2-aminobenzenethiol-functionalized silver-decorated nanoporous silicon photo-electrodes for selective CO_2 reduction *Angew. Chem. Int. Ed.* **59** 11462–9

[77] Gurudayal , Beeman J W, Bullock J, Wang H, Eichhorn J, Towle C, Javey A, Toma F M, Mathews N and Ager J W 2019 Si photocathode with Ag-supported dendritic Cu catalyst for CO_2 reduction *Energy Environ. Sci.* **12** 1068–77
[78] Ding P *et al* 2019 Controlled chemical etching leads to efficient silicon–bismuth interface for photoelectrochemical CO_2 reduction to formate *Mater. Today Chem.* **11** 80–5
[79] Gong Q *et al* 2019 Structural defects on converted bismuth oxide nanotubes enable highly active electrocatalysis of carbon dioxide reduction *Nat. Commun.* **10** 2807
[80] Kempler P A, Richter M H, Cheng W H, Brunschwig B S and Lewis N S 2020 Si microwire-array photocathodes decorated with cu allow CO_2 reduction with minimal parasitic absorption of sunlight *ACS Energy Lett.* **5** 2528–34
[81] Roh I, Yu S, Lin C K, Louisia S, Cestellos-Blanco S and Yang P 2022 Photoelectrochemical CO_2 reduction toward multicarbon products with silicon nanowire photocathodes interfaced with copper nanoparticles *J. Am. Chem. Soc.* **144** 8002–6
[82] Rao K R, Pishgar S, Strain J, Kumar B, Atla V, Kumari S and Spurgeon J M 2018 Photoelectrochemical reduction of CO_2 to HCOOH on silicon photocathodes with reduced SnO_2 porous nanowire catalysts *J. Mater. Chem.* A **6** 1736–42
[83] Jiang X X, De H X, Tarek M, Saravanan P, Alqadhi R, Chin S Y and Rahman Khan M M 2020 Tailoring the properties of g-C_3N_4 with CuO for enhanced photoelectrocatalytic CO_2 reduction to methanol *J. CO_2 Util.* **40** 101222
[84] Wang Q, Wang X, Yu Z, Jiang X, Chen J, Tao L, Wang M and Shen Y 2019 Artificial photosynthesis of ethanol using type-II g-C_3N_4/ZnTe heterojunction in photoelectrochemical CO_2 reduction system *Nano Energy* **60** 827–35
[85] Sahara G, Kumagai H, Maeda K, Kaeffer N, Artero V, Higashi M, Abe R and Ishitani O 2016 Photoelectrochemical reduction of CO_2 coupled to water oxidation using a photocathode with a Ru(II)–Re(I) complex photocatalyst and a CoO_x/TaON photoanode *J. Am. Chem. Soc.* **138** 14152–8
[86] Kamata R, Kumagai H, Yamazaki Y, Sahara G and Ishitani O 2019 Photoelectrochemical CO_2 reduction using a Ru(II)–Re(I) supramolecular photocatalyst connected to a vinyl polymer on a NiO electrode *ACS Appl. Mater. Interfaces* **11** 5632–41
[87] Sekizawa K, Sato S, Arai T and Morikawa T 2018 Solar-driven photocatalytic co_2 reduction in water utilizing a ruthenium complex catalyst on p-type Fe_2O_3 with a multiheterojunction *ACS Catal.* **8** 1405–16
[88] Kuk S K *et al* 2020 CO_2-reductive, copper oxide-based photobiocathode for Z-scheme semi-artificial leaf structure *Chem. Sus. Chem.* **13** 2940–4
[89] Madhusudan P, Wageh S, Al-Ghamdi A A, Zhang J, Cheng B and Yu Y 2020 Graphene–$Zn_{0.5}Cd_{0.5}S$ nanocomposite with enhanced visible-light photocatalytic CO_2 reduction activity *Appl. Surf. Sci.* **506** 144683
[90] Zhang Z, Shu M, Jiang Y and Xu J 2021 Fullerene modified $CsPbBr_3$ perovskite nanocrystals for efficient charge separation and photocatalytic CO_2 reduction *Chem. Eng. J.* **414** 128889
[91] Olowoyo J O, Kumar M, Jain S L, Babalola J O, Vorontsov A V and Kumar U 2019 Insights into reinforced photocatalytic activity of the CNT–TiO_2 nanocomposite for CO_2 reduction and water splitting *J. Phys. Chem.* C **123** 367–78
[92] Tan L L, Ong W J, Chai S P, Goh B T and Mohamed A R 2015 Visible-light-active oxygen-rich TiO_2 decorated 2D graphene oxide with enhanced photocatalytic activity toward carbon dioxide reduction *Appl. Catal.* B **179** 160–70

[93] Xing M, Shen F, Qiu B and Zhang J 2014 Highly-dispersed boron-doped graphene nanosheets loaded with TiO_2 nanoparticles for enhancing CO_2 photoreduction *Sci. Rep.* **4** 6341
[94] Bie C, Zhu B, Xu F, Zhang L and Yu J 2019 *In situ* grown monolayer N-doped graphene on CdS hollow spheres with seamless contact for photocatalytic CO_2 reduction *Adv. Mater.* **31** 1902868
[95] Wang X, Li Q, Zhou C, Cao Z and Zhang R 2019 ZnO rod/reduced graphene oxide sensitized by α-Fe_2O_2 nanoparticles for effective visible-light photoreduction of CO_2 *J. Colloid Interface Sci.* **554** 335–43
[96] Fu Z C *et al* 2018 Highly efficient photocatalytic system constructed from cop/carbon nanotubes or graphene for visible-light-driven CO_2 reduction *Chem. Eur.* J **24** 4273–8
[97] Ma B, Blanco M, Calvillo L, Chen L, Chen G, Lau T C, Dražić G, Bonin J, Robert M and Granozzi G 2021 Hybridization of molecular and graphene materials for CO_2 photocatalytic reduction with selectivity control *J. Am. Chem. Soc.* **143** 8414–25
[98] Wang Y, Cai Q, Yao M, Kang S, Ge Z and Li X 2018 Easy synthesis of ordered mesoporous carbon-carbon nanotube nanocomposite as a promising support for CO_2 photoreduction *ACS Sustain. Chem. Eng.* **6** 2529–34
[99] Yadav D, Yadav R K, Kumar A, Park N J and Baeg J O 2016 Functionalized graphene quantum dots as efficient visible-light photocatalysts for selective solar fuel production from CO_2 *Chem. Cat. Chem.* **8** 3389–93
[100] Yan Y, Chen J, Li N, Tian J, Li K, Jiang J, Liu J, Tian Q and Chen P 2018 Systematic bandgap engineering of graphene quantum dots and applications for photocatalytic water splitting and CO_2 reduction *ACS Nano* **12** 3523–32
[101] Hasan M R, Hamid S B A and Basirun W J 2015 Charge transfer behavior of graphene-titania photoanode in CO_2 photoelectrocatalysis process *Appl. Surf. Sci.* **339** 22–7
[102] Hasan M R, Abd Hamid S B, Basirun W J, Meriam Suhaimy S H and Che Mat A N 2015 A sol–gel derived, copper-doped, titanium dioxide-reduced graphene oxide nanocomposite electrode for the photoelectrocatalytic reduction of CO_2 to methanol and formic acid *RSC Adv.* **5** 77803–13
[103] Razzaq A, Grimes C A and Il S 2016 Facile fabrication of a noble metal-free photocatalyst: TiO_2 nanotube arrays covered with reduced graphene oxide *Carbon* **98** 537–44
[104] Zhang M, Cheng J, Xuan X, Zhou J and Cen K 2016 CO_2 synergistic reduction in a photoanode-driven photoelectrochemical cell with a Pt-modified TiO_2 nanotube photoanode and a Pt reduced graphene oxide electrocathode *ACS Sustain. Chem. Eng.* **4** 6344–54
[105] De Souza M K R, Cardoso E D S F, Fortunato G V, Lanza M R V, Nazário C E, Zanoni M V B, Maia G and Cardoso J C 2021 Combination of Cu–Pt–Pd nanoparticles supported on graphene nanoribbons decorating the surface of TiO_2 nanotube applied for CO_2 photoelectrochemical reduction *J. Environ. Chem. Eng.* **9** 105803
[106] Kecsenovity E, Endrodi B, Pápa Z, Hernádi K, Rajeshwar K and Janáky C 2016 Decoration of ultra-long carbon nanotubes with Cu_2O nanocrystals: a hybrid platform for enhanced photoelectrochemical CO_2 reduction *J. Mater. Chem.* A **4** 3139–47
[107] Kecsenovity E, Endrödi B, Tóth P S, Zou Y, Dryfe R A W, Rajeshwar K and Janáky C 2017 Enhanced photoelectrochemical performance of cuprous oxide/graphene nanohybrids *J. Am. Chem. Soc.* **139** 6682–92
[108] Szaniawska E, Rutkowska I A, Seta E, Tallo I, Lust E and Kulesza P J 2020 Photoelectrochemical reduction of CO_2: stabilization and enhancement of activity of

copper(I) oxide semiconductor by over-coating with tungsten carbide and carbide-derived carbons *Electrochim. Acta* **341** 136054

[109] Yuan J, Wang P, Hao C and Yu G 2016 Photoelectrochemical reduction of carbon dioxide at $CuInS_2$/graphene hybrid thin film electrode *Electrochim. Acta* **193** 1–6

[110] Andrei V, Jagt R A, Rahaman M, Lari L, Lazarov V K, MacManus-Driscoll J L, Hoye R L Z and Reisner E 2022 Long-term solar water and CO_2 splitting with photoelectrochemical BiOI–$BiVO_2$ tandems *Nat. Mater.* **21** 864–8

[111] Zhang H, Chen Y, Wang H, Wang H, Ma W, Zong X and Li C 2020 Carbon encapsulation of organic–inorganic hybrid perovskite toward efficient and stable photo-electrochemical carbon dioxide reduction *Adv. Energy Mater.* **10** 2002105

[112] Wang Q, Tao L, Jiang X, Wang M and Shen Y 2019 Graphene oxide wrapped $CH_3NH_3PbBr_3$ perovskite quantum dots hybrid for photoelectrochemical CO_2 reduction in organic solvents *Appl. Surf. Sci.* **465** 607–13

[113] Liu Z, Wang Z, Qing S, Xue N, Jia S, Zhang L, Li L, Li N, Shi L and Chen J 2018 Improving methane selectivity of photo-induced CO_2 reduction on carbon dots through modification of nitrogen-containing groups and graphitization *Appl. Catal.* B **232** 86–92

[114] Aoi S, Mase K, Ohkubo K, Suenobu T and Fukuzumi S 2017 Selective CO production in photoelectrochemical reduction of CO_2 with a cobalt chlorin complex adsorbed on multiwalled carbon nanotubes in water *ACS Energy Lett.* **2** 532–6

[115] Tang L, Jia C-t, Xue Y-C, Li L, Wang A-Q, Xu G, Liu N and Wu M-h 2017 Fabrication of compressible and recyclable macroscopic g-C_3N_4/GO aerogel hybrids for visible-light harvesting: a promising strategy for water remediation *Appl. Catal.* B **219** 241–8

[116] Zhang Y, Zhang D, Zhou L, Zhao Y, Chen J, Chen Z and Wang F 2018 Polypyrrole/reduced graphene oxide aerogel particle electrodes for high-efficiency electro-catalytic synergistic removal of Cr(VI) and bisphenol A *Chem. Eng. J.* **336** 690–700

[117] Jiang L, Zhang Y, Zhou M, Liang L and Li K 2018 Oxidation of rhodamine B by persulfate activated with porous carbon aerogel through a non-radical mechanism *J. Hazard. Mater.* **358** 53–61

[118] Hu J, Chen D, Li N, Xu Q, Li H, He J and Lu J 2018 3D aerogel of graphitic carbon nitride modified with perylene imide and graphene oxide for highly efficient nitric oxide removal under visible light *Small* **14** 1800416

[119] Liu Q, Shen J, Yang X, Zhang T and Tang H 2018 3D reduced graphene oxide aerogel-mediated Z-scheme photocatalytic system for highly efficient solar-driven water oxidation and removal of antibiotics *Appl. Catal.* B **232** 562–73

[120] Zhao H, Chen Y, Peng Q, Wang Q and Zhao G 2017 Catalytic activity of MOF(2Fe/Co)/carbon aerogel for improving H_2O_2 and [rad] OH generation in solar photo–electro–Fenton process *Appl. Catal.* B **203** 127–37

[121] Zhang M, Cheng J, Xuan X, Zhou J and Cen K 2017 Pt/graphene aerogel deposited in Cu foam as a 3D binder-free cathode for CO_2 reduction into liquid chemicals in a TiO_2 photoanode-driven photoelectrochemical cell *Chem. Eng. J.* **322** 22–32

[122] Yan S, Mahyoub S A, Lin J, Zhang C, Hu Q, Chen C, Zhang F and Cheng Z 2022 Au aerogel for selective CO_2 electroreduction to CO: ultrafast preparation with high performance *Nanotechnology* **33** 125705

[123] Zhang R, Huang Z, Li C, Zuo Y and Zhou Y 2019 Monolithic g-C_3N_4/reduced graphene oxide aerogel with *in situ* embedding of Pd nanoparticles for hydrogenation of CO_2 to CH_4 *Appl. Surf. Sci.* **475** 953–60

[124] Tong Z, Yang D, Shi J, Nan Y, Sun Y and Jiang Z 2015 Three-dimensional porous aerogel constructed by g-C_3N_4 and graphene oxide nanosheets with excellent visible-light photocatalytic performance *ACS Appl. Mater. Interfaces* **7** 25693–701
[125] Yang M, Wang P, Li Y, Tang S, Lin X, Zhang H, Zhu Z and Chen F 2022 Graphene aerogel-based NiAl-LDH/g-C_3N_4 with ultratight sheet-sheet heterojunction for excellent visible-light photocatalytic activity of CO_2 reduction *Appl. Catal.* B **306** 121065
[126] Jiang G, Wang J, Li N, Hübner R, Georgi M, Cai B, Li Z, Lesnyak V, Gaponik N and Eychmüller A 2022 Self-supported three-dimensional quantum dot aerogels as a promising photocatalyst for CO_2 reduction *Chem. Mater.* **34** 2687–95
[127] Wen M, Ren B, Ye X, He J, Jiang H and Xiong C 2022 Hierarchical loading of CuO on SiO_2 aerogel@high crystalline TiO_2 nanofibers for efficiently photocatalytic reduction of CO_2 without sacrificial agent *Nano Res.* **15** 3872–9

IOP Publishing

Materials Technology for the Energy and Environmental Nexus, Volume 2

R V Mangalaraja, Arunachalam Arulraj, Radhamanohar Aepuru and Chan Siew Hwa

Chapter 5

Two-dimensional advanced material nanocomposites for flexible energy storage devices

B V Bhaskara Rao, P Durga Prasad, Radhamanohar Aepuru, Manda Sathish, R V Mangalaraja and Rodrigo Espinoza-González

In recent years, the requirement for flexible and high-performance energy storage devices has increased substantially because of the rapidly increasing use of wearable electronic devices. In this context, many attempts have been made to boost the use of advanced 2D electrode materials, namely graphene, MXene, etc. The nanocomposites with novel 2D materials such as MXene and graphene using polymers are useful in making flexible device applications such as flexible supercapacitors, batteries, and nanogenerators. Novel 2D materials such as titanium carbide and nitrides and their composite forms with carbon nanostructures, such as graphene oxides and carbon nanotubes, using polymers with high volumetric capacitance and energy densities, could meet the current need for flexible energy storage devices such as lithium-ion batteries (LIBs) and supercapacitors. Flexible devices using 2D materials with high capacity and energy density are vital in flexible electronics because of their high strength, flexibility, bendability, low density, light weight, and ultra-thin properties. These 2D material composite flexible ultra-thin papers can be prepared using the cost-effective spray coating, solvent casting, and vacuum filtration methods. The 2D material MXenes/graphene in nanocomposites not only serve as an electronic conduction pathway for fast electron transport, which results in high charge storage, but also provides high strength free-standing papers that do not need the extra support of current collectors in energy storage devices. Further, these 2D material flexible electrodes can lead to cost effective flexible electronics such as wearable devices and implantable medical devices. This chapter

doi:10.1088/978-0-7503-5729-6ch5

focuses on the use of 2D materials for flexible electrode fabrication and the energy storage performance enhancement of flexible LIBs. It gives a brief overview of the various fabrication methods for flexible materials and the methods to improve the capacity and density of flexible storage devices, and discusses the mechanisms for understanding energy storage. Finally, we summarize the recent advances in this field and describe future opportunities for improving energy storage performance and flexible devices in the increasingly common wearable technology.

5.1 Introduction

The electronics industry has a rapidly increasing demand for portable and flexible energy storage devices. Wearable technological devices and electronic textiles are a key development area for research into thin, flexible film fabrication and flexible batteries. The standard batteries can meet the energy requirements of wearable devices, but they struggle to achieve flexibility, thinness, and light weight. This market needs to open the window for flexible energy storage devices with new fabrication aspects. High-power thin-film flexible batteries have the highest potential followed by printed rechargeable lithium/sodium/zinc batteries (figure 5.1) [1].

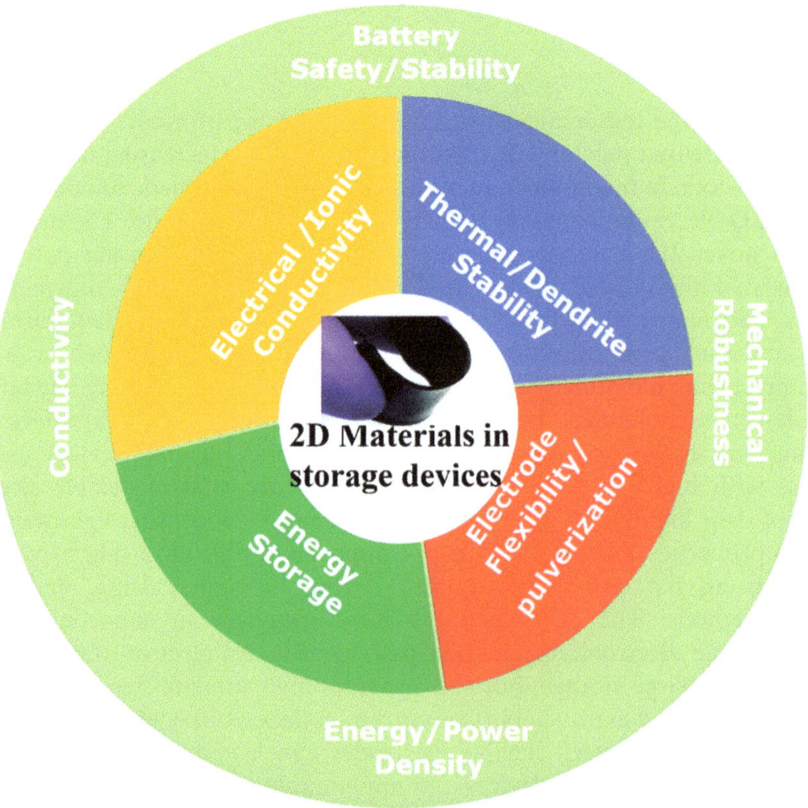

Figure 5.1. The use of 2D material properties in storage device applications.

This chapter discusses the role of two-dimensional materials and composites in flexible battery applications, addressing their challenges and opportunities. The 2D materials, namely MXene and graphene derivatives, are used widely in flexible battery applications because of their unique properties such as a layered structure, flexibility, electrical, ionic and thermal conductivity, and dendrite prevention, leading to their use in flexible electrode fabrication.

5.1.1 Scientific limitations of existing electrode materials for batteries

Flexible batteries, as a new form of power source, still have some big challenges and prospects for research [1]. The challenge is to design and develop novel flexible electrode materials with improved mechanical and electrochemical properties. In a lithium-ion cell, the anode materials are the negative electrodes and are paired with cathode materials of positive electrodes. These anode materials act as the host which reversibly admits lithium-ion intercalation/de-intercalation during charge/discharge cycles. The limitations of existing anode materials include high first cycle irreversible loss, low coulombic efficiency, low cycle life, slow lithium-ion diffusion in and out, low ionic and electronic conductivity, fast degradation (serious volume changes due to large structural changes upon charge and discharge), low specific capacity (mAh g^{-1}), and non-uniform solid electrolyte interface (SEI) layer formation upon cycling [3–5].

The design and fabrication of flexible anodes is crucial for flexible LIBs with improved electrochemical performance. Several inorganic materials such as RuO_2, Fe_3O_4, MnO_2, SnO_2, etc, and alloys such as Si, Ge, etc, and carbon compounds have been studied as anode materials, but most of them are brittle, have a low elastic deformation limit, and thus are not useful individually in fabricating flexible anodes for flexible LIBs [9, 10]. To overcome the flexibility issue while maintaining high electrochemical properties, we need intrinsically flexible conductive supports, such as MAX/MXene ($Ti_3C_2T_x$) papers, carbon nanotubes, porous carbon papers, and carbon nanofibers. Hybrid nanocomposites such as the 2D materials MAX/MXenes with MWCNT/rGO free-standing thin, flexible papers and some other nonconductive substrates, such as polymers, cellulose paper, or nano-dimensional active materials with high capacity and high elastic limit, are proposed to be important for preparing thin, cost-effective, and flexible anodes [2–4, 11].

Energy storage applications with low discharge specific capacity and power density have been explored well with conventional battery anode materials such as those reported in the literature. For instance, Staffolani *et al* reported a specific capacity of 980 mAh g^{-1} at 4C for Fe_3O_4@graphene compared to that of benchmarking active material (i.e. RuO_2) with a capacitance of 1000–1500 F cm^{-3}. However, the cost of RuO2 is high compared to that of the reported material which limits the development of flexible devices with high storage performance. [14]. In this regard, a nanocomposite of high capacitive materials, such as 2D materials (MXenes), and high aspect, conducting carbon-based systems, namely CNTs and rGO carbon nanostructures, are the solutions to explore [15–17]. Efficient energy storage devices should permit a large amount of charge storage per unit volume (high energy density) that can be released rapidly (high power density).

The currently available commercial batteries have either very low power or energy density and low discharge capacity, i.e. a few hundreds of mAh g^{-1} or 1/10 of hybrid battery to sustain with emerging needs of power storage and are very expensive to manufacture [16]. Zhao *et al* reported a capacitance of 350 F cm^{-3} at a current density of 5 A g^{-1} using an anode based on an MXene/carbon nanotube sandwich-like paper composite [23]. In this endeavor, flexible LIB batteries with 2D anode materials of optimized MXene@ carbon nanostructures have better electrical conductivity, a larger surface area available to ions, and mechanical flexibility, suggesting their suitability for applications in flexible electronics, load stabilization, power back-up, motorized industry, and wearable electronic applications. The electrochemical performance evaluation of the materials was analyzed using the below formulation and measurement methods [18–25].

The performance evaluation of MXene-CNT/rGO flexible papers as the anode of an LIB battery with polymer gel electrolyte paste containing a liquid electrolyte of 1 M $LiPF_6$ in ethylene carbonate (1:1 v/v), separator celgard (2320), and a standard cathode $LiCoO_2$ in the form of a sealed pouch cell and tested using a battery analyzer (Galvanostat, MTI-5V 1mA) and cyclic voltammetry (SP 300) to determine the electrochemical properties such as discharge capacity, charge–discharge, cycle life, rate performance, coulombic efficiency, etc.

The diffusion of Li-ions occurs due to small changes in the electrolyte concentration (*C*), which is a function of time (*t*) and distance (*x*) and is governed by

$$\frac{\partial C}{\partial t} = -\frac{D}{\varepsilon}\frac{\partial C}{\partial x}, \tag{5.1}$$

where D is the diffusion constant (7.5 × 10–^{10} m s^{-1} Li-ions in $LiPF_6$ electrolyte) and ϵ is the porosity of the electrolyte (0.7724).

$$\text{Discharge capacity}(Q) = \text{Applied current}(i) \times \text{Discharge time}(t)\,, \tag{5.2}$$

which is specific for the material weight or area.

$$\text{Coulombic efficiency} = (\text{Capacity discharged/Capacity charged}) \times 100, \tag{5.3}$$

where capacity discharged means the charge obtained in de-lithiation and capacity charged means the charge provided in lithiation.

5.1.2 Advantages of 2D material (MXene) intercalated CNT/rGO nanocomposites as anode materials for LIBs

MXenes are a recently emerging class of 2D materials discovered in 2011 by Dr Yury Gogotsi and his research team at Drexel University. The MXenes are transition metal carbides, carbonitrides, and nitrides. Gogotsi's team transformed the three-dimensional titanium–aluminum carbide, called the MAX phase, into a two-dimensional structure (Ti_3C_2) with different properties called MXene using ball-milling and chemical etching methods, namely hydrofluoric (HF) treatment (1.3 g LiF in 20 ml of 12 M HCl). The common formula of MXene is $M_{n+1}X_nT_x$ (n = 1–3), where M denotes an early transition metal (Sc, Ti, V, Cr, Mo, Ta, Nb, or Zr), and X is carbon/nitrogen, and T_x signifies for the surface terminating groups (OH, O, or F). Some of the widely used MXene examples are Ti_2CT_x, $Ti_3C_2T_x$, $Nb_4C_3T_x$, and $Mo_2TiC_2T_x$. The understanding of the layer nature and control of the ion dynamics between MXene sheets can enable the development of flexible, high-performance batteries [4, 23–30].

The MXene intercalated carbon nanotubes or graphene oxide anode materials in lithium-ion cells act as the host, which reversibly admit lithium-ion intercalation and de-intercalation during the charge and discharge cycles. The advantages of the layer structured 2D material of MXene intercalated with carbon nanostructured materials of rGO/CNT as the anode include: (i) low irreversible loss in the first cycle; (ii) high coulombic efficiency; (iii) fast lithium-ion diffusion into and out of the anode; (iv) high ionic and electronic conductivity; (v) low structural changes upon charge and discharge meaning low volume change; (vi) high theoretical specific capacity; (vii) the capability to form and maintain a stable, uniform SEI layer upon cycling; and (viii) a long cycle life (table 5.1) [4, 31–45].

The new batteries are categorized using numerous aspects including:

i. *Footprint or size of battery*: Micro-batteries or large-area batteries.
ii. *Thickness of battery*: Thin-film batteries or bulky batteries.
iii. *Manufacturing methods*: Printing, coating, or injecting.
iv. *Mechanical properties*: Flexibility, bendability, rollability, foldability, and stretchability.
v. *Technologies*: Solid-state, lithium–polymer batteries, carbon–zinc batteries, etc.

Table 5.1. Electrode designs for flexible batteries. (Adapted from [1]. CC BY 4.0.)

Electrode design	Advantages	Disadvantages	References
Thin film, stacked	Advanced fabrication	Low cell capacity, and complex fabrication	[8, 9]
Thin film, coplanar	Good flexibility	Low capacity and complicated fabrication	[10, 11]
Island-bridge	Fairly high cell capacity, superb flexibility and durability	Incompatible with liquid electrolytes, complex fabrication	[12–17]
Miura fold	Outstanding flexibility	Complicated fabrication and limited durability	[18, 19]
Wavy electrodes	Superb stretchability	Complicated fabrication	[20, 21]
Percolating network	Superb flexibility, self-healing	Complex fabrication and moderate conductivity of current collectors	[22–26]
Porous electrodes/ current collectors	Flexible designs and reasonably high capacity	Poor uniformity of the porous structures and lack of reliable characterizations	[27–35]
One-dimensional and textile	Superb flexibility and versatile designs	Low capacity and complicated fabrication	[36–45]

5.1.3 Evalution of 2D MXenes from MAX phases

The MXenes, mostly titanium carbide and nitrides from their respective MAX phases, are synthesized by metal-ion etching methods, namely the acid (HF/LiF +HCl) or alkali (NaOH) etching routes (figure 5.2) [1, 4].

5.1.4 Graphene with 2D layers and derivative composites

Graphene and it derivatives, such as graphene oxide (GO), reduced GO, graphene ribbons, and porous graphene, are synthesized by a modified Hummer's method from a graphite source. Among these graphene derivatives, for example, reduced GO is explored for flexible electrode fabrication for flexible supercapacitor devices because of its outstanding mechanical, electrical, thermal, and structural properties (figure 5.3) [2].

5.1.5 2D material nanocomposites in flexible batteries

2D materials such as MXenes and graphene derivatives are used widely in nanocomposites because of their usage in flexible electrode fabrication. The MXenes/graphene in nanocomposites will not only serve as an electronic conduction pathway for fast electron transport, which results in high charge storage, but also allow high strength free-standing papers that do not need the extra support of current collectors in energy storage devices (figure 5.4).

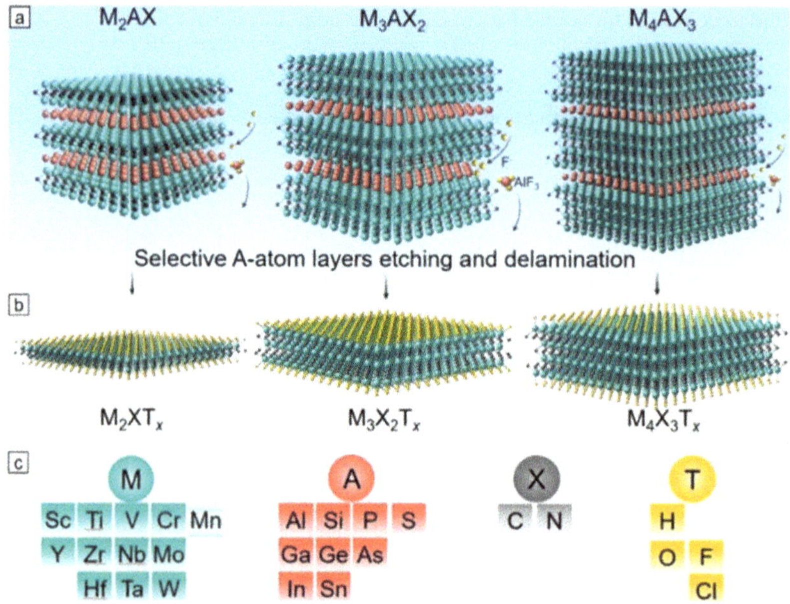

Figure 5.2. MXenes from MAX phases. (Reproduced with permission from [4]. Copyright 2020 Springer Nature and The Materials Research Society.)

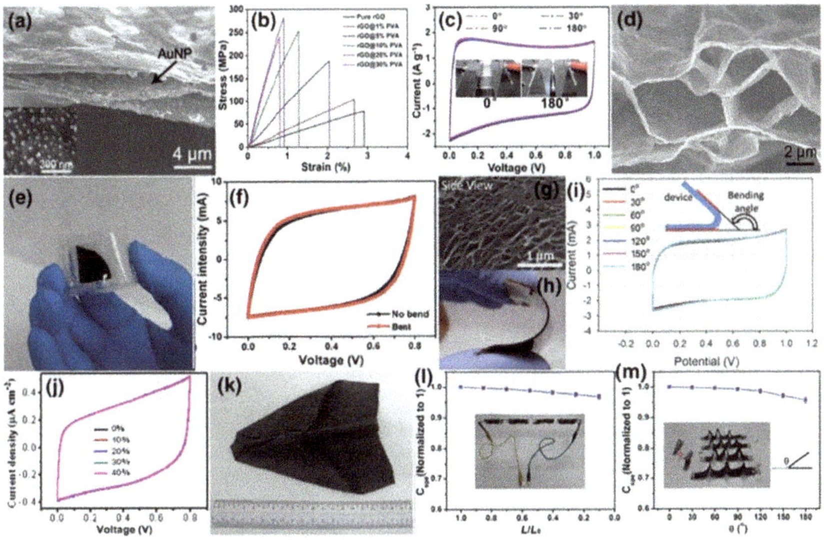

Figure 5.3. Graphene and composites use in supercapacitor applications. (Reproduced with permission from [2]. Copyright 2018 Elsevier.)

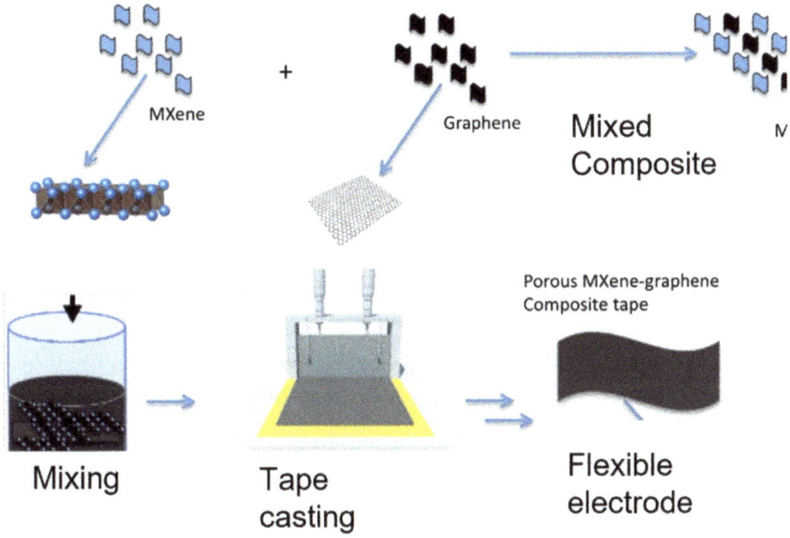

Figure 5.4. Preparation of 2D material based nanocomposites.

5.2 Evolution of flexible batteries and fabrication methods

Flexible batteries have recently been used increasingly in wearable electronics, portable devices, and bioelectronics, and the evolution of flexible batteries is detailed in figure 5.5.

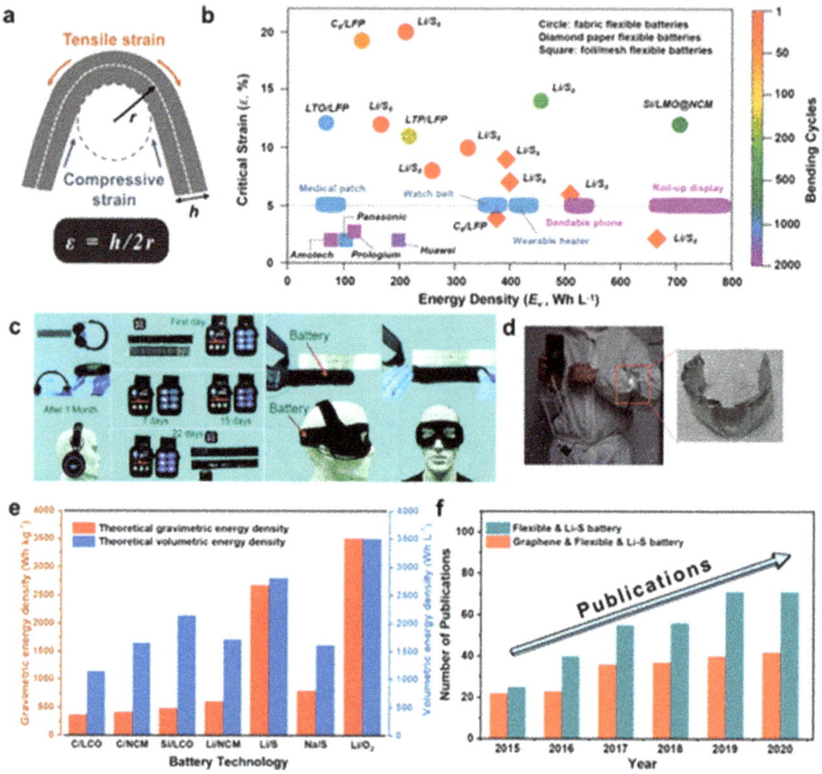

Figure 5.5. Evolution of flexible batteries. (Reproduced with permission from [3]. Copyright 2021 American Chemical Society.)

Flexible batteries are classified into various categories based on the fabrication method, components, and design, which are detailed in table 5.2.

5.2.1 Fabrication methods for flexible batteries

There are various methods for fabricating flexible batteries. First, a flexible electrode and flexible substrate are key in making a flexible battery.

5.2.2 Fabrication methods for flexible electrodes

The flexible electrode substrates have been fabricated using different designs and casting methods which are illustrated in figure 5.6.

The electrodes in figure 5.6 were tested under different conditions such as stretching, bending, and twisting while monitoring performance evolution. The results show that the performance of the electrodes was not much affected while applying these conditions. Thus these electrodes are effective in the final fabricated flexible battery without loss of performance.

Table 5.2. Categories of flexible batteries based on fabrication methods.

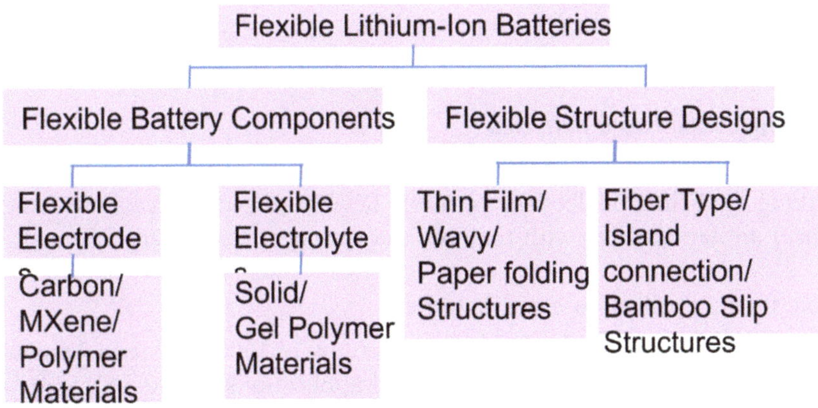

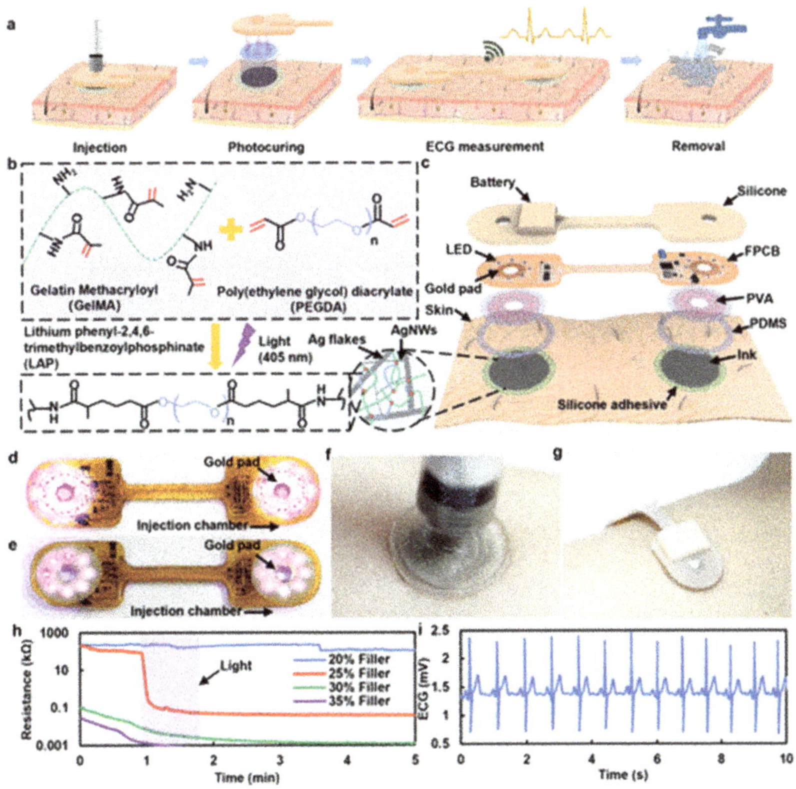

Figure 5.6. Flexible electrode fabrication processes by injection. (Reproduced with permission from [5]. Copyright 2023 American Chemical Society.)

5.3 Application based flexible batteries

There are varieties of flexible batteries based on their usage in medical and wearable electronics/textile applications, which are detailed in the following sections.

5.3.1 Medical application-based batteries

In the medical field there is a tremendous need for batteries in the range of 10 mAh for use in making medicinal devices for portable flexible transdermal patches and portable medical implant devices with the required specifications (figure 5.7).

5.3.2 Wearable textile application-based batteries

Wearable textiles with energy storage batteries with a range of 100 mAh are in demand because of their use in wearable electronics, medical applications (figure 5.8).

5.3.3 Wearable electronic device-based batteries

The rapid development and innovation in electronics technology requires flexible batteries with a range of 1 Ah for use in making the electronic devices that are flexible, bendable, foldable, lightweight, and small volume, with long working hours to meet user-friendly requirements (figure 5.9).

For example, Samsung launched the SDI band battery in 2015 for wearable electronic devices with a bending radius equal to the size of the human wrist. The global market for flexible batteries was valued at USD 69.5 million in 2015 and is expected to increase exponentially [1]. Other high-power and high-capacity

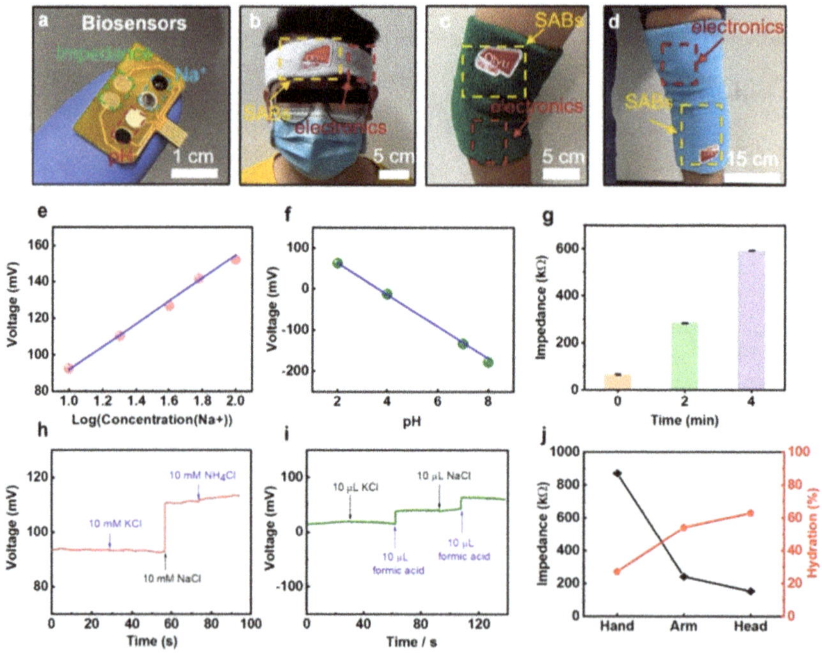

Figure 5.7. Flexible batteries in smart medicinal patch application. (Reproduced from [6]. CC BY 4.0.)

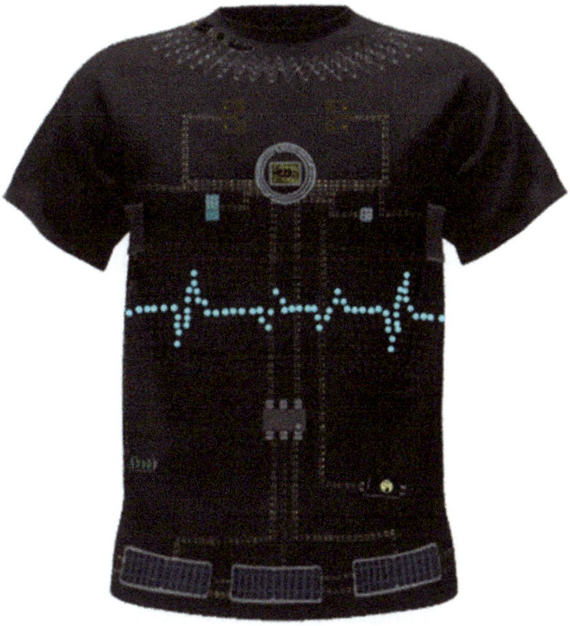

Figure 5.8. Flexible battery in a wearable textile application. (Reproduced with permission from [7]. Copyright 2022 American Chemical Society.)

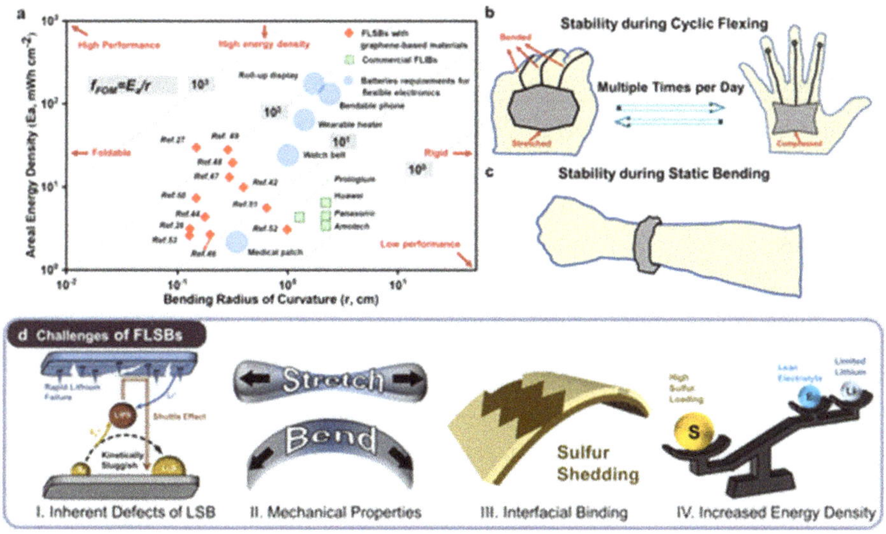

Figure 5.9. Flexible 2D material electrodes of batteries with flexible electronic devices. (Reproduced with permission from [3]. Copyright 2021 American Chemical Society.)

applications, such as transport in electronic-vehicles and large-scale storage systems (grids), with desirable designs would be allowed by flexible batteries with a capacity range from 100 Ah to more than 1 kAh.

5.4 Conclusions and future scope

Two-dimensional materials and their composites are types of flexible electrode materials that are used widely in energy storage applications because of their outstanding versatility, low cost, high strength, strong mechanical properties, simple processing, and superior performance and stability. In this chapter we summarize the recent progress in flexible electrodes with 2D materials, namely graphene and MXene based composites, focusing primarily on the electrode fabrication process, flexible design, incorporation of these electrodes in flexible batteries, enhancing the crystalline phase and piezoelectric coefficient, and energy harvesting applications. In addition, the application of these storage devices in various fields of study was discussed briefly. Through significant efforts, the output power of 2D material based electrodes is sufficient to power high-capacity lightweight and wearable electronic devices such as flexible cellphones and laptops, as well as electric watches, smart medicinal patches and so on. However, more work is needed to address specific issues, and the following research aspects must be considered in the future:

i. The incorporation of other functional layered materials could significantly improve the output performance of flexible lithium battery storage devices. As a result, a thorough understanding of the coupling mechanisms of various components would then aid in their coordination and provide a better understanding of storage devices with high capacity and light weight.
ii. Advanced polymers with high mechanical strength, bending, twisting, and stretching properties could enhance the flexibility of wearable devices.

Acknowledgments

The author acknowledges FONDECYT 2021 Project No. 3200452, ANID Chile for financial support.

References

[1] Zhao Y and Guo J 2020 Development of flexible Li-ion batteries for flexible electronics *Info. Mat.* **2** 866–78
[2] Arulraj A and Nishanthi S T 2020 2D materials–based flexible supercapacitors for high energy storage devices *Nanostructured, Functional, and Flexible Materials for Energy Conversion and Storage Systems* (Amsterdam: Elsevier) pp 417–36
[3] Tian Y *et al* 2021 Graphene-based materials for flexible lithium–sulfur batteries *ACS Nano* **15** 13901–23
[4] Hong W, Wyatt B, Nemani S K and Anasori B 2020 Double transition metal MXenes: atomistic design of two-dimensional carbides and nitrides *MRS Bull.* **45** 850–61
[5] Wan C *et al* 2023 *In situ* formation of conductive epidermal electrodes using a fully integrated flexible system and injectable photocurable ink *ACS Nano* **17** 10689–700
[6] Huang X *et al* 2022 Garment embedded sweat-activated batteries in wearable electronics for continuous sweat monitoring *npj Flex. Electron.* **10** 1–8
[7] Chen G *et al* 2022 Electronic textiles for wearable point-of-care systems *Chem. Rev.* **122** 3259–91

[8] Koo M *et al* 2012 Bendable inorganic thin-film battery for fully flexible electronic systems *Nano Lett.* **12** 4810–6

[9] Gockeln M, Glenneberg J, Busse M, Pokhrel S, Mädler L and Kun R 2018 Flame aerosol deposited $Li_4Ti_5O_{12}$ layers for flexible, thin film all-solid-state Li-ion batteries *Nano Energy* **49** 564–73

[10] Zheng S *et al* 2018 All-solid-state planar integrated lithium-ion micro-batteries with extraordinary flexibility and high-temperature performance *Nano Energy* **51** 613–20

[11] Zheng S *et al* 2018 All-solid-state flexible planar ithium ion micro-capacitors *Energy Environ. Sci.* **11** 2001–9

[12] Qian G *et al* 2018 Bioinspired, spine-like, flexible, rechargeable lithium-ion batteries with high energy density *Adv. Mater.* **30** e1704947

[13] Liao X *et al* 2019 High-energy-density foldable battery enabled by zigzag-like design *Adv. Energy Mater.* **9** 1802998

[14] Antunes S *et al* 2021 Fe_3O_4/graphene composite anode material for fast-charging Li-ion batteries *Molecules.* **26** 4316

[15] Park M, Cha H, Lee Y, Hong J, Kim S Y and Cho J 2017 Postpatterned electrodes for flexible node-type lithium-ion batteries *Adv. Mater.* **29** 1605773

[16] Xu S *et al* 2013 Stretchable batteries with self-similar serpentine interconnects and integrated wireless recharging systems *Nat. Commun.* **4** 1543

[17] Arulraj A, Ilayaraja N, Rajeshkumar V and Ramesh M 2019 Direct synthesis of cubic shaped Ag2S on Ni mesh as binder-free electrodes for energy storage applications *Sci. Rep.* **9** 10108

[18] Song Z *et al* 2014 Origami lithium-ion batteries *Nat. Commun.* **5** 3140

[19] Song Z *et al* 2015 Kirigami-based stretchable lithium-ion batteries *Sci. Rep.* **5** 10988

[20] Weng W *et al* 2015 A gum-like lithium-ion battery based on a novel arched structure *Adv. Mater.* **27** 1363–9

[21] Wang C, Zheng W, Yue Z, Too C O and Wallace G G 2011 Buckled, stretchable polypyrrole electrodes for battery applications *Adv. Mater.* **23** 3580–4

[22] Minsu G W J S *et al* 2019 Stretchable batteries with gradient multilayer conductors *Sci. Adv.* **5** eaaw1879

[23] Meng-Qiang Z *et al* 2015 Flexible MXene/carbon nanotube composite paper with high volumetric capacitance *Adv. Mater.* **27** 339–45

[24] Zhao Y *et al* 2016 A self-healing aqueous lithium-ion battery *Angew. Chem. Int. Ed. Engl.* **55** 14384–8

[25] Park S, Thangavel G, Parida K, Li S and Lee P S 2019 A stretchable and self-healing energy storage device based on mechanically and electrically restorative liquid-metal particles and carboxylated polyurethane composites *Adv. Mater.* **31** e1805536

[26] Sun Y *et al* 2016 A stretchable graphitic carbon/Si anode enabled by conformal coating of a self-healing elastic polymer *Adv. Mater.* **28** 2455–61

[27] Arulraj A, Mangalaraja R V and Khalid M 2022 MXene: Pioneering 2D Materials *Fundamental Aspects and Perspectives of MXenes* ed M Khalid, A N Grace, A Arulraj and A Numan (Berlin: Springer) pp 1–16

[28] Zhao S, Guo J, Jiang F, Su Q and Du G 2016 Porous $CoFe_2O_4$ nanowire arrays on carbon cloth as binder-free anodes for flexible lithium-ion batteries *Mater. Res. Bull.* **79** 22–8

[29] Amin K *et al* 2018 A carbonyl compoundbased flexible cathode with superior rate performance and cyclic stability for flexible lithium-ion batteries *Adv. Mater.* **30** 1703868

[30] Ni L, Jia L, Chen G, Wang F, Liu X and Ma R 2018 Facile synthesis of porous $FeCo_2O_4$ nanowire arrays on flexible carbon cloth with superior lithium storage properties *J. Phys. Chem. Solid* **122** 261–7

[31] Balogun M S *et al* 2018 Achieving high gravimetric energy density for flexible lithium-ion batteries facilitated by core–double-shell electrodes *Energy Environ. Sci.* **11** 1859–69

[32] Wang X *et al* 2014 Ultralong-life and high-rate weblike $Li_4Ti_5O_{12}$ anode for high-performance flexible lithiumion batteries *Nano Res.* **7** 1073–82

[33] Shen L, Uchaker E, Zhang X and Cao G 2012 Hydrogenated $Li_4Ti_5O_{12}$ nanowire arrays for high rate lithium ion batteries *Adv. Mater.* **24** 6502–6

[34] Xue L, Savilov S V, Lunin V V and Xia H 2018 Self-standing porous $LiCoO_2$ nanosheet arrays as 3D cathodes for flexible Li-ion batteries *Adv. Funct. Mater.* **28** 1705836

[35] Xia H, Xia Q, Lin B, Zhu J, Seo J K and Meng Y S 2016 Self-standing porous $LiMn_2O_4$ nanowall arrays as promising cathodes for advanced 3D microbatteries and flexible lithium-ion batteries *Nano Energy* **22** 475–82

[36] Kwon Y H *et al* 2012 Cable-type flexible lithium-ion battery based on hollow multi-helix electrodes *Adv. Mater.* **24** 5192–7 5145

[37] Ren J *et al* 2013 Twisting carbon nanotube fibers for both wire-shaped micro-supercapacitor and micro-battery *Adv. Mater.* **25** 1155–9 1224

[38] Weng W *et al* 2014 Winding aligned carbon nanotube composite yarns into coaxial fiber full batteries with high performances *Nano Lett.* **14** 3432–8

[39] Ren J *et al* 2014 Elastic and wearable wire shaped lithium-ion battery with high electrochemical performance *Angew. Chem. Int. Ed. Engl.* **53** 7864–9

[40] Zhang Y *et al* 2016 A fiber-shaped aqueous lithium-ion battery with high power density *J. Mater. Chem.* A **4** 9002–8

[41] Wang X *et al* 2019 Stretchable fiber-shaped lithium metal anode *Energy Storage Mater.* **22** 179–84

[42] Zhang Y *et al* 2014 Super-stretchy lithium-ion battery based on carbon nanotube fiber *J. Mater. Chem.* A **2** 11054–9

[43] Zhang Y *et al* 2014 Flexible and stretchable lithium-ion batteries and supercapacitors based on electrically conducting carbon nanotube fiber springs *Angew. Chem. Int. Ed. Engl.* **53** 14564–8

[44] Liu K *et al* 2018 Stretchable lithium metal anode with improved mechanical and electrochemical cycling stability *Joule* **2** 1857–65

[45] Wang Y *et al* 2017 3D-printed all-fiber Li-ion battery toward wearable energy storage *Adv. Funct. Mater.* **27** 1703140

Chapter 6

Recent advances in cobalt-based cathode materials for intermediate solid oxide fuel cells

Mudaliar Mahesh Margoni, Ali Akbari-Fakhrabadi and R V Mangalaraja

The solid oxide fuel cell (SOFC) is distinguished by its superior efficiency when compared to other types of fuel cells. The SOFC's operating temperature ranges between 800 °C and 1000 °C and remains a challenge for its use. The primary aim is to reduce the SOFC's operating temperature to below 1000 °C in order to enable more applications and improve system longevity. It is necessary to make improvements to the SOFC's primary components, i.e. the air/oxygen electrode referred to as a cathode, in order to increase the electrochemical performance of the SOFC. The materials known to be employed as SOFC cathodes include spinel-based cobaltite, ferro-cobaltite, and perovskite-based compounds. However, the main hurdles for the SOFC cathode are to identify a suitable material with good electrocatalytic capability at lower operational temperatures of 400 °C–800 °C. However, at lower temperatures, the oxygen reduction reaction (ORR) at the cathode is slow and its exposure to Cr-based interconnects causes chromium (Cr) poisoning. Also the degradation of materials in humid and CO_2 environments and the suitability of current intermediate temperature (IT-SOFC) electrolytes is still a problem for their widespread development. Further, the performance of SOFCs is also greatly influenced by the composition and microstructure of the cathode materials. It is possible to improve the ionic and electronic conductivities as well as the catalytic characteristics for oxygen reduction in the cathode by designing the composition of materials rationally through regulated oxygen non-stoichiometry and defect features. This chapter examines current advancements in SOFC cathodes with a focus on cobalt-based cathode material selection. Finally, some insights on the potential outlook of research will be provided as a conclusion to this overview.

doi:10.1088/978-0-7503-5729-6ch6

6.1 Introduction

Fuel cells use oxidation and reduction reactions, or redox reactions, to transform fuels into electrical energy [1]. A fuel cell has a higher electrical efficiency (up to 60%) than a standard generator, when hydrogen is utilised as a fuel it produces only water, and it operates more quietly since it has fewer moving parts [2]. The distinctive characteristic of a solid oxide fuel cell (SOFC) using a solid oxide as its electrolyte is that it must be run at a high temperature (550 °C–1000 °C) to obtain adequate performance. Within this temperature range, it is now feasible to employ a less expensive catalyst, bringing down the overall cost of the fuel cell. As a more affordable alternative to hydrogen, it is also possible to use biofuels and hydrocarbon fuels [3, 4]. The slower kinetic oxygen reduction reaction (ORR) in the cathode component of the cell has a significant impact on the performance of the cell. As a result, selecting the cathode is far more crucial than selecting the anode. Typically, the primary cathode electrode reactions take place at the electrode, electrolyte, and gas, known as the triple phase boundaries (TPBs) [5, 6]. The following elementary reactions are proposed for composite electrode or mixed ion–electron conductor (MIEC) materials: (i) the process of reducing oxygen (O_2) molecules includes adsorption, reduction, dissociation, and incorporation of the oxygen anion into the materials' lattice; (ii) the ion hopping into the lattice of the electrolyte; and (iii) ionic transfer toward the electrolyte through a porous cathode [7]. A few steps among the three basic reactions could be the rate-limiting steps in the oxygen reduction process. It is not always possible to foresee which elementary reaction will be the rate-limiting step; it may depend on regional factors such as temperature and oxygen partial pressure as well as microstructural circumstances. The oxygen reduction process accounts for the majority of the overall cell resistance, therefore enhancements to the cathode's catalytic activity have a significant influence on the performance of the cell as a whole. In addition to the ORR, the polarisation effect degrades the SOFC's performance [8]. Polarisation refers to voltage losses caused by the choice of the fuel cell's materials and microstructure. There are three forms of polarisation: ohmic polarisation, concentration polarisation, and activation polarisation [9]. Consequently, a low value of polarisation resistance often denotes a fuel cell's performance [10]. Numerous studies have been conducted to increase the kinetics of ORR and minimise polarisation resistance by better understanding the ORR process. If the elements are combined properly, an electronic conductor or MIEC can be produced that has a reduced activation energy for ORR in the cathode as well as retaining its power density and high electrical conductivity [11, 12]. The materials' increased porosity ensures that gases diffuse quickly, reducing concentration polarisation and increasing interaction between the cathode and oxygen gas [13]. Degradation is less likely if the cathode, anode, and electrolyte materials are combined properly. These materials should have less interfacial reactivity with one another and should also be impervious to contaminants. To avoid mechanical damage at high temperatures, there must not be much of a difference in the thermal expansion coefficient (TEC) between the electrolyte and cathode layers [14, 15].

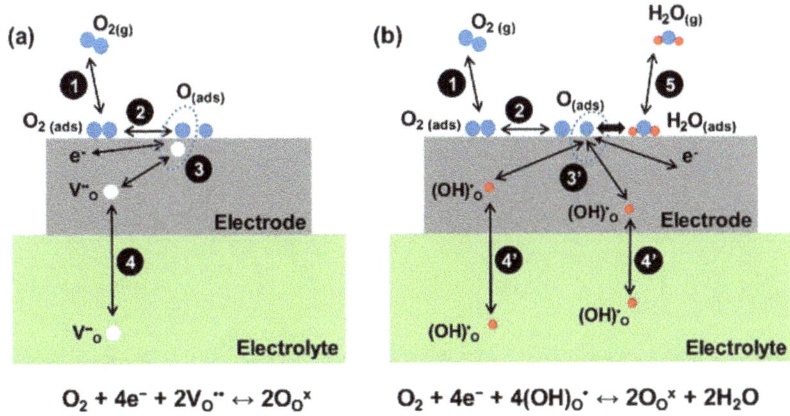

Figure 6.1. Schematic illustration of the operating mechanism of SOFC. Reaction mechanism for (a) ion-oxide-based and (b) ideal 'pure' proton-conducting SOFC. (Reproduced from [16]. Copyright 2016 The Authors. Published by The Electrochemical Society and Institute of Physics Publishing.)

A schematic of an SOFC with electrode reactions is shown in figure 6.1 [16]. The oxygen reduction process, in which O_2 is reduced to O^{2-} ions at the cathode by consuming electrons (ORR), occurs when an external load is applied. The oxygen ion conducting solid electrolyte allows these O^{2-} ions to diffuse through it. The cathode reaction is expressed as follows:

$$\frac{1}{2}O2 + 2e^{'} \leftrightarrows O^{2-}. \tag{6.1}$$

As seen from the ORR (equation (6.1)), it includes a complicated interaction between mobile electronic (e′), ionic (O_2), and molecular species (O_2) at the TPB, as illustrated in figure 6.2 [17]. Hence the cathode material needs high electronic conductivity whereas high conductivity is required for the solid electrolyte (SE) for oxygen ions to diffuse to the anode. Furthermore, the cathode should have adequate porosity to allow gas penetration to and from the reaction sites [18]. The oxide ions appear at the anode after migrating through the electrolyte and react with fuel (H_2 or CO) as follows:

$$H_2 + O^{2-} \leftrightarrows H_2O + 2e'. \tag{6.2}$$

Electrons generated during the reactions are transported by the current collector to an external load through an electrical circuit that is linked externally [19].

As a result, the whole cell response may be stated as follows:

$$H_2 + \frac{1}{2}O^2 \leftrightarrows H_2O. \tag{6.3}$$

The following Nernst equation describes the open-circuit voltage (OCV) associated with the reaction mentioned in equation (6.3) [20]:

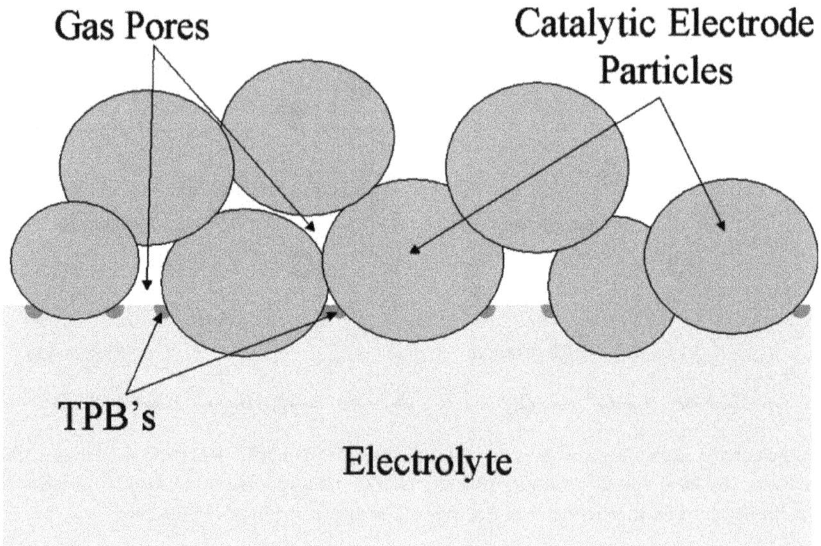

Figure 6.2. A simplified schematic diagram of the electrode/electrolyte interface in a fuel cell, where the catalytically active electrode particles, electrolyte phase, and gas pores intersect. (Reproduced with permission from [17]. Copyright 2005 The Electrochemical Society. Published by Institute of Physics Publishing.)

$$E = E^0 + \frac{RT}{2F} \ln \frac{\left(p_{H_2} p_{O_2}{}^{0.5}\right)}{p_{H_2O}}, \tag{6.4}$$

where R is the universal gas constant, F is the Faraday constant, p is the gas pressure, T is the absolute temperature, and E is the equilibrium potential. The denominator's coefficient '2' indicates how many electrons are transported for every mole of oxygen that reacts in the cell. As a result, the mass action expression is

$$K_{eq} = \frac{P_{H_2O}}{(P_{H_2})(P_{O_2})^{1/2}}, \tag{6.5}$$

where the fuel gas reaction's equilibrium constant is denoted by K_{eq}. Following that, equations (6.6) and (6.7) may be used to represent the Gibbs free energy change (G) and oxygen partial pressure (p_{O_2}) at the anode side, assuming no water condensation happens in the gas supply system [19]:

$$\Delta GT = \Delta H_T - T\Delta S_T = -RT \ln\left(K_{eq}\right). \tag{6.6}$$

Here, ΔH_T is heat change, ΔS_T is entropy change, and

$$p_{O_2} = \left(\frac{p_{H_2O}}{p_{H_2}}\right)^2 \exp\left[\frac{2(T\Delta S_T - \Delta H_T)}{RT}\right]. \tag{6.7}$$

With the aid of equations (6.4) and (6.7), the theoretical Nernst potential can be calculated at any SOFC operating temperature. Equation (6.4) illustrates the connection between the ideal standard potential E_0, for the SOFC under standard circumstances for the processes, and the ideal or open-circuit potential E at any other given temperature and pressure of the reactants and products. When the current is drawn, the actual cell potential is lower than its equilibrium potential. Figure 6.3 depicts an I–V curve schematic with different polarisation losses [21]. The decrease in cell potential is related to a variety of losses or polarisations depending on the electrode and the electrolyte materials, their microstructures, cell design, and working temperatures [22–24]. Hence, a cell's total polarisation (η), which may be expressed as

$$\eta = \eta_{\text{ohm}} + \eta_{\text{act}} + \eta_{\text{conc}}. \tag{6.8}$$

As a result, the actual cell potential can be estimated using the following equation:

$$E_{\text{cell}} = E_{\text{Nernst}} - \eta_{\text{ohm}} - \eta_{\text{act}} - \eta_{\text{conc}}.$$

The overall performance of the SOFC at lower temperatures is highly influenced by polarisations, which are closely correlated with operating temperatures [25]. Cathode polarisation appears to be a key challenge in lowering the operating temperature of SOFCs in the 400 °C–600 °C range, since the activation energy (E_a) required for ORR is very high [26]. Furthermore, the materials used and their microstructure have an important effect in determining the working temperature of an SOFC. The different polarisations that have an impact on the performance of an low temperature (LT-SOFC) are discussed below.

The ohmic polarisation is the voltage loss caused by the contact resistance between coupling parts and other cell components, the electrode–electrolyte

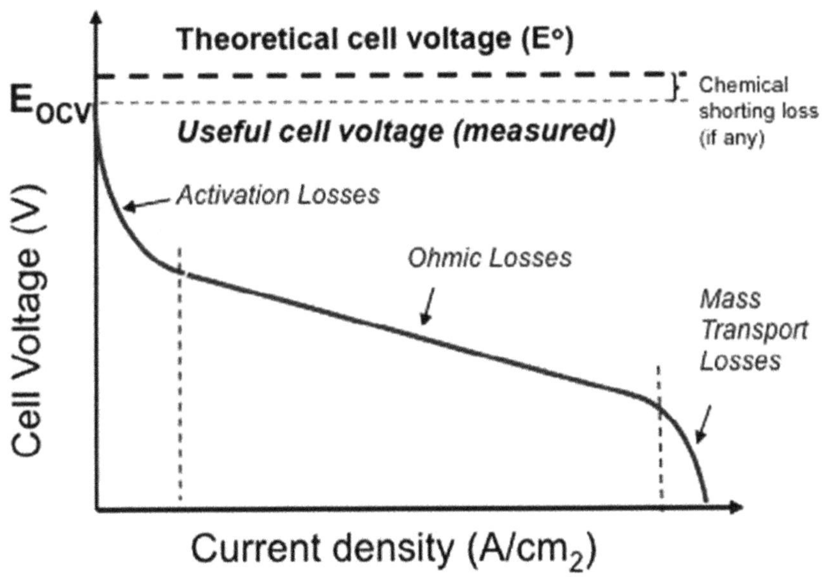

Figure 6.3. An illustration of the I–V characteristics of a typical SOFC. (Reproduced with permission from [21]. Copyright 2016 Elsevier.)

interface, and ionic resistance of the electrode, whereas for the low operating temperature of 600 °C, ohmic losses are primarily introduced by the transport resistance of the ionic current as it moves through the electrolyte [27]. This overpotential complies with Ohm's rule, which is $\eta_{\text{ohm}} = I \times R_i$ where where I stands for current drawn from the cell (A), η_{ohm} stands for ohmic polarisation (V), and R_i stands for the electrolyte's ionic resistance [28, 29]. Second, the voltage loss caused by the ORR, indicating that the substantial loss in the cell potential requires the reacting species to get past a potential energy barrier that prevents electron transfer, is said to be the activation polarisation. Charge transport is a factor in activation polarisation which in turn hangs on the electrode–electrolyte interfaces [30]. The electrochemical process is limited at low currents by the sluggish oxidation and reduction reaction rates at the cell electrodes; as a result, activation losses predominate. The well-known Butler–Volmer formula [31] can be used to express

$$i = i_0\left[\exp\left(\frac{\alpha_1 F\eta_a}{RT}\right) - \left(\frac{-\alpha_2 F\eta_a}{RT}\right)\right], \tag{6.9}$$

where current density is denoted by i (A cm^{-2}), exchange current density by i_0 (A cm^{-2}), activation polarisation by η_a, absolute temperature by T, universal gas constant by R, charge-transfer coefficients by α_1 and α_2, and the Faraday constant by F. Finally, concentration polarisation is the voltage loss brought on by the passage of product (e^-) and reactant (O^{2-}) species to and from the reaction sites, which indicates the loss due to the transport of gaseous fuel that comes in and goes out.

The movement of reactant (O_2^- ions) and product (e^-) species, respectively, into and out of the reaction sites is limited at high current densities, which causes concentration polarisation. Concentration polarisation can be expressed as

$$\eta_{\text{conc.}} = \frac{RT}{nF}\ln\left(1 - \frac{i}{i_L}\right), \tag{6.10}$$

where limiting current density is denoted by i_L and the quantity of electrons involved in the process is denoted by n [31]. Therefore, to maximise the performance of LT-SOFCs, it has been crucial to reduce ohmic, activation, and concentration polarisation at intermediate temperatures (<600 °C) and to handle material characteristics such as transport properties, exchange kinetics, and microstructure.

The basic requirements of the cathode material in SOFCs are as follows [3, 32–34]. The cathode in SOFCs serves as site for the oxygen reduction electrochemical reaction. Under oxidising conditions, a high electronic conductivity of >100 S cm^{-1} is required. The cathode should be chemically compatible with the electrolyte employed, and the thermal expansion coefficient should be matched and compatible with the interconnect material. The cathode material should possess enough porosity for the diffusion of the oxygen between the cathode–electrolyte interface. The cathode material should possess high stability under operational conditions and during the fabrication process. The cathode material should show high catalytic activity during the OR and the cost of preparation should be low.

The mechanism of the ORR of the cathode material [35] is

$$\frac{1}{2}O2\ (g) + 2e'(\text{cathode}) \rightleftharpoons O2 - (\text{electrolyte}).$$

The ORR routes in a porous cathode microstructure on a solid electrolyte are mainly determined by the cathode materials used. The concomitant existence of oxygen vacancy, electrons, and molecule oxygen should be on or close to the surface of the cathode. For the ORR, the cathode materials act as a catalysts as well as a means of delivering charged species to the solution and promote the ORR-related electrical charge. The oxygen reduction process on the cathode side can be broken down into a number of simple stages [36]. The stages are:

1. $O_2(g)$ diffusion on the cathode.
2. $(O_2(g) \rightarrow O_{2,ad})$.
3. $(O_{2,ad} \rightarrow 2O_{ad})$.
4. $(O_{ad} + 2e' + V_{O}^{\cdot\cdot} \rightarrow O_{O}^{\times})$.
5. Diffusion of O-ions on the surface of the electrode and in the bulk electrolyte.

When investigating novel cathode materials for SOFCs, a number of parameters need to be taken into account. A suitable cathode material for SOFCs should have a variety of characteristics in order to reduce ohmic losses. The cathode materials should have enough electrical conductivity and enough catalytic activity for the oxygen reduction process. It is possible to prevent thermomechanical stresses by choosing an electrolyte material with a desirable thermal expansion coefficient, often in the range of 10×10^{-6}–12.5×10^{-6} K^{-1}. The cathode material should be structurally stable throughout manufacture and operation, have good chemical compatibility with the cell components which include the electrolyte and other elements in the stack, have a suitable porous shape to reduce tortuosity and the losses of mass transportation, and be effective in terms of cost. This chapter includes a brief overview of cobalt-based cathode materials that support high cell power efficiency in addition to several tactics used to lower cathode polarisation and, as a result, lower the operating temperature of SOFCs to between 400 °C and 600 °C. We review the cobalt-based cathode materials with spinel and perovskite structure.

6.1.1 Spinel cathodes

In SOFCs the protective layers between the cathode and interconnections are typically investigated as spinel oxides with the general formula AB_2O_4 (A, B = Co, Fe, Cu, Mn, Ni, and so on) because of their high thermal compatibility and chemical durability, suitable electrical conductivity, and compliance with SOFC components' thermal expansion requirements [37]. In recent years, researchers have begun to explore the use of spinel oxides as cathode materials in SOFCs [38]. In the chemical composition of spinel oxides, alkaline earth and rare earth elements were used such as $La_{0.5}Sr_{0.5}Co_{0.5}Cu_{0.5}O_{3d}$, $Pr_{1-x}Ca_xBaCo_2O_{5+\delta}$, $Sm_{0.5}Sr_{0.5}CoO_3$–$La_{0.6}Sr_{0.4}FeO_3$, $BaBi_{0.05}Zr_{0.1}Co_{0.85-x}Nb_xO_{3-\delta}$, and $La_{0.9}Sr_{0.1}Ga_{0.8}Mg_{0.2}O_{3-\delta}$. This inherent property ensures better chemical compatibility with neighbouring components in SOFCs, as well as long-term durability in CO_2 containing environments. The spinel oxides, however, have minimal oxygen ion conductivity and are basically pure electronic conductors. The oxygen ion conductivity of spinel oxides is extremely low and the catalytic active sites for the ORR on spinel cathodes are primarily confined to the TPBs at the interface of the cathode/electrolyte [39]. In this regard, at intermediate temperatures, the majority of spinel oxides have weaker catalytic activity than the perovskite-related cathodes.

6.1.2 Perovskite cathodes

Perovskite materials have been a popular choice for SOFC cathodes [45]. It is essential to first comprehend the foundations of the perovskite structure in order to properly develop and optimise the cathode materials. The typical formula for a perovskite-type oxide is ABO_3, where A and B are cations with a combined charge of +6. The lower valence A cations (such as, La, Ba, Sr, Ca, and Pb, etc) are larger and coordinated to 12 oxygen anions while the B cations (such as, Ti, Cr, Ni, Fe, Co, and Zr, etc) occupy a much smaller space and are coordinated to six oxygen anions. It is possible to completely or partially substitute A or B cations with cations of differing valence. When the combined valence of the A- and B-site cations (n + m) is less than six, the missing charge is compensated for by creating vacancies at the oxygen lattice sites [46, 47]. It is feasible to completely or partially replace A or B cations with cations of a different valence. When the sum of the valences of the A-site and B-site cations is less than six, the number of charges that are less will be covered by vacancies at oxygen lattice sites [44, 45]. Doping has the ability to modify the TEC of the perovskite substance as well as greatly improve its electrical conductivity. For SOFCs, La-based perovskites are employed because of their well-defined crystal structure, high thermal and outstanding structural stability, superior ionic conductivity, redox behaviour, TEC compliance, and alloy chemical stability. It is anticipated that perovskite oxide will increase the durability of the metal interconnect and reduce or even prevent Cr from volatilising in the oxide layer. Lanthanum cobaltate ($LaCoO_3$), lanthanum strontium cobaltate ($La_{1-x}Sr_xCoO_3$), lanthanum chromate ($LaCrO_3$), lanthanum strontium manganate ($La_{1-x}Sr_xMnO_3$), and other elements are common perovskite components [46–49].

6.2 Review of spinel- and perovskite-based cobaltite cathode materials

Cathode materials are being studied extensively by researchers for various SOFCs. When compared to other cathode materials, significant performance was observed with cobalt-based cathode materials. Thaheem *et al* [50] synthesised cathodic $NiCo_2O_4$ spinel oxide using a glycine-nitrate process. Addition of Ni improves the electrical conduction by small polaron hopping observed from x-ray photoelectron spectroscopy (XPS) analysis. The electrical conductivity observed is 36.9 S cm^{-1} at 800 °C. Further the nickel cobaltite shows excellent compatibility at 900 °C for 10 h with $((Sc_2O_3)0.1(CeO_2)0.01(ZrO_2)0.89)$ (ScCeSZ) electrolyte. The low polarisation resistance analysed using EIS showed 0.16 Ω cm^2 at 800 °C, 0.25 Ω cm^2 at 750 °C,

0.46 Ω cm^2 at 700 °C, and 0.93 Ω cm^2 at 650 °C. Finally the maximum power density obtained for single cells ($NiCo_2O_4$ cathode, electrolyte, and anode) achieved 0.73 W cm^2 at 750 °C whereas it was 1.04 W cm^2 at 800 °C, thereby showing that the $NiCo_2O_4$ spinel oxide is a suitable cathode material for IT-SOFCs. Shao *et al* [51] synthesised a $CuCo_2O_4$ cathode by employing a citrate-ethylenediaminetetraacetic acid (EDTA) acid process reveals good chemical compatibility and good electrocatalysis with SSZ. Also, the $CuCo_2O_4$ spinel structure has the capability of activating oxygen molecules and the obtained polarisation resistance of $CuCo_2O_4$ is obtained as 0.12 Ω cm^2 at 800 °C. Further, a $CuCo_2O_4$ cathode supported with anode-supported single cells achieved a high power output of 972 mW cm^{-2}. Liu *et al* [37] employed the citrice EDTA acid process to synthesise spinel oxide nanoparticles. When the addition of Co is decreased, the $(Mn,Co)_3O_4$ shows a phase transition from the cubic to tetragonal phase. Sufficient electrical conductivity is obtained with the increase in Co content in order to be employed as a cathode. Also, the electrochemical performances was improved with the addition of Co in $(Mn,Co)_3O_4$ spinel oxides which makes them a suitable candidate for IT-SOFCs. Liu *et al* [52] reported the preparation of yttria stabilised zirconia (YSZ) and $Mn_{1.5}Co_{1.5}O_4$ (MCO) composite cathodes via the liquid phase impregnation technique by infiltration of the MCO oxides into porous YSZ backbones using aqueous solutions of the corresponding nitrate salts. The optimised volume of MCO ($V_{MCO} = 13.5\%$) shows high phase connectivity and high porosity for the MCO-YSZ composite as well as a polarisation resistance of 0.43 Ω cm^2 in air at 800 °C. The impedance measurements of the MCO-YSZ cathode reveal the rate-limiting step due to double ionisation of adsorbed oxygen atoms or charge transfer at triple phase boundaries in varied oxygen partial pressure during electrochemical reduction of oxygen. Kim *et al* synthesised $NdBa_{0.5}Sr_{0.5}Co_{2-x}Mn_xO_{5+\delta}$ ($x = 0$, 0.25, and 0.5) by the Pechini process as a cathode material for intermediate temperature SOFCs and reported that the Co-based materials with stronger Mn–O bonds and suppressed Co^{3+} spin state transitions resulted in lower TEC for the substitution of Mn for Co; the electrical conductivity and electrochemical performance were affected negatively by the trapping of effective mobile charge with higher Mn concentration. Liu *et al* [54] discovered that the spinel oxide $Mn_{1.5}Co_{1.5}O_4$ (MCO) alone is astonishingly capable of activating oxygen molecules and a single cell with it. Since, the $Mn_{1.5}Co_{1.5}O_4$ cathode exhibits high output power when compared to that of the conventional $La_{1-x}Sr_xMnO_3$–YSZ (LSM–YSZ) composite cathode. A composite cathode with a current collector was created, which significantly increased the output power, exhibiting 912 mW cm^2 at 800 °C which is relatively higher compared to the LSM–YSZ composite cathode. The findings show that a novel class of possible cathode materials compatible with the YSZ electrolyte may be used in intermediate temperature SOFCs. Thaheem *et al* [55] prepared $Mn_{1.3}Co_{1.3}Cu_{0.4}O_4$ (MCCO) spinel oxide by a facile glycine-nitrate process and studied their physicochemical and electrical properties. The addition of Cu in the MCO structure reveals that the Mn ions of MCO moved to octahedral positions from tetrahedral with higher oxidation states, enabling modest polaron hopping of valence electrons. Moreover, the cathodic area-specific resistance (ASR) values for the MCCO–ScSZ composite cathode were low, measuring 0.03 Ω cm^2 for 800 °C and 0.06 Ω cm^2 and 750 °C, respectively.

Finally, the maximum power density of 1.92 W cm^{-2} at 800 °C and 1.74 W cm^{-2} at 750 °C, respectively, was attained by the anode-supported SOFC utilising a MCCO–ScSZ cathode, which is the best performance to date for all SOFCs using spinel-based cathodes, making it a potential cathode material for solid oxide fuel cells applications. Li *et al* [56] synthesised $FeCo_2O_{4-x}Ce_{0.8}Sm_{0.2}O_{1.9}$ (x = 0, 30, 40, 50, 60, 70 wt%), catalyst/electrolyte composite oxides as a cathode material for SOFCs. Among the prepared composites, the lowest polarisation resistance of about 0.094 Ω cm^2 at 800 °C in air was obtained for the $FeCo_2O_4$–$60Ce_{0.8}Sm_{0.2}O_{1.9}$ composite cathode, whereas for the $FeCo_2O_4$ cathode it was 4.18 Ω cm^2. In comparison to the $FeCo_2O_4$ cathode, based on an electrolyte supported single-cell arrangement, the $FeCo_2O_4$–$60Ce_{0.8}Sm_{0.2}O_{1.9}$ composite cathode exhibited higher electrocatalytic activity and a higher power density of 397 mW cm^{-2} at 800 °C. $CuCo_2O_4$ spinel nanoparticles were coated by Shao *et al* [57] via a solution impregnation technique on a porous 10 mol% scandia stabilised zirconia (SSZ) framework. The porous SSZ backbones' surface is evenly covered in $CuCo_2O_4$ nanoparticles, as a result lengthening is observed in the TPBs. As anticipated, the impregnated nanostructured $CuCo_2O_4$ obtained low polarisation resistance (0.087 Ω cm^2) in air at 800 °C and produced 1136 mW cm^2 of high peak power density. Wang *et al* [58] developed multiphase $(Mn_{0.6}Mg_{0.4})_{0.8}Sc_{0.2}Co_2O_4$-Pr (MMSCO-Pr) and $(Mn_{0.6}Mg_{0.4})_{0.8}Sc_{0.2}Co_2O_4$-Ce (MMSCO-Ce) spinel oxide nanocomposites via a co-precipitation method. The nanocomposite consists of 84 wt% MMSCO spinel oxide, 11.3 wt% perovskite phase of orthorhombic crystal structure and PrO_2 spontaneous combination, and 4.7 wt% of Sc_2O_3 phase, whereas in MMSCO-Ce there was 87.6 wt % MMSCO spinel oxide, 8.2 wt% CeO_2, and 4.2 wt% Sc_2O_3 phase. The ORR kinetics are clearly accelerated due to many heterogeneous interfaces (the $PrCoO_3$-based perovskite and Sc_2O_3 phase are anchored at the surface of the spinel phase), originated by the self-assembled nanocoating which activates the surface of the MMSCO-Pr phase, where the area-specific resistance (ASR) of 0.11 Ω cm^2 is reached at 750 °C whereas for MMSCO-Ce, the ASR is 0.24 Ω cm^2 at 750 °C. Furthermore, compared to a cell with a $MnCo_2O_4$ cathode (456.2 mW cm^2), a single cell with the cathode MMSCO-Pr exhibits a peak power density (PPD) of 1144.1 mW cm^2 at 750 °C. Due to the decreased thermal expansion coefficient (13.9×10^{-6} K^{-1}), an improved stability of 120 h (0.8 A cm2, 750 °C) is also attained. The development of heterointerfaces, an increase in the mechanical compatibility, and the refining of grains are all responsible for enhancement in ORR kinetics and stability. Fulgêncio *et al* [59] reported the preparation of Misfit Ca-cobaltites ($Ca_3Co_4O_{9-\delta}$) by solid state synthesis as a cathode material for solid oxide fuel cells. Impedance spectroscopy was used to study the electrochemical behaviour of screen-printed $Ca_3Co_4O_{9-\delta}$ porous electrodes that were deposited on both sides of gadolinia-doped ceria substrate. The ASR for $Ca_3Co_4O_{9-\delta}$ was 1.47 Ω cm^2 at 800 °C in air. Natile *et al* [60] synthesised $LaCo_{0.5}Cu_{0.5}O_{3d}$ and $La_{0.5}Sr_{0.5}Co_{0.5}Cu_{0.5}O_{3d}$ by the citrate method. X-ray diffraction confirms the formation of the preferred phase and no impurity phases are observed. According to results obtained from the oxygen-temperature programmed oxygen desorption (O_2-TPD), one can observe that the mobility of oxygen ions is significantly increased by doping Sr and Cu in $LaCoO_3$. This increase in oxygen ion mobility shows a decrement in the ASR values as observed from the EIS measurement done for half the cells ($LaCo_{0.5}Cu_{0.5}O_3/Ce_{0.9}Gd_{0.1}O_{1.95}/LaCo_{0.5}Cu_{0.5}O_3$ and $La_{0.5}Sr_{0.5}Co_{0.5}Cu_{0.5}O_3/Ce_{0.9}Gd_{0.1}O_{1.95}/La_{0.5}Sr_{0.5}Co_{0.5}Cu_{0.5}O_3$) where $LaCo_{0.5}Cu_{0.5}O_3$ based cells showed better performance when compared to $La_{0.5}Sr_{0.5}Co_{0.5}Cu_{0.5}O_3$, especially at the lower temperature of 500 °C–600 °C. Develos-Begarinao *et al* [61] deposited single layers of $(La_{0.6}Sr_{0.4})CoO_{3-\delta}$ (LSC) and gadolinia-doped ceria (GDC) (10 mol % $GdO_{1.5}$ doped CeO_2) thin films on YSZ single crystal substrates with (100) oriention by a employing pulsed laser deposition (PLD) technique. At a low temperature of 500 °C–700 °C, the oxygen surface exchange coefficient was found to be in the order of 10^{-5} cm s^{-1} which shows an increase in the kinetics of oxygen exchange. The presence of Ce^{3+} at the interface of the LSC and GDC multilayer generates high density oxygen vacancies resulting in enhanced oxygen exchange kinetics. ASR values were obtained from electrochemical impedance spectroscopy (EIS) measurements carried out using LSC-GDC multilayers and show 0.03 Ω cm^2 at 700 °C and 0.9 Ω cm^2 at 500 °C. Further cell configuration and layer thickness optimisations are predicted to minimise the ASR and fully exploit the interface-induced improvement in cathode characteristics. Wang *et al* [62] reported the development of Cu and Mo ions doped separately with $Ca_3Co_4O_{9-\delta}$ by taking the position of the Co ion to alter its electrical conductivity and electrochemical behaviour. The addition of Cu in $Ca_3Co_4O_{9-\delta}$ significantly improved the densification, increased grain size, and triggered the formation of $Ca_3Co_2O_6$, whereas Mo doped $Ca_3Co_4O_{9-\delta}$ showed no impurity phases even at a higher temperature of 1050 °C. Enhanced electrical conductivity of 218.8 S cm^{-1} was obtained for a Cu doped $Ca_3Co_4O_{9-\delta}$ cathode that was employed to construct a single-cell SOFC with an SDC electrolyte layer (20 μm), showing 0.367 W cm^{-2} power density at 750 °C. Samson *et al* [63] synthesised $Ca_3Co_4O_{9+\delta}$ and $Ce_{0.9}Gd_{0.1}O_{1.95}$ (CGO/$Ca_3Co_4O_{9+\delta}$) by a solid state reaction and observed that the CGO electrolyte can reduce the cathode's polarisation resistance to 0.64 Ω cm^2. Yu *et al* [64] reported the formation of $Ca_3Co_4O_9$ by calcinating Co_3O_4 and $CaCO_3$. When the $Ca_3Co_4O_9$ and $Ca_9Co_{12}O_{28}$ were, respectively, formed at 800 °C and 900 °C where the average valence Co ion in $Ca_3Co_4O_9$ is increased with increasing temperature from 750 °C to 900 °C. The oxidation state of cobalt ions has a considerable impact on the performance of calcium cobaltite cathodes in IT-SOFCs. The $Ca_9Co_{12}O_{28}$ shows lower area-specific resistance and high peak power density due to the higher average oxidation state of cobalt $Co^{3+/4+}$ when compared to $Ca_3Co_4O_9$. A new Co based cathode material ($Ca_3Co_2O_6$) was studied by Wei *et al* [65]. The unique coplanar-sharing chain structure is due to the mixed oxide-ion/electron conductive behaviour, matching thermal and mechanical properties with the electrolyte, and the thermoelectric effect are all factors that contribute to the $Ca_3Co_2O_6$ cathode's exceptional electrochemical performance. Guo *et al* [66] studied the effect of Sm on $Sm_xSr_{1-x}CoO_{3-\delta}$ (SSC) ($0.2 \leqslant x \leqslant 0.8$) using the sol–gel process and reported that the SSC6 and SSC5 showed high electrical conductivity of 1885 and 1731 S cm^{-1} in air atmosphere whereas in nitrogen atmosphere, SSC5 showed higher electrical conductivity of 176 S cm^{-1} at 300 °C and 436 S cm^{-1} at 900 °C. Also the electrochemical studies revealed that the SSC5 exhibits low ASR values of 0.42 cm^2 and increased current density of 60 mA cm^{-2} at 600 °C. Tatko *et al* [67] prepared $Sm_{0.5}Sr_{0.5}CoO_3$–$La_{0.6}Sr_{0.4}FeO_3$ (SSC–LSF) composites by screen printing technique where the SSC and

LSF mixed in the weight ratios of 3:7 named as SSC3LSF7, 5:5 named as SSC5–LSF5, and 7:3 named as SSC7–LSF3. The SSC5–LSF5 composite showed the lowest area-specific resistance and thermal expansion coefficient of 17.7×10^{-6} K^{-1}. Si *et al* [68] reported that the cathodic $Sr_{0.9}Y_{0.1}CoO_{2.5+\delta}$ (SYC) nanoparticles were infiltrated into $Ce_{0.8}Gd_{0.2}O_{2-\delta}$ (GDC) to catalyse the ORR for cathode applications as well as in $La_{0.6}Sr_{0.4}Co_{0.2}Fe_{0.8}O_{3-\delta}$ (LSCF) to promote and enhance the ORR of LSCF. It also achieved the lowest area-specific polarisation resistance at 650 °C, loading of 30 wt%, and calcination temperatures of 350 °C for SYC10. The NPs were rigorously optimised and obtained an ASR of 0.733 Ω cm^2 at 500 °C. Zhou *et al* [69] prepared $BaBi_{0.05}Zr_{0.1}Co_{0.85-x}O_{3-\delta}$ (BBZC) and $BaBi_{0.05}Zr_{0.1}Co_{0.85-x}Nb_xO_{3-\delta}$ (BBZCN) oxides ($x = 0$ and 0.05) using a solid state reaction method where bismuth, zirconium, and niobium were doped. A lowered thermal expansion coefficient and improved phase structure were observed for BBZCN due to the addition of 5% Nb in BBZCN. With $La_{0.9}Sr_{0.1}Ga_{0.8}Mg_{0.2}O_{3-\delta}$ (LSGM) electrolyte, the BBZC and BBZCN cathodes showed ASR values of 0.021 Ω cm^2 and 0.023 Ω cm^2 at 800 °C in air atmosphere. The peak power density obtained from a single cell for the BBZC and BBZCN cathodes are, respectively, 515 and 566 mW cm^{-2}. Jiang *et al* [70] synthesised cathodic $Pr_{1+x}Ba_{1-x}Co_2O_{5+\delta}$ (PBC, $x = 0$–0.30) oxides by varying the Pr/Ba ratios using the sol–gel process. XPS analysis revealed the existence of mixed Co^{3+}/Co^{4+} ions and Pr^{3+}/Pr^{4+} and observed three distinct oxygen chemical states. With rising temperatures, electrical conductivity drops. At 800 °C every PBC sample exhibits an abruptly high conductivity (>550 S cm^{-1}). The performance is enhanced by replacing Ba with a small quantity of Pr. The best electrochemical performance is shown by the $x = 0.10$ sample with a power density of 732 mW cm^{-2}. It has the highest conductivity and lowest polarisation resistance making it a suitable candidate for SOFCs. Fu *et al* [71] employed a conventional solid state reaction to prepare $Pr_{1-x}Ca_xBaCo_2O_{5+\delta}$ ($x = 0.1$–0.4; PCBCO) double perovskites where the samples with $x = 0.1$ and 0.2 show pure tetragonal structure and $x > 3$ shows impurity phases. As the Ca content increases, the intensity of the impurity peaks also increased. The TEC value of the cathode is decreased by Ca doping whereas the increase in Ca reduces the conductivity which may be due to increase in the concentration of oxygen vacancy. At 950 °C for 10 h, the PCBCO materials showed high chemical compatibility using the SDC electrolyte. As the Ca concentration increased, more and more oxygen vacancies formed, leading to the creation of defect clusters, which reduced the electrochemical performance of the PCBCO cathodes. For $x = 0.1$ and 0.3, the ASR values increased from 0.081 Ω cm^2 to 0.089 Ω cm^2 at 700 °C while the corresponding maximum power density values decreased from 646.5 mW cm^{-2} to 620.6 mW cm^{-2} at 800 °C. Lim *et al* [72] prepared Ln doped $LnBaCo_2O_{5+\delta}$ (where Ln = Pr, Nd, Sm, and Gd) using the Pechini method and found it was a suitable cathode material for SOFCs due to its surface exchange coefficient and enhanced oxygen diffusion. The layered perovskite structure was changed to a simple perovskite structure by increasing the amount of Ca in $PrBa_{1-x}Ca_xCo_2O_{5+\delta}$ (PBCCO) where $x = 0$, 0.1, 0.2, 0.3, and 0.4. The simple and layered perovskite structures were observed for $x = 0.3$ in $PrBa_{1-x}Ca_xCo_2O_{5+\delta}$. Ca doping can prevent mobile oxygen loss from the lattice owing to increased oxygen bonding strength. With an increase in the Ca concentration up to 30 wt%, the conductivity increases from

400 to 1100 S cm^{-1} and TEC values decreased, which may be due to an increase in the mobile oxygen concentration. Li *et al* [73] reported perovskite $SrCo_{0.8}Nb_{0.1}Ta_{0.1}O_{3-\delta}$ co-substituted with niobium and tantalum with excellent electrochemical activity. In a symmetrical cell configuration, the $SrCo_{0.8}Nb_{0.1}Ta_{0.1}O_{3-\delta}$ cathode shows an ASR value of 0.16 Ω cm^2 at 500 °C and 0.68 Ω cm^2 at 450 °C. The co-doping of Nb^{5+} and Ta^{5+} enhanced the bulk oxygen ionic conductivity. Further, the density functional theory calculations reveal that the co-doping of Nb^{5+} and Ta^{5+} in the $SrCo_{0.8}Nb_{0.1}Ta_{0.1}O_{3-\delta}$ cathode results in high performance of the $SrCo_{0.8}Nb_{0.1}Ta_{0.1}O_{3-\delta}$ cathode. Kalpana *et al* [74] reported the composition of $Gd_{0.5}Pr_{0.5}BaCo_2O_{5+\delta}$ (GPBC) on $Ba_{0.5}Sr_{0.5}Ce_{0.6}Zr_{0.2}Gd_{0.1}Y_{0.1}O_{3-\delta}$ (BSCZGY) electrolyte in order to obtain electrical conductivity and oxygen content at a balanced level. The ASR value for the above composition measured at 700 °C is 2.4 Ω cm^2 in air and 1.9 Ω cm^2 in wet air for the ORR with an activation energy of 1.2 eV, whereas ASR is greater and the activation energy is 1.6 eV in the symmetrical cell of the GPBC with Y_2O_3-stabilised zirconia (8YSZ), which may be the result of an interfacial interaction. Sun and Cheng prepared $BaZr_{0.1}Ce_{0.7}Y_{0.1}Yb_{0.1}O_{3-\delta}$ (BZCYYb) electrolyte material and $Ba_{0.5}Sr_{0.5}Co_{0.8}Fe_{0.2}O_{3-\delta}$ cathode material synthesised using a combustion method followed by heat treatment, whereas the $SrCo_{0.8}Nb_{0.1}Ta_{0.1}O_{3-\delta}$ (SCNT) leading cathode material was prepared using a solid state reaction and investigated and compared with the BSCF cathode at 750 °C–450 °C. Because of the restricted proton production, pure SCNT as well as BSCF do not perform well under working conditions. A composite cathode was created by combining SCNT and BZCYYb to increase the performance by balancing the demands of oxygen exchange, water absorption, and proton conduction.

6.3 Summary and future outlook

Researchers are aiming to decrease the working temperature of SOFCs (to $\leqslant$ 600 °C). The problem with reducing the temperature is that it will increase the electrode polarisation and also decrease the ionic conductivity of the electrolyte, which will lead to reduced overall performance of the fuel cell. Recently various methods, such as preparing highly porous structures, reducing the size of th electrolyte layer, or mixing the electrolyte with cathode material during the synthesis process were used to reduce the ohmic resistance and obtain high power density at low temperatures. Also, polarisation of the cathode material which possesses a high activation energy at low temperatures needs to be addressed in other ways by developing hierarchical microstructures that are highly nanoporous in nature to achieve low polarisation. Based on the review undertaken here, one can say that $SrCo_{0.8}Nb_{0.1}Ta_{0.1}O_{3-\delta}$ co-substituted ferroactive highly charged cations such as niobium and tantalum with excellent electrochemical activity. In a symmetrical cell configuration, the $SrCo_{0.8}Nb_{0.1}Ta_{0.1}O_{3-\delta}$ cathode shows an ASR value of 0.16 Ω cm^2 at 500 °C and 0.68 Ω cm^2 at 450 °C. The co-doping of Nb^{5+} and Ta^{5+} enhanced the bulk oxygen ionic conductivity. Further, density functional theory calculations reveal that the co-doping of Nb^{5+} and Ta^{5+} in the $SrCo_{0.8}Nb_{0.1}Ta_{0.1}O_{3-\delta}$ cathode results in high performance of the $SrCo_{0.8}Nb_{0.1}Ta_{0.1}O_{3-\delta}$ cathode. Several improvements and techniques have been executed during the study to address issues with the cathode components of SOFCs for low temperature operation. Recently, an attractive technique to enhance the performance of the cathode material has been mixing the cathode material with the electrolyte material which will improve the stability, reduce the cost, and enhance the performance at low temperatures. Electrolyte-free fuel cells will lead the SOFCs toward commercialisation by developing a fuel cell with only a cathode and anode, and this will also be more sustainable. Despite the fact that several materials have been discovered in recent years, none of them can be regarded as the ideal cathode material since none of them performs as well electrocatalytically at low temperatures. The technology for component manufacturing and material processing still needs to be improved.

References

[1] Sharaf O Z and Orhan M F 2014 An overview of fuel cell technology: fundamentals and applications *Renew. Sustain. Energy Rev.* **32** 810–53

[2] Chubbock S and Clague R 2016 Comparative analysis of internal combustion engine and fuel cell range extender *SAE Int. J. Altern. Powertrains* **5** 175–82

[3] Hussain S and Yangping L 2020 Review of solid oxide fuel cell materials: cathode, anode, and electrolyte *Energy Transitions* **4** 113–26

[4] Lohmann F P, Schulze P S C, Wagner M, Naumov O, Lotnyk A, Abel B and Varga A 2017 The next generation solid acid fuel cell electrodes: stable, high performance with minimized catalyst loading *J. Mater. Chem.* A **5** 15021–5

[5] He S and Jiang S P 2021 Electrode/electrolyte interface and interface reactions of solid oxide cells: recent development and advances *Prog. Nat. Sci.: Mater. Int.* **31** 341–72

[6] Jiang S P 2015 Thermally and electrochemically induced electrode/electrolyte interfaces in solid oxide fuel cells: an AFM and EIS study *J. Electrochem. Soc.* **162** F1119–28

[7] Malti A *et al* 2016 An organic mixed ion–electron conductor for power electronics *Adv. Sci.* **3** 1500305

[8] Zouhri K and Lee S Y 2016 Exergy study on the effect of material parameters and operating conditions on the anode diffusion polarization of the SOFC *Int. J. Energy Environ. Eng.* **7** 211–24

[9] Padinjarethil A K, Bianchi F R, Bosio B and Hagen A 2021 Electrochemical characterization and modelling of anode and electrolyte supported solid oxide fuel cells *Front. Energy Res.* **9** 668964

[10] Sangeetha T, Yan W M, Chen P T, Yang C J and Huan K D 2020 Electrochemical polarization analysis for optimization of external operation parameters in zinc fuel cells *RSC Adv.* **10** 28807–18

[11] Anasuhah W, Yusoff W, Nur W, Samat A, Somalu M, Muchtar A and Baharuddin N A 2018 Fabrication process of cathode materials for solid oxide fuel cells *J. Adv. Res. Fluid Mech. Therm. Sci.* **50** 153–60

[12] Neelima M, Banerjee A, Alka G, Shobit O and Kantesh B 2015 Progress in material selection for solid oxide fuel cell technology: a review *Prog. Mater Sci.* **72** 141–337

[13] Shimada H, Suzuki T, Yamaguchi T, Sumi H, Hamamoto K and Fujishiro Y 2016 Challenge for lowering concentration polarization in solid oxide fuel cells *J. Power Sources* **302** 53–60

[14] Park K, Yu S, Bae J, Kim H and Ko Y 2010 Fast performance degradation of SOFC caused by cathode delamination in long-term testing *Int. J. Hydrogen Energy* **35** 8670–7

[15] Golkhatmi S Z, Imran Asghar M and Lund P D 2022 A review on solid oxide fuel cell durability: latest progress, mechanisms, and study tools *Renew. Sustain. Energy Rev.* **161** 112339
[16] Sun S and Cheng Z 2017 Effects of H_2O and CO_2 on electrochemical behaviors of BSCF cathode for proton conducting IT-SOFC *J. Electrochem. Society* **164** F81–8
[17] O'Hayre R, Barnett D M and Prinz F B 2005 The triple phase boundary a mathematical model and experimental investigations for fuel cells *J. Electrochem. Soc.* **152** A439–44
[18] Adler S B 1996 Electrode kinetics of porous mixed-conducting oxygen electrodes *J. Electrochem. Soc.* **143** 3554
[19] Mahato N, Banerjee A, Gupta A, Omar S and Balani K 2015 Progress in material selection for solid oxide fuel cell technology: a review *Prog. Mater Sci.* **72** 141–337
[20] Brett D J L, Atkinson A, Brandon N P and Skinner S J 2008 Intermediate temperature solid oxide fuel cells *Chem. Soc. Rev.* **37** 1568
[21] Gür T M 2016 Comprehensive review of methane conversion in solid oxide fuel cells: prospects for efficient electricity generation from natural gas *Prog. Energy Combust. Sci.* **54** 1–64
[22] Xing L, Hao J, Li X, Zhang Y, Hu Z and Gao Y 2017 Polarization modeling and performance optimization of a molten sodium hydroxide direct carbon fuel cell (MHDCFC) *J. Power Sources* **363** 428–41
[23] Severson H and Assadi M 2011 Modeling of overpotentials in an anode-supported planar SOFC using a detailed simulation model *J. Fuel Cell. Sci. Tech.* **8** 051021
[24] Subramanyan K and Diwekar U M 2007 Optimizing model complexity with application to fuel cell based power systems *Chem. Eng. Process.* **46** 1116–28
[25] Yamamoto O and O 2000 Solid oxide fuel cells: fundamental aspects and prospects *Electrochim. Acta* **45** 2423–35
[26] Gao Z, Mogni L V, Miller E C, Railsback J G and Barnett S A 2016 A perspective on lowtemperature solid oxide fuel cells *Energy Environ. Sci.* **9** 1602–44
[27] Steele B C H and Heinzel A 2001 Materials for fuel-cell technologies *Nature* **414** 345–52
[28] Sahli Y, Zitouni B and Ben-Moussa H 2017 Solid oxide fuel cell thermodynamic study *Çankaya Univ. J. Sci. Eng.* **14** 134–51
[29] Sezer N, Khalid F, Biçer Y and Koç M 2020 Electrochemical modeling and performance assessment of H_2S/air solid oxide fuel cell *Energy Technol.* **8** 1–8
[30] Virkar A V, Chen J, Tanner C W and Kim J W 2000 Role of electrode microstructure on activation and concentration polarizations in solid oxide fuel cells *Solid State Ion.* **131** 189–98
[31] Srinivasan S, Krishnan L and Marozzi C 2006 Fuel cell principles *Fuel Cells* (Boston, MA: Springer) pp 189–233
[32] Sun C, Hui R and Roller J 2010 Cathode materials for solid oxide fuel cells: a review *J. Solid State Electrochem.* **14** 1125–44
[33] Chen Y, Zhou W, Ding D, Liu M, Ciucci F, Tade M and Shao Z 2015 Advances in cathode materials for solid oxide fuel cells: complex oxides without alkaline earth metal elements *Adv. Energy Mater.* **5** 1500537
[34] Peng R, Wu T, Liu W, Liua X and Meng G 2010 Cathode processes and materials for solid oxide fuelcells with proton conductors as electrolytes *J. Mater. Chem.* **20** 6218–25
[35] Ma R, Lin G, Zhou Y, Liu Q, Zhang T, Shan G, Yang M and Wang J 2019 A review of oxygen reduction mechanisms for metal-free carbon-based electrocatalysts *npj Comput. Mater.* **5** 78

[36] Shi Y, Yong Lee W and Ghoniem A F 2010 Elementary reaction models for CO electrochemical oxidation on an Ni/YSZ patterned anode *Proc. of the ASME 2010 8th Int. Fuel Cell Science, Engineering and Technology Conf. (New York)* (New York: ASME International)
[37] Liu H Y, Zhu X F, Cheng M J, Cong Y and Yang W S 2013 Electrochemical performances of spinel oxides as cathodes for intermediate temperature solid oxide fuel cells *Int. J. Hydrogen Energy* **38** 1052–7
[38] Zhen S Y *et al* 2016 High performance cobalt-free $Cu_{1.4}Mn_{1.6}O_4$ spinel oxide as an intermediate temperature solid oxide fuel cell cathode *J. Power Sources* **315** 140–4
[39] Li N, Sun L, Li Q, Xia T, Huo L and Zhao H 2021 Electrode properties of $CuBi_2O_4$ spinel oxide as a new and potential cathode material for solid oxide fuel cells *J. Power Sources* **511** 230447
[40] Paramvir Kaur K 2020 Review of perovskite-structure related cathode materials for solid oxide fuel cells *Ceram. Int.* **46** 5521–35
[41] Kalpana S, Kannan R and Thangadurai V 2019 Perspective of perovskite-type oxides for proton conducting solid oxide fuel cells *Solid State Ion.* **339** 114951
[42] Afroze S, Karim A, Cheok Q, Eriksson S and Azad A K 2019 Latest development of double perovskite electrode materials for solid oxide fuel cells: a review *Front. Energy Res.* **13** 770–97
[43] Ding P, Wenlu L, Zhao H, Wu C, Zhao L, Dong B and Wang S 2021 Review on Ruddlesden–Popper perovskites as cathode for solid oxide fuel cells *J. Phys.: Mater.* **4** 022002
[44] Pelosato R, Cordaro G, Stucchi D, Cristiani C and Dotelli G 2015 Cobalt based layered perovskites as cathode material for intermediate temperature solid oxide fuel cells: a brief review *J. Power Sources* **298** 46–67
[45] Li Z, Li M and Zhu Z 2022 Perovskite cathode materials for low-temperature solid oxide fuel cells: fundamentals to optimization *Electrochem. Energy Rev.* **5** 263–311
[46] Kim J, Chen X, Shih P C and Yang H 2017 Porous perovskite-type lanthanum cobaltite as electrocatalysts toward oxygen evolution reaction *ACS Sustain. Chem. Eng.* **5** 10910–7
[47] Acchar W, Sousa C R C and Mello-Castanho S R H 2012 Mechanical performance of $LaCrO_3$ doped with strontium and cobalt for SOFC interconnect *Mater. Sci. Eng.* A **550** 76–9
[48] Manishanma S and Dutta A 2023 Synthesis and characterization of nickel doped LSM as possible cathode materials for LT-SOFC application *Mater. Chem. Phys.* **297** 127438
[49] Ren H, Lee Y H, Wu E A, Chung H, Meng Y S, Fullerton E E and Minh N Q 2020 Nano-ceramic cathodes via Co-sputtering of Gd–Ce alloy and lanthanum strontium cobaltite for low-temperature thin-film solid oxide fuel cells *ACS Appl. Energy Mater.* **3** 8135–42
[50] Thaheem I, Ali S, Waqus M, Hussain A, Soomro A M, Bhutto Z, Shah S S R, Muhammad W and Shah J 2022 Electrochemical performance of $NiCo_2O_4$ spinel cathodes for intermediate temperature solid oxide fuel cells *Phys. Status Solidi* A **219** 2100542
[51] Shao L, Wang Q, Fan L, Wang P, Zhang N and Sun K 2016 Copper cobalt spinel as a high performance cathode for intermediate temperature solid oxide fuel cells *Chem. Commun.* **52** 8615–8
[52] Liu X, Han D, Wu H, Meng X, Zeng F and Zhan Z 2013 $Mn_{1.5}Co_{1.5}O_{4-\delta}$ infiltrated yttria stabilized zirconia composite cathodes for intermediate-temperature solid oxide fuel cells *Int. J. Hydrogen Energy* **38** 16563–8

[53] Kim J, Choi S, Park S, Kim C, Shin J and Kim G 2013 Effect of Mn on the electrochemical properties of a layered perovskite $NdBa_{0.5}Sr_{0.5}Co_{2-x}Mn_xO_{5+\delta}$ (x = 0, 0.25, and 0.5) for intermediate-temperature solid oxide fuel cells *Electrochim. Acta* **112** 712–8

[54] Liu H, Zhu X, Cheng M, Cong Y and Yang W 2011 Novel $Mn_{1.5}Co_{1.5}O_4$ spinel cathodes for intermediate temperature solid oxide fuel cells *Chem. Commun.* **47** 2378–80

[55] Thaheem I, Joon Kim K, Jun Lee J, Woo Joh D, Jeong I and Taek Lee K 2019 High performing $Mn_{1.3}Co_{1.3}Cu_{0.4}O_4$ spinel based composite cathodes for intermediate temperature solid oxide fuel cells *J. Mater. Chem.* A **7** 19696–703

[56] Li H, Su C, Wang C and Lü Z 2020 Electrochemical performance evaluation of $FeCo_2O_4$ spinel composite cathode for solid oxide fuel cells *J. Alloys Comp.* **829** 154493

[57] Shao L, Wang P, Zhang Q, Fan L, Zhang N and Sun K 2017 Nanostructured $CuCo_2O_4$ cathode for intermediate temperature solid oxide fuel cells via an impregnation technique *J. Power Sources* **343** 268–74

[58] Wang H, Zhang W, Meng J, Pei Y, Qiu X, Meng Z and Liu X 2021 Effectively promoting activity and stability of a $MnCo_2O_4$ based cathode by *in situ* constructed heterointerfaces for solid oxide fuel cells *ACS Appl. Mater. Interfaces* **1313** 24329–40

[59] Fulgêncio E B G A, Melo K P V, Silva R M, Grilo J P F, Caignaert V, Cesário M R, Campos L F A and Macedo D A 2017 Misfit Ca-cobaltite from a mixture of mollusk shells and cobalt oxide as potential SOFC cathode material *Ceram. Int.* **43** 9564–7

[60] Natile M M, Eger G, Batocchi P, Mauvy F and Glisenti A 2017 Strontium and copper doped $LaCoO_3$: new cathode materials for solid oxide fuel cells? *Int. J. Hydrogen Energy* **42** 1724–35

[61] Develos-Bagarinao K, Vero J D, Kishimoto H, Ishiyama T, Yamaji K, Horita T and Yokokawa H 2018 Multilayered LSC and GDC: an approach for designing cathode materials with superior oxygen exchange properties for solid oxide fuel cells *Nano Energy* **52** 369–80

[62] Wang S F, Hsu Y F, Chang J H, Cheng S and Lu H C 2016 Characteristics of Cu and Mo-doped $Ca_3Co_4O_{9-\delta}$ cathode materials for use in solid oxide fuel cells *Ceram. Int.* **42** 11239–47

[63] Samson A J, Søgaard M, Nong N V, Pryds N and Bonanos N 2011 Enhanced electrochemical performance of the solid oxide fuel cell cathode using $Ca_3Co_4O_{9+\delta}$ *J. Power Sources* **196** 10606–10

[64] Yu S, He S, Chen H and Guo L 2015 Effect of calcination temperature on oxidation state of cobalt in calcium cobaltite and relevant performance as intermediate temperature solid oxide fuel cell cathodes *J. Power Sources* **280** 581–7

[65] Wei T, Rui Zeng Y H H, Yuan L X, Hu X L, Zhang W X, Jiang L, Yang J Y and Zhang Z L 2013 Evaluation of $Ca_3Co_2O_6$ as cathode material for high-performance solid-oxide fuel cell *Sci Rep.* **3** 1125

[66] Guo Y, Chen D, Shi H, Ran R and Shao Z 2011 Effect of Sm^{3+} content on the properties and electrochemical performance of $Sm_xSr_{1-x}CoO_{3-\delta}$ ($0.2 \leq x \leq 0.8$) as an oxygen reduction electrodes on doped ceria electrolytes *Electrochim. Acta* **56** 2870–6

[67] Tatko M, Mosiałek M, Dudek M, Nowak P, Kędra A and Bielanska E 2015 Composite cathode materials $Sm_{0.5}Sr_{0.}5CoO_3$–$La_{0.6}Sr_{0.4}FeO_3$ for solid oxide fuel cells *Solid State Ion.* **271** 103–8

[68] Si F, Shao L, Kang X, Qin C, Huang K, Fu X and Luo J 2019 Infiltrated $Sr_{0.9}Y_{0.1}CoO_{2.5+\delta}$ nanoparticles as a cathode material for solid oxide fuel cells operated at 450–650 °C *Int. J. Hydrogen Energy* **44** 31305–11

[69] Zhou Q, He Y, Bu F, Zhang T, Wang M, Li Y and An D 2020 Comparative studies of $BaBi_{0.05}Zr_{0.1}Co_{0.85-x}Nb_xO_{3-\delta}$ ($x = 0$ and 0.05) as cathodes for intermediate-temperature solid oxide fuel cells *J. Mater. Sci., Mater. Electron.* **31** 11819–24

[70] Jianga L, Li F, Wei T, Zenga R and Huang Y 2014 Evaluation of $Pr_{1+x}Ba_{1-x}Co_2O_{5+\delta}$ ($x = 0$–0.30) as cathode materials for solid-oxide fuel cells *Electrochim. Acta* **133** 364–72

[71] Fu D, Jin F and He T 2016 A-site calcium-doped $Pr_{1-x}Ca_xBaCo_2O_{5+\delta}$ double perovskites as cathodes for intermediate-temperature solid oxide fuel cells *J. Power Sources* **313** 134–41

[72] Lim C, Jun A, Jo H, Min Ok K, Shin J, Ju Y W and Kim G 2016 Influence of Ca-doping in layered perovskite $PrBaCo_2O_{5+\delta}$ on the phase transition and cathodic performance of a solid oxide fuel cell *J. Mater. Chem.* A **4** 6479–86

[73] Li M, Zhao M, Li F, Zhou W, Peterson V K, Xu X, Shao Z, Gentle I and Zhu Z 2017 A niobium and tantalum co-doped perovskite cathode for solid oxide fuel cells operating below 500 °C *Nat. Commun.* **8** 13990

[74] Kalpana S, Baral A K and Thangadurai V 2016 Electrochemical studies of $Gd_{0.5}Pr_{0.5}BaCo_2O_{5+\delta}$ (GPBC) cathode for oxide ion and proton conducting solid oxide fuel cells *Solid State Ion.* **288** 351–6

[75] Sun S and Cheng Z 2020 $SrCo_{0.8}Nb_{0.1}Ta_{0.1}O_{3-\delta}$ based cathodes for electrolyte-supported proton-conducting solid oxide fuel cells: comparison with $Ba_{0.5}Sr_{0.5}Co_{0.8}Fe_{0.2}O_{3-\delta}$ based cathodes and implications *J. Electrochem. Soc.* **167** 024514

Chapter 7

The role of sacrificial agents in water splitting

Mano Ganapathy, Shankar Mani, Viswanathan Alagan and Chang Tang Chang

Water splitting is the process of decomposition of water into hydrogen and oxygen in the presence of semiconductor photocatalysts. In the photocatalyst based water splitting process, the substance (semiconductor) will absorb photons which accelerates the electrons from the valence band to the conduction band. These photogenerated electron–hole pairs undergo a reduction reaction which leads to the formation of hydrogen and oxygen. Water splitting has several advantages, such as being clean, producing a carbon-neutral fuel, and thus providing green energy. However, the recombination of photogenerated charge carriers limits the efficiency of water splitting. To overcome this problem, several strategies have been adopted such as doping, semiconductor coupling, sacrificial agents, etc. Among them, the use of sacrificial agents has been improving the efficiency of water splitting efficiently. The sacrificial agents act as electron and hole donors, which limit the charge carrier recombination and increase the mobility of the charge carriers as well as their lifetime. This promotes the reactive oxygen species which enhances the efficiency of water splitting. Different kinds of sacrificial agents such as organic and/or inorganic materials have been employed in water splitting applications. The selection of sacrificial agents depends on the hydrogen or oxygen production. Many organic compounds have been used as sacrificial agents such as alcohols, organic acids, hydrocarbons, etc, which act as hole scavengers. Similarly, inorganic molecules such as IO_3^-/I^-, Ce^{4+}/Ce^{3+}, SO_3^{2-}/S^{2-}, etc, act as hole scavengers. These hole scavengers enhance the efficiency of hydrogen production. Electron scavengers such as silver nitrate, ferric ions, silver cations, etc, promote oxygen production. This chapter discusses the role of the sacrificial agents in the water splitting process.

doi:10.1088/978-0-7503-5729-6ch7

7.1 Introduction

Water splitting is the process of decomposing water into hydrogen (H_2) and oxygen (O_2). Different energy sources such as thermal, electrical, and light are utilized to cleave the H–O–H bonds [1, 2]. Water splitting has several advantages such as being clean, producing carbon-neutral fuel, and the providing green energy [2–4]. The water splitting process is classified based on the energy sources utilized such as thermolysis (heat), electrolysis (current), and photolysis (light). In the process of electrolysis, water is split into hydrogen and oxygen by utilizing electrical energy which is converted into chemical energy [3, 5–9]. The water reacts with the anode and cathode to form oxygen and hydrogen, respectively. In thermolysis, hydrogen and oxygen are derived from the conversion of fossil fuels, coal, biomass, etc [10–12]. Photolysis is the process of water conversion into hydrogen and oxygen under the solar or artificial light irradiation [13, 14]. In most of these methods, electron–hole pairs were generated during the application of external energy.

These electron–hole pairs undergo the reduction reaction which yields hydrogen and oxygen simultaneously, as shown in figure 7.1 [12]. In ideal conditions, the observed water splitting efficiency is poor. Moreover, the recombination of photo-generated electrons–holes limits the water splitting efficiency [15]. Also, there is the possibility of the formation of a reversible reaction (the formation of H_2O) due to the large driving force of the backward reaction. To overcome these drawbacks, several approaches such as doping, semiconductor coupling, and sacrificial agents have been adopted to improve the efficiency of water splitting [16–18]. Among them,

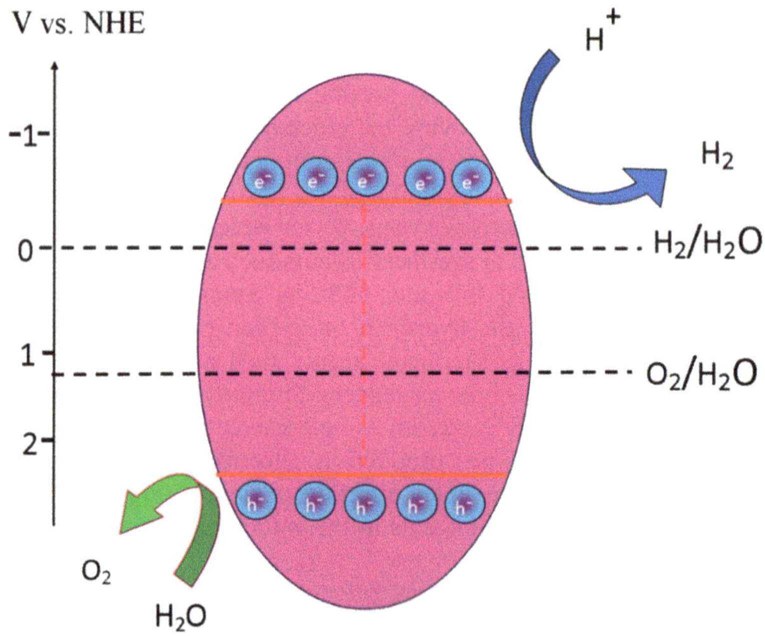

Figure 7.1. Schematic representation of water splitting.

the use of sacrificial agents improves the efficiency of the water splitting efficiently. The sacrificial agents will act as electron and hole donors, which limit the charge carrier recombination, increase the charge carrier's mobility as well as lifetime, and inhibit the backward reaction [19, 20]. This promotes the formation of reactive oxygen species (ROS) which enhances the water splitting efficiency [21]. Different kinds of sacrificial agents, including organic and/or inorganic materials, have been employed in water splitting applications [22–24]. The selection of sacrificial agents depends upon the hydrogen or oxygen production. The sacrificial agent properties such as electron donating ability, chemical stability, and polarity strongly influence the efficiency of the photocatalytic water splitting [18, 25].

7.2 Hydrogen production

The increased human population and rapid industrialization have led to the consumption of a large amount of fuel and electricity. This had led to the current major energy crisis of the present and also of the future [6, 26]. The combustion of fossil fuel energy sources, such as oil and natural gas, produces harmful by-products such as carbon dioxide (CO_2) [2, 11, 27, 28]. CO_2 pollution is a matter of concern and has a remarkable impact on ecosystems and global warming. Therefore, there is an urgent need to search for alternative energy sources. In recent years, green hydrogen as a source of energy carrier has emerged as a carbon-neutral fuel for next-generation transportation and industrial applications [29]. In recent years, water splitting has received more attention for hydrogen production than other conventional methods due to its promising properties. Hydrogen is extensively used in many applications such as ammonia synthesis, the petrochemical industry, electronic chip manufacturing, and metallurgy [27, 30]. The use of sacrificial agents has accelerated the hydrogen production in the water splitting process [31].

7.2.1 Organic sacrificial agents

The organic materials such as alcohols, organic acids, and hydrocarbons are used widely in hydrogen production due to their strong reducibility and ease of oxidation by photogenerated holes. In several attempts, methanol and ethanol have been employed widely as the sacrificial agent, however, these sacrificial agents contributes in the exceeded loss [32, 33]. Thus other organic electron donors such as glucose, glycerol, and triethanolamine are being used as the sacrificial agents, and are derived from organic waste, renewable biomass, and alcohols [34–36]. These kinds of materials promote hydrogen production and preclude environmental pollution, since all these organic materials are non-toxic [37].

7.2.1.1 Methanol

In recent years, methanol has been used frequently as a sacrificial agent in hydrogen production due to its effective electron donors and consuming holes that inhibit oxidation (figure 7.2). Methanol yields high-purity hydrogen due to its high H/C ratio and its low boiling point. The methanol oxidation potential is 0.02 V which is

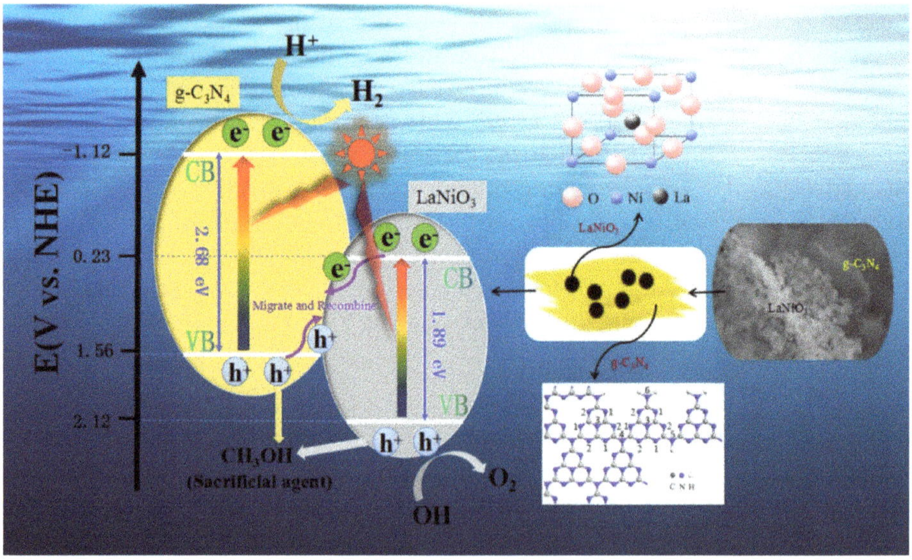

Figure 7.2. Schematic reaction mechanism of photocatalytic hydrogen production using a methanol sacrificial agent. (Reproduced from [72]. CC BY 4.0.)

less than water's oxidation potential $E_oO/H_2O = 1.23$ V at pH = 0 [38]. The reaction mechanism of hole oxidation using methanol is classified into two types:

1. Direct oxidation of photogenerated holes.
2. Photogenerated hole oxidation through the –OH radicals which were produced from $OH^{\bullet}$ radicals.

Methanol donates the electrons which interact directly with the photogenerated holes which lead to hole oxidation. Similarly, radicals were formed through an interaction with the photogenerated electron and hole pairs of the catalyst. These kinds of reactions were effectively scavenging the hole which promotes the efficient production of hydrogen.

A mixture of water and alcohol shows greater efficiency than the others. By increasing the methanol concentration in the water mixture, the production efficiency of H_2 is increased. Yang *et al* prepared $SrTiO_3/TiO_2$ heterostructures by hydrothermal and *in situ* deposition methods, where platinum (Pt) was used as the co-catalyst and methanol (25%) was used as the sacrificial agent. The $SrTiO_3/TiO_2$ nanowires produced 111.26 mmol g^{-1} h^{-1} hydrogen gas at room temperature, which was superior to bare TiO_2 nanowires (44.18 mmol g^{-1} h^{-1}) and P25 (35.77 mmol g^{-1} h^{-1}) [39].

Similarly, ZnO/Bi_2S_3 heterostructures were prepared using *in situ* generation methods to improve photocatalytic organic pollutant degradation and hydrogen production under solar light irradiation. Methanol (25%) has been used as a sacrificial agent to act as a hole scavenger and produced free radicals that accelerate hydrogen production. The measured amount of H_2 produced was found to be 2791 μmol [40].

Likewise, the TiO_2 surface was modified through a sacrificial coating of nickel oxide (NiO) quantum dots. The strategy of maintaining a short distance was adopted to reduce the electron–hole pair recombination. The NiO/TiO_2 heterojunction was prepared using a temperature-programmed method, 10% methanol was used as a sacrificial agent, and the measured hydrogen production was 1.35 mmol h^{-1} g^{-1}, which was 37 times higher than for NiO/TiO_2, and also 56 times higher than that of bare TiO_2 [41]. RuO_2/TiO_2 nanoporous heterostructures were prepared with different RuO_2 loading in the range of 1–10 wt% to improve photocatalytic hydrogen production. The following reaction mechanism was proposed in hydrogen production using methanol as the sacrificial agent:

$$RuO_2/TiO_2 + h\gamma \rightarrow e^-(TiO_2) + h^+(RuO_2)$$

$$h^+ + CH_3OH \rightarrow CH_2OH + h^+$$

$$CH_2OH \rightarrow HCHO + h^+ + e^-$$

$$2H_2O + 2e^- \rightarrow H_2 + 2OH^-$$

$$CH_3OHh\gamma, \xrightarrow{\text{catalyst}} HCHO + H_2$$

$$HCHO + H_2Oh\gamma, \xrightarrow{\text{catalyst}} HCOOH + H_2$$

$$HCOOHh\gamma, \xrightarrow{\text{catalyst}} CO_2 + H_2.$$

The methanol based sacrificial agent produces $\bullet CH_3OH$ and has negative oxidation potential (−0.74 V). It further reacts and produces H_2 and formaldehyde. Further, H_2 was produced when formaldehyde changed into formic acid. The 5% RuO_2/TiO_2 shows superior hydrogen production efficiency of 618 μmol h^{-1} [42].

7.2.1.2 Ethanol

As mentioned above, ethanol is the second most used sacrificial agent in hydrogen production. Ethanol belongs to an alkyl group with two carbon atoms and it is attached to the –OH group. Also, it is miscible with water and mixtures have a smaller volume than the sum of the two individual volumes of the water and ethanol. Moreover, ethanol and water are exothermic and the surface tension of water is reduced upon the addition of ethanol. Ethanol shows a lower oxidation potential of 0.08 V versus normal hydrogen electrode (NHE) which is less than that of the water oxidation potential. The electrons were donated by the ethanol which is directly interacting with the photogenerated holes which leads to the oxidation of the holes. Ethanol is a most promising sacrificial agent, like methanol, due to its bio-renewable origin and is derived from common industrial wastes. This may support reducing the cost of H_2 production and also promotes the effective use of the waste products. These kinds of reactions effectively scavenge the holes which promotes efficient hydrogen production.

The ethanol sacrificial reaction mechanism of the photocatalytic hydrogen production is as follows:

$$CH_3CH_2OH + TiO_2 \rightarrow CH_3CH_2O - Ti^{4+} + OH$$

$$TiO_2 + UVlight \rightarrow 2e^- + 2h^+$$

$$CH_3CH_2O - Ti^{4+} + 2h + \rightarrow CH_3CHO + Ti^{4+}$$

$$2OH + e^- \rightarrow H_2 + 2O^{2+}.$$

The Zn(O,S)/GO nanocomposite was prepared by feasible synthesis for photocatalytic hydrogen production. Different kinds of sacrificial agents, such as ethanol, methanol, isopropanol, and ethylene glycol, were used as hole scavengers and the photocatalytic hydrogen production efficiency was studied. The photocurrent response is 85, 42, 30, and 28 μA corresponding to ethanol, methanol, isopropanol, and ethylene glycol, respectively. Among the different sacrificial agents used, ethanol shows a superior photocurrent response compared to the others due to effective hole oxidation which promotes hydrogen production efficiency. The photocatalytic hydrogen production rate of different sacrificial agents is ethanol > methanol > isopropanol > ethylene glycol. Photocatalytic hydrogen production of 2840 μmol of H_2 was measured in the presence of an ethanol sacrificial agent and UV light [43]. Zn $(OH)_2/Cd_{0.3}Zn_{0.7}S$ photocatalyst was prepared by a precipitation method and the photocatalytic water splitting was analyzed under the influence of organic and inorganic sacrificial agents. The different kinds of organic and inorganic sacrificial agents, such as Na_2SO_3/Na_2S, lactic acid, and ethanol, are used to measure photocatalytic hydrogen production. Among the organic sacrificial agents ethanol (10 ml) shows superior photocatalytic hydrogen production efficiency of 3200 μmol h^{-1} g^{-1} [23]. Likewise, the Au/TiO_2 catalysts tuned their photonic band gap and electronic band gap which resulted in improved photocatalytic hydrogen production under direct sunlight. Here, ethanol was used as a sacrificial agent (0.5 vol. %). The ethanol-donating electrons had a fast reaction with holes within a fraction of a nanosecond. Moreover, the rate of charge carrier disappearing occurs at multiples of nanoseconds in TiO_2 [44]. Pd-TiO_2/ZSM-5 catalysts were prepared and studied for photocatalytic water splitting efficiency by varying the catalyst concentration, pH, and sacrificial agent concentration. Ethanol was used as a sacrificial agent and its concentration was varied from 4% to 16%. With an improved concentration of sacrificial agent, the efficiency of the water splitting was enhanced with a maximum production of 712 μmol h^{-1} for the 16% sacrificial agents [45].

7.2.1.3 Isopropanol

The isopropanol sacrificial reaction mechanism of the photocatalytic hydrogen production is as follows:

$$\left[{}^{S_2^{\ddagger}}_{Cd_2}{>}(CdS)\right]_2 + H_2O \rightarrow {}^{S_2^{\ddagger}}_{Cd_2}{>}Cd(II)SH + {}^{S_2^{\ddagger}}_{Cd_2}{>}S(-II)Cd(II)OH$$

$$ {}^{S_2}_{Cd_2}\!\ddagger > CdSH + 2 \rightarrow {}^{S_2}_{Cd_2}\!\ddagger > CdSH \ + \ H^+ $$

$$ {}^{S_2}_{Cd_2}\!\ddagger > CdSH \ \rightarrow {}^{S_2}_{Cd_2}\!\ddagger > CdS \ - \ + H^+ $$

$$ {}^{S_2}_{Cd_2}\!\ddagger > CdOH + 2 \rightarrow {}^{S_2}_{Cd_2}\!\ddagger > CdOH \ + \ H^+ $$

$$ {}^{S_2}_{Cd_2}\!\ddagger > CdOH \ \rightarrow {}^{S_2}_{Cd_2}\!\ddagger > CdO \ - \ + H^+ $$

$$ Cd2 + S2 > {}^{S_2}_{Cd_2}\!\ddagger > Cd(+II)S(0) + \ + C_3H_7OH \ \rightarrow {}^{S_2}_{Cd_2}\!\ddagger > Cd(+II)S(-I)H + + C_3H \cdot 6OH $$

$$ {}^{S_2}_{Cd_2}\!\ddagger > Cd(+II)S(-I)H + +C_3H_7OH \ \rightarrow {}^{S_2}_{Cd_2}\!\ddagger > Cd(+II)S(-II)H + 2 + C_3H \cdot 6OH $$

$$ 2H \cdot \rightarrow H_2 $$

$$ 2C_3H_6OH \rightarrow 2C_3H_5O + H_2 $$

$$ {}^{S_2}_{Cd_2}\!\ddagger > Cd(+II)S(-II)H^{+2} \rightarrow {}^{S_2}_{Cd_2}\!\ddagger > Cd(+II)S(-II)H + H^+ $$

$$ {}^{S_2}_{Cd_2}\!\ddagger > CdOH \ + \ e - CB \rightarrow {}^{S_2}_{Cd_2}\!\ddagger > CdO \ - \ + H^+. $$

The Al/$SrTiO_3$ were prepared using a solid state reaction method and the homogeneous precipitation method was adopted for gold (Au) loading over Al/ $SrTiO_3$ and used as photocatalysts in the water splitting process. Methanol, ethanol, and isopropyl alcohol were used as sacrificial agents with different weight percentages, such as 10%, 20%, and 30% respectively. Among the various sacrificial agents used, 30% isopropyl alcohol shows better performance than the others. The H_2 production efficiency increased gradually when the isopropyl alcohol concentration increased from 10% to 30%, and ethanol shows the lowest water splitting efficiency. The measured photocatalytic H_2 production was found to be 347 mmol $h^{-1}.gcat^{-1}$ [46].

7.2.1.4 Ethylene diamine tetra acetic acid (EDTA)

EDTA has had a great deal of attention for use as a sacrificial agent in water splitting applications. EDTA has four and two groups of carboxyls and amines, respectively. Moreover, EDTA is a Lewis based compound which can donate electron pairs. This electron undergoes a hole oxidation process which influences the efficiency of the water splitting.

Core–shell TiO_2/Fe_2O_3 were prepared and used in photocatalytic water splitting, where methanol, ethanol, triethanolamine, acetic acid, lactic acid, and EDTA were used as sacrificial agents. Among the different sacrificial agents, EDTA exhibited superior H_2 production efficiency due to the efficient separation of photogenerated holes which leads to a long photogenerated electron lifetime. Moreover, EDTA can bind with the centre metal ions which produce electron pairs. All these parameters helped to improve the photocatalytic H_2 production efficiency of 2700 μmol h^{-1} under optimized conditions [47].

TiO_2 sensitized with Au nanoclusters as single electrode systems were prepared to improve the photocatalytic H_2 production. Methanol and EDTA were used as sacrificial agents and Na_2SO_4 was used as an electrolyte. Among the different sacrificial agents, EDTA showed better photocatalytic H_2 production efficiency with four-fold increased current density compared to methanol. Moreover, negligible backward current was detected in the comparison of methanol and EDTA donating electrons which oxidized the holes in order to enhance the water splitting efficiency (figure 7.3) [48].

The coastal deposits of black sand contain mixtures of oxides (such as Fe_2O_3, TiO_2, SiO_2, ZrO_2, MnO, Al_2O_3, etc) and metals (V, Cr, Ni, Sr, Ce, etc) which have been used as photocatalysts. 10 mM EDTA was used as the sacrificial agent with a presence of 0.5 g l^{-1} and the amount of H_2 produced was observed to be 35,479.58 μmol h^{-1} g^{-1} [49].

Porphyrin–pyrene conjugate functionalized Pt nancomposites were prepared and used as photocatalysts in photocatalytic water splitting and EDTA was used as a sacrificial agent. The conjugate forms were excited moieties under visible light irradiation and electron transfer took place from the excited porphyrin to Pt which produced the hydrogen. The H_2 production efficiency was increased during the increment of pH which indicated a fast reduction reaction. The electron donating ability of EDTA was decreased when the pH value was decreased due to its protonation. In this regard, the photocatalysis parameters were optimized [50].

7.2.1.5 Triethanolamine (TEOA)

TEOA contains both triol and tertiary amine and its viscous organic compound. The molecules can coordinate with metal ions in the solution which occupy the surface adsorption sites of semiconductor-like graphitic carbon nitride (g-C_3N_4). This exhibited a negative effect on the adsorption of metal ions subsequent to the photodeposition. The sacrificial agents limit the surface affinity of g-C_3N_4 of metal ions:

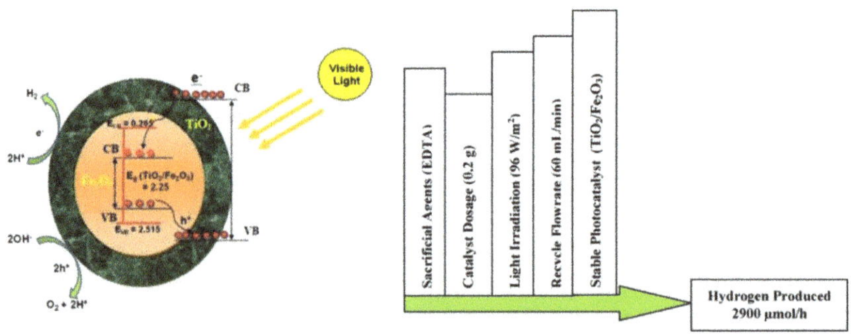

Figure 7.3. Schematic reaction mechanism of photocatalytic hydrogen production using EDTA. (Reproduced with permission from [47]. Copyright 2018 Elsevier.)

$$6H_{15}NO_3 \rightarrow C_6H_{15}NO_3^+ + e^-$$

$$C_6H_{15}NO_3^+ \rightarrow C_6H_{14}NO_3 + H^+$$

$$C_6H_{14}NO_3 \rightarrow C_6H_{14}NO_3^+ + e^-$$

$$C_6H_{14}NO_3^+ + H_2O \rightarrow C_4H_{11}NO_3 + CH_3CHO + H^+.$$

CoS_2@SCN was prepared by modified thermal oxidation and doping of sulphur using thermal oxidation etching to improve the photocatalytic H_2 production. Different kinds of sacrificial agents such as triethanolamine, methanol, ethanol, ethylene glycol, and polyethylene glycol were employed. The charge transfer resistance was decreased upon the introduction of sacrificial agents indicates that the charge transfer ability was increased after hole was oxidized. The measured H_2 production is 223.6 μmol g^{-1} h^{-1} with the presence of 10% triethanolamine [32].

$Ni_3(C_3N_3S_3)_2$ coordination polymer was prepared by the simple wet-chemical method for photocatalytic water splitting applications, where the triethanolamine was used as a sacrificial agent. Poor photocatalytic H_2 production efficiency was observed without the addition of co-catalyst and sacrificial agent. After the introduction of a sacrificial agent (triethanolamine), there was a two-fold increase in the water splitting observed with H_2 production of 112.6 μmol [51].

TiO_2/Pt sensitized with eosin Y have been studied for photocatalytic water splitting under different eosin, sacrificial agent, Pt, and catalyst concentrations. The dye absorbs more photons and the oxidized dye molecules are regenerated with the help of triethanolamine. Moreover, small interactions were observed between eosin Y–Pt and eosin Y–TEOA pairs. Superior hydrogen production was observed under the optimized concentration of eosin Y, TEOA, Pt, and pH values corresponding 2.5×10^3 M, 0.275 M, 0.25%, and 7.0, respectively [36]. The interaction effects plot exhibited that the visible light supports in yielding more H_2 production.

Covalent triazine frameworks (CTFs) were synthesized from a benzylamine-functionalized monomer which enhanced the hydrophilicity and crystallinity. Triethanolamine is used as a sacrificial agent in this report. The wetting ability helps in better dispersibility of the photocatalyst in water which enhanced better interaction between water and the photocatalyst. The combination of hydrophilicity, crystallinity, and triethanolamine was favourable for charge carrier generation and separation which enhanced the photocatalytic water splitting efficiency with a value of 9200 μmol g^{-1} h^{-1} under visible light irradiation [52].

7.2.1.6 Lactic acid

A two-dimensional (2D) cadmium sulphide/tungsten disulphide (CdS/WS_2) heterostructure was prepared with different weight percentages of WS_2, and tested against the hydrogen evolution reaction (HER). Lactic acid (1 ml) was used as a sacrificial agent and it scavenged the holes in the CdS valance band. Among the different WS_2 weight percentages, 10% WS_2 showed superior hydrogen production of 14.1 mmol g^{-1} h^{-1} under visible light irradiation [53].

Molybdenum disulphide (MoS_2) quantum dots (QDs) were prepared by the hydrothermal method followed by liquid exfoliation and the CdS co-catalyst was sensitized into MoS_2, and the homogeneous solution was added into the sacrificial agent (lactic acid). The small particle size of the MoS_2 QDs-L provided more active sites and the heterostructure had a large specific surface area which enhanced the light absorption. Moreover, the co-catalysts help to improve the photogenerated charge carrier lifetime and the production efficiency. The measured hydrogen production efficiency was found to be 1032.1 μmol^{-1} [54].

The photocatalytic hydrogen production and photocorrosion properties of CdS were investigated in the medium of lactic acid. During the reaction the cobalt lactate complex was formed due to the chemical reactions between the lactic acid and cobalt ions. By increasing the photocorrosion resistance the H_2 production efficiency was enhanced. Moreover, the cobalt lactate complex was consumed in the CdS photogenerated holes in addition to the Au which was photodeposited over the CdS facilitating the electron transaction. The observed photocatalytic hydrogen production efficiency of CdS/Au/-Co is about 86.46 μmol h^{-1} [55].

Nanostructured antimony and cerium co-doped with tin dioxide (Ce/ATO) were prepared through sol–gel method. During the light irradiation Ce^{4+} gains electrons from Ce^{3+} and also Ce^{4+} gains electrons from the lactic acid (which was used as the sacrificial agent). The concentration of Ce increased by 6 mol% which promotes the Ce^{4+} and decreased the Ce^{3+} ions. The observed H_2 production of Ce/ATO was 8.5 mmol g^{-1} h^{-1} under optimized conditions [56].

MoS_2/g-C_3N_4 composites were prepared by different techniques such as *in situ* photodeposition, thermal decomposition, and sonochemical methods. The 2H MoS_2 phase was formed during the photodeposition due to the reduction of $[MoS_4]^{2-}$. The 2H MoS_2 formed as an intimate interfacial contact with CN which is favourable for the photogenerated charge separation and promotes the H_2 production efficiency. The Pt and lactic acid were used as the co-catalyst and sacrificial agents, respectively. The 2H phase MoS_2/g-C_3N_4 composites show superior photocatalytic hydrogen production efficiency [57].

7.2.2 Inorganic sacrificial agents

The sulphite and sulphide materials are widely employed as inorganic sacrificial agents for H_2 production due to the efficient acceptors of photogenerated holes, as shown in figure 7.4. The sulphite and sulphide are more easily oxidizable than alcohols which leads to reduced photoanodic corrosion. Another advantages of sulphide involves the reactions with certain sulphide materials which leads to rebuilding the catalyst materials [38].

Zinc oxide (ZnO) was doped with other non-metals such as nitrogen and boron by a mechanochemical combustion technique for photocatalytic water splitting. Na_2S–Na_2SO_3 was used as the sacrificial agent. An electron–hole pair was generated in the ZnO during light illumination over the catalyst surface. The photogenerated electrons undergoes a reduction reaction thereby generating the H_2 gas. The hydroxyl radicals and H^+ ions are formed during the reaction of photogenerated

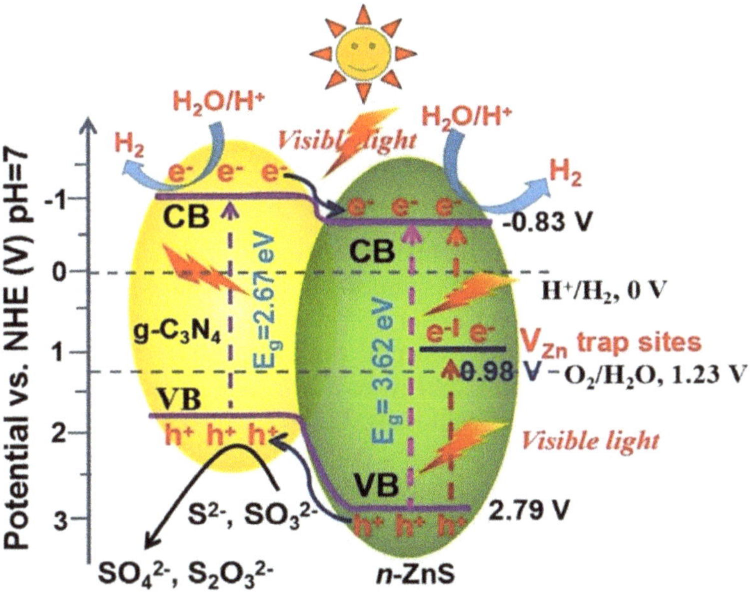

Figure 7.4. Schematic reaction mechanism of photocatalytic hydrogen production using the Na_2S–Na_2SO_3 sacrificial agent. (Reproduced with permission from [73]. Copyright 2018 Elsevier.)

holes with the water molecule. The Na_2S–Na_2SO_3 effectively harvests the photogenerated holes by three routes. The S_2^{2-} act as optical filters and reduces the proton reduction. The ZnS formed on the ZnO surface creates the p–n heterojunction. The observed photocatalytic H_2 production efficiency is about 1730 μmol g^{-1} for 3 h [24].

CdS@MoS_2 core–shell structures were prepared using the hydrothermal method and Na_2S–Na_2SO_3 was used as the sacrificial agent. The core–shell structure promotes the interfacial contact between the CdS and MoS_2. The pristine CdS has high photogenerated charge carrier recombination which exhibited a low photocurrent. The formation of the core–shell structure (CdS@MoS_2) inhibited the charge carrier recombination and the sacrificial agent harvested the photogenerated holes effectively, which enhanced the photocatalytic H_2 production with a rate of 17 203.7 μmol g^{-1} h^{-1} under visible light irradiation.

CdS and ZnS were prepared using the precipitation method and the photocatalytic H_2 production was analysed under sacrificial agents such as methanol (organic) and Na_2S–Na_2SO_3 (inorganic). ε-$Zn(OH)_2$ and ZnS intermediates were observed for the Na_2S–Na_2SO_3 and methanol, respectively. The sacrificial agents influence the composite transformations strongly and the Na_2S–Na_2SO_3 show superior photocatalytic H_2 production compared to methanol in the presence of $Zn(OH)_2$/$Cd_{0.3}Zn_{0.7}S$ photocatalyst. The observed amount of H_2 was 8900 μmol g^{-1} h^{-1} under visible light irradiation [23].

The enriched zinc vacancy mediated S-scheme ZnCdS/ZnS composite was derived for the photocatalytic H_2 production through the *in situ* ZnCdS/MOF-5 method and Na_2S–Na_2SO_3 was used as the sacrificial agent. The vacancy defects in the synthesized material improved the charge carrier separation process. The measured photocatalytic H_2 rate was 12.31 mmol g h^{-1} for the ZnCdS/ZnS(20) photocatalyst under optimized conditions [58].

7.3 The role of sacrificial agent in oxygen evolution

7.3.1 Introduction to photocatalytic oxygen production

In recent years, the increase in the human population and industrial development have led to an increase in energy consumption and environmental pollution [59, 60]. As mentioned above, the current major energy needs are met by fossil fuels, which cause climatic change and environmental pollution [60]. In order to reduce the pollution and meet the current energy requirements, it is necessary to find an alternative to the conventional fossil fuels. Photocatalytic water splitting paves the way for renewable energy generation and reduces the environmental pollution in an ecofriendly manner. It has attracted the attention of researchers across the globe owing to its merits. Photocatalytic water splitting is a thermodynamic process which results in the evolution of hydrogen and oxygen [61]. Photocatalysis based on the semiconductor material takes place in three steps: (i) the generation of holes and electrons by absorbing a photon of energy greater than or equal to the band gap (E_g) of the material; (ii) charge separation (electron and holes) in the semiconductor material; and (iii) a chemical reaction takes place at the surface between the photogenerated charge carriers and other compounds [62].

7.3.2 Sacrificial agents in photocatalysis

Initiating the photocatalytic water splitting reaction is a tedious process, because the stable H_2O molecule resists charge dissociation. This leads to the recombination of the charge carriers produced and results in decreased efficiency of the photocatalytic reaction [63]. The oxidation and reduction of water is a complicated reaction which involves four electrons [38]. Sacrificial agents are those with a decreased tendency of recombination of the charge carriers (electrons and holes) and increased efficiency of the photocatalytic reaction [31]. Several sacrificial agents have been employed to increase the efficiency of the reaction and they also result in good H_2 and O_2 evolution rates. Several sacrificial agents such as methanol, sodium sulphide, EDTA, butyl alcohol, silver nitrate, and Fe^{3+} ions play key roles in better oxygen evolution. Not only the sacrificial agents, but also the photocatalyst plays a vital role in the oxygen evolution reaction (OER). The yield of oxygen generation varies depending on the photocatalyst and sacrificial agent used. The role of sacrificial agents in the oxygen generation is less explored when compared with H_2 generation. The upcoming sections will discuss the different sacrificial agents and their role in the OER. Table 7.1 and Table 7.2 provides a comparative analysis of the sacrificial agent employed in the photocatalytic hydrogen and oxygen generation with their evolution rate.

Table 7.1. Comparison of different sacrificial agents and photocatalysts in the hydrogen evolution reaction.

S. No.	Photocatalysts	Catalyst dosage	Light source	Sacrificial agent	Evolved H_2	References
1.	$SrTiO_3/TiO_2$	10 mg	UV	25% methanol	111.26 mmol g^{-1} h^{-1}	[39]
2.	Bi_2S_3/ZnO	50 mg		25% methanol	2791 μmol g^{-1} h^{-1}	[40]
3.	NiO QDs/TiO_2			10% methanol	1.35 mmol	[41]
4.	RuO_2/TiO_2		UV	10 ml	618 μmol g^{-1} h^{-1}	[42]
5.	$SrTiO_3$–TiO_2	3(2×3) thin films	UV	40 ml distilled water and 5 ml methanol	386.6 μmol $(g\ h)^{-1}$	
6.	Zn(O,S)/GO nanocomposite	0.225 g	UV	50 ml ethanol	2840 μmol	[43]
7.	$Zn(OH)_2$/ $Cd_{0.3}Zn_{0.7}S$	50 mg	Visible light	10 ml ethanol and 90 ml water	3200 μmol h^{-1} g^{-1}	[23]
8.	Au/TiO_2	10–25 mg	Sunlight	0.5 vol. % of ethanol	1 mmol H_2/ gcatal h	[44]
9.	Pd-TiO_2/ZSM-5	1.5 g l^{-1}	UV	16% ethanol	712 μmol g^{-1} h^{-1}	[45]
10.	Au-Al/$SrTiO_3$	2.5gcat L^{-1}	Visible light	30% isopropyl alcohol	347 mmol h. gcat.	[46]
11.	TiO_2/Fe_2O_3	0.2 g	Visible light	30 ml EDTA	2700 μmol h^{-1}	[47]
12.	CoS_2@SCN	50 mg	Visible light	10 Vol % TEOA	223.6 μmol g^{-1} h^{-1}	
13.	$Ni_3(C_3N_3S_3)_2$	300 mg	UV	TEOA	112.6 μmol	[51]
14.	Eosin *Y*-TiO_2/Pt		Visible light	0.275 M TEOA	87.44%	[36]
15.	CTF-HUST-A1		Visible light	TEOA	9200 μmol g^{-1} h^{-1}	[52]
16.	CdS/WS_2	3 mg	Visible light	1 ml lactic acid	14.1 mmol g^{-1} h^{-1}	[53]

7.3.3 Sodium iodate

TiO_2 is widely used as a catalyst because of its band gap energy (3.2 eV). Due to its fast electron–hole recombination, TiO_2 can split H_2 and O_2 from water [64, 65]. However, by using a sacrificial agent we can reduce the recombination of electron–hole pairs and enhance the evolution of H_2 and O_2. Galinska and her team reported the role of sacrificial agents in the water splitting of Pt-TiO_2 [66]. Photocatalytic reactions were carried out for different sacrificial agents (such as methanol, $NaIO_3$, NaI, EDTA, and Na_2S) in a closed reactor using a mercury lamp as a source. $NaIO_3$

Table 7.2. Comparison of the different sacrificial agents and photocatalysts in the oxygen evolution reaction.

S. No.	Photocatalysts	Catalyst dosage	Light source	Sacrificial agent	Evolved O_2	References
1.	Pt-TiO_2	0.3 g	Mercury lamp	$NaIO_3$	0.410 mmol/30 min	[66]
2.	Pt-TiO_2	0.3 g	Mercury lamp	NaI	0.121 mmol/30 min	[66]
3.	$SrTiO_3$:Mn	0.3 g	Xenon lamp	$AgNO_3$	2.7 μmol h^{-1}	[67]
4.	$SrTiO_3$:Ru	0.3 g	Xenon lamp	$AgNO_3$	3.9 μmol h^{-1}	[67]
5.	In-doped GaON	500 mg	Visible light	$AgNO_3$	around 50 μmol h^{-1}	[68]
6.	Sr_2CoTaO_6-F (cubic shaped) Sr_2CoTaO_6-S (irregular shaped)	0.2 g	Visible light	$AgNO_3$	31 μmol h^{-1} g^{-1}20 μmol h^{-1} g^{-1}	[69]
7	FX-$LaTiO_2N$ (2 wt% CoO_x as co-catalyst)	0.2 g	Xenon lamp	$AgNO_3$	736 μmol h^{-1}	[70]
8	PC-$LaTiO_2N$ (2 wt% CoO_x as co-catalyst)	0.2 g	Xenon lamp	$AgNO_3$	489 μmol h^{-1}	[70]
9	Pt-TiO_2	0.3 g	Mercury lamp	EDTA	0.118 mmol/30 min	[66]
10	Ni@NiO-CdS-3	2 mg	300 W Xenon arc lamp	$Na_2S_2O_8$	55 120 μmol g^{-1} h^{-1}	[71]

exhibited a high oxygen evolution rate of 0.410 mmol/30 min when compared to NaI, EDTA, and Na_2S, whereas the methanol yielded a high hydrogen evolution rate of 11.50 mmol/90 min. The high evolution rate contributed to the sacrificial agents which act as scavengers for the electron and holes.

7.3.4 Silver nitrate ($AgNO_3$)

Silver nitrate is a well-known sacrificial agent to enhance the rate of oxygen evolution in the photocatalytic reaction. The results on the photocatalytic behaviour of $SrTiO_3$ doped with noble metal ions (Mn, Ru, Rh, and Ir) was reported by Ryoko Konta *et al* which show a higher oxygen evolution rate of 3.9 μmol h^{-1} for the photocatalyst $SrTiO_3$:Rh in the presence of $AgNO_3$ as the sacrificial agent [67], whereas the oxygen evolution rate of $SrTiO_3$:Mn was found to be 2.7 μmol h^{-1}. The difference in oxygen evolution rate arises due to the active sites generated by the dopants. Very small and negligible oxygen evolution rates were observed for the Rh and Ir doped $SrTiO_3$ photocatalysts.

Indium (In) doped gallium oxynitrides (GaON) with Pt as the co-catalyst were found to exhibit a good oxygen evolution rate in the presence of a sacrificial agent ($AgNO_3$) [68]. The oxygen evolution rate of GaON with respect to different percentages of In doping (0%, 0.1%, 0.5%, 1%, and 3%) were studied and reported

[68], in which 0.5% In doping shows a higher oxygen evolution rate due to its enhanced dispersion of hybridized orbitals.

The photocatalytic OER and HER of the cubic shaped Sr_2CoTaO_6-F and irregular shaped Sr_2CoTaO_6-S synthesized by flux and solid state reaction methods have been studied [69]. The photocatalytic OER activity was carried out in visible light in the presence of aqueous $AgNO_3$ as the sacrificial agent. Photocatalytic OER activity of 31 μmol h^{-1} g^{-1} for Sr_2CoTaO_6-F and 20 μmol h^{-1} g^{-1} for Sr_2CoTaO_6–S were observed. The higher oxygen evolution rate in the regular cubic structure (Sr_2CoTaO_6–F) is due to its photochemical stability [69].

Single crystalline cobalt (Co) modified $LaTiO_2N$ was prepared by the impregnation method for efficient oxidation of water under visible light [70]. The photocatalysts were prepared using the flux method (FX-LTON) and polymerized complex method (PC-LTON), then they were subjected to the photocatalytic reaction with different weight percentages of cobalt and iron oxide as co-catalysts in the presence of aqueous silver nitrate solution. A good oxygen evolution rate was observed in both FX-LTON (736 μmol h^{-1}) and PC-LTON (489 μmol h^{-1}) with CoO_x as the co-catalyst.

7.3.5 Ethylene diamine tetra acetic acid

As mentioned in section 7.2.1.5, EDTA is an organic acid which is being used as a scavenger of electrons and holes in photocatalytic water splitting. It supports reducing the recombination of charge carriers thereby enhancing the oxygen evolution. Pt-TiO_2 based photocatalysts were prepared using a dry impregnation method [66]. In this report, 0.3 g of photocatalyst was kept in an enclosed photoreactor illuminated using a mercury lamp, and 2 mmol of EDTA was used as the sacrificial agent for the oxygen evolution. The obtained oxygen evolution rate was found to be 0.118 mmol/30 min, and the same photocatalyst was subject to hydrogen evolution in 3 mmol of EDTA which yielded 2.209 mmol/90 min. These results indicate that EDTA acts as a hole scavenger.

7.3.6 Sodium per sulphate ($Na_2S_2O_8$)

Qiao *et al* studied the photocatalytic behaviour of Ni@NiO spheres modified by CdS nanoparticles [71]. The photocatalytic water splitting of Ni@NiO-Cd-3 was studied under the illumination of a 300 W xenon lamp in the presence of $Na_2S_2O_8$. There was no oxygen evolution observed in the absence of light and upon illuminating the light, a good oxygen generation rate of 55 120 μmol g^{-1} h^{-1} was obtained (figure 7.5) [71], which indicates that the modified Ni@NiO vacancies had good photocatalytic activity towards oxygen evolution.

7.4 Summary

In this chapter insight into the importance of photocatalytic water splitting in the modern era and the role of the sacrificial agents is provided. The chapter also discusses the use of different sacrificial agents, including organic and inorganic materials, employed with different photocatalysts in the water splitting process. The

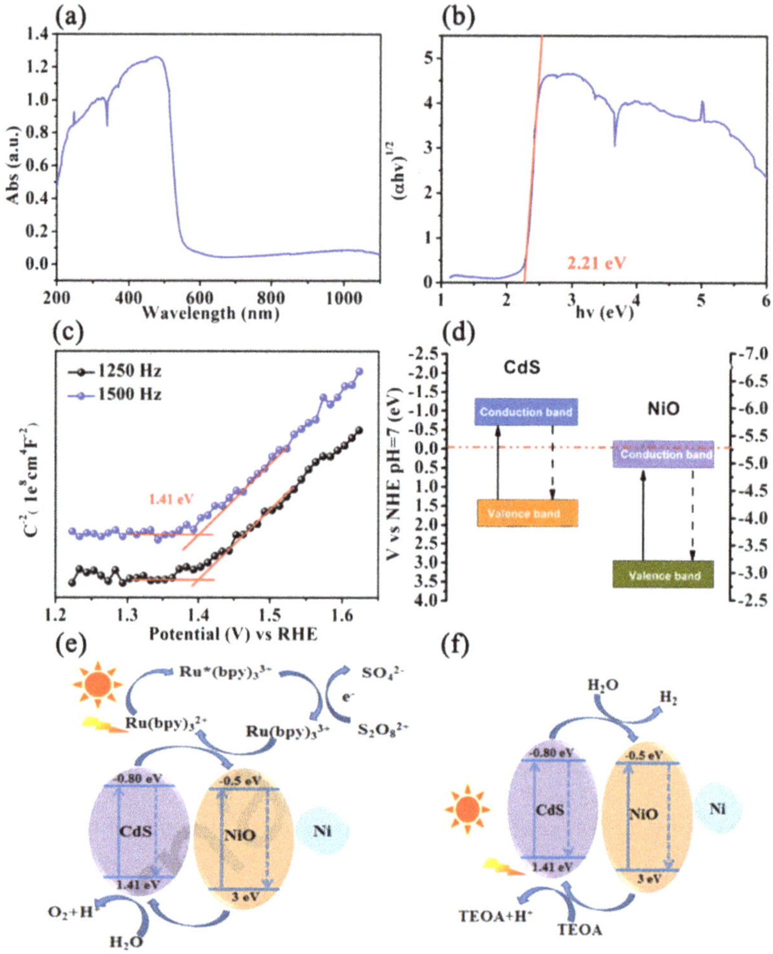

Figure 7.5. (a) UV–vis diffuse reflectance spectra, (b) Kubelka–Munk spectra, (c) Mott–Schottky plots, (d) band positions, and (e) and (f) schematic diagrams of the O_2 production of CdS composites under visible light radiation. (Reproduced with permission from [71]. Copyright 2020 Elsevier.)

sacrificial agents showed an excellent contribution in the water splitting efficiency due to their potential properties. However, the reaction mechanism of the hole oxidation and electron reduction is yet to be explored. Therefore, it is essential to find a suitable probe technique to analyse the reaction mechanism of the sacrificial agents and also find suitable sacrificial materials to support further improvement of the efficiency of water splitting.

References

[1] Basile A, Centi G, De Falco M and Iaquaniello G 2019 The vision of future sustainable energy, catalyst, and chemistry: opportunities for innovation and business *Stud. Surf. Sci. Catal.* **179** 7–20

[2] Moniz S J A *et al* 2015 Visible-light driven heterojunction photocatalysts for water splitting-a critical review *Energy Environ. Sci.* **8** 731–59
[3] Kahng S, Yoo H and Kim J H 2020 Recent advances in earth-abundant photocatalyst materials for solar H_2 production *Adv. Powder Technol.* **31** 11–28
[4] Wang Y *et al* 2021 Oxygen vacancies in actiniae-like Nb_2O_5/Nb_2C MXene heterojunction boosting visible light photocatalytic NO removal *Appl. Catal.* B **299** 120677
[5] Arulraj A *et al* 2023 Nanoarchitectonics of layered metal chalcogenides-based ternary electrocatalyst for water splitting *Energies* **16** 1669
[6] Basile A and Iulianelli A 2014 *Advances in Hydrogen Production, Storage and Distribution* (Amsterdam: Elsevier) pp 1–546
[7] Praveen Kumar M *et al* 2022 NiFe layered double hydroxide electrocatalyst prepared via an electrochemical deposition method for the oxygen evolution reaction *Catalysts* **12** 1470
[8] Arulraj A, Thachnatharen N, Rajkumar C and Mangalaraja R V 2022 Engineering 2D chalcogenides for energy and environmental remediation *Nanostructured Materials for Sustainable Energy and Environmental Remediation* (Bristol: IOP Publishing) pp 2-1–2-28
[9] Wang W, Tadé M O and Shao Z 2015 Research progress of perovskite materials in photocatalysis- and photovoltaics-related energy conversion and environmental treatment *Chem. Soc. Rev.* **44** 5371–408
[10] Arulraj A *et al* 2021 Chalcogenides nanocrystals and its applications ed N M Mubarak, M Khalid, R Walvekar and A Numan Contemporary Nanomaterials in Material Engineering Applications. Engineering Materials (Cham: Springer) pp 201–27
[11] de Klerk A 2020 *Transport Fuel: Biomass-, Coal-, Gas- and Waste-to-Liquids Processes* (Amsterdam: Elsevier)
[12] Patial S *et al* 2020 Tunable photocatalytic activity of $SrTiO_3$ for water splitting: strategies and future scenario *J. Environ. Chem. Eng.* **8** 103791
[13] Hou H and Zhang X 2020 Rational design of 1D/2D heterostructured photocatalyst for energy and environmental applications *Chem. Eng. J.* **395** 125030
[14] Hisatomi T and Domen K 2019 Reaction systems for solar hydrogen production via water splitting with particulate semiconductor photocatalysts *Nat. Catal.* **2** 387–99
[15] Ganapathy M *et al* 2021 Preparation of $SrTiO_3/Bi_2S_3$ heterojunction for efficient photocatalytic hydrogen production *Energy Fuels* **35** 14995–5004
[16] Wei Y *et al* 2021 Hollow multishelled structured $SrTiO_3$ with La/Rh co-doping for enhanced photocatalytic water splitting under visible light *Small* **17** 2005345
[17] Rahman M Z, Kibria M G and Mullins C B 2020 Metal-free photocatalysts for hydrogen evolution *Chem. Soc. Rev.* **49** 1887–931
[18] Ahmad H, Kamarudin S K, Minggu L J and Kassim M 2015 Hydrogen from photo-catalytic water splitting process: a review *Renew. Sustain. Energy Rev.* **43** 599–610
[19] Sayama K *et al* 2002 A new photocatalytic water splitting system under visible light irradiation mimicking a Z-scheme mechanism in photosynthesis *J. Photochem. Photobiol.* A **148** 71–7
[20] Wang J *et al* 2020 Few-layer $BiVO_4$ nanosheets decorated with $SrTiO_3$:Rh nanoparticles for highly efficient visible-light-driven overall water splitting *Appl. Catal.* B **279** 119377
[21] Nosaka Y and Nosaka A Y 2017 Generation and detection of reactive oxygen species in photocatalysis *Chem. Rev.* **117** 11302–36
[22] Bai Y, Hippalgaonkar K and Sprick R S 2021 Organic materials as photocatalysts for water splitting *J. Mater. Chem.* A **9** 16222–32

[23] Markovskaya D V *et al* 2020 The influence of the sacrificial agent nature on transformations of the $Zn(OH)_2/Cd0._3Zn_0.7S$ photocatalyst during hydrogen production under visible light *RSC Adv.* **10** 1341–50
[24] Gomathisankar P *et al* 2013 Photocatalytic hydrogen production from aqueous Na_2S + Na_2SO_3 solution with B-doped ZnO *ACS Sustain Chem. Eng.* **1** 982–8
[25] Kumar P V, Bernardi M and Grossman J C 2013 The impact of functionalization on the stability, work function, and photoluminescence of reduced graphene oxide *ACS Nano* **7** 1638–45
[26] Lubitz W and Tumas W 2007 Hydrogen: an overview *Chem. Rev.* **107** 3900–3
[27] Ávila-Neto C N *et al* 2009 Hydrogen production from methane reforming: thermodynamic assessment and autothermal reactor design *J. Nat. Gas Sci. Eng.* **1** 205–15
[28] Megia P J, Vizcaino A J, Calles J A and Carrero A 2021 Hydrogen production technologies: from fossil fuels toward renewable sources. A mini review *Energy Fuels* **35** 16403–15
[29] Abdin Z *et al* 2020 Hydrogen as an energy vector *Renew. Sustain. Energy Rev.* **120** 109620
[30] Yao Q, DIng Y and Lu Z H 2020 Noble-metal-free nanocatalysts for hydrogen generation from boron- and nitrogen-based hydrides *Inorg. Chem. Front.* **7** 3837–74
[31] Singh R and Dutta S 2021 Integrated photocatalytic hydrogen production and pollutants or wastes treatment: prospects and challenges *Sustainable Fuel Technologies Handbook* (Amsterdam: Elsevier) pp 541–9
[32] Wang L *et al* 2022 Effects of various alcohol sacrificial agents on hydrogen evolution based on CoS_2@SCN nanomaterials and its mechanism *Chemosphere* **286** 131558
[33] Majumdar P, Bera M K, Pant D and Patra S 2018 Enzymatic electrocatalysis of CO_2 reduction *Encyclopedia of Interfacial Chemistry: Surface Science and Electrochemistry* 5.2 pp 577–89
[34] Geng R *et al* 2021 Biomimetic nanozymes based on coassembly of amino acid and hemin for catalytic oxidation and sensing of biomolecules *Small* **17** 2008114
[35] Kohlsdorf A and Taffa D H 2018 Microwave assisted synthesis of Ta_2O_5 nanostructures for photocatalytic hydrogen production *J. Photochem. Photobiol.* A **366** 41–7
[36] Chowdhury P, Gomaa H and Ray A K 2011 Factorial design analysis for dye-sensitized hydrogen generation from water *Int. J. Hydrogen Energy* **36** 13442–51
[37] Vimal Michael R J *et al* 2015 Cu_2S-incorporated ZnS nanocomposites for photocatalytic hydrogen evolution *RSC Adv.* **5** 30175–86
[38] Schneider J and Bahnemann D W 2013 Undesired role of sacrificial reagents in photocatalysis *J. Phys. Chem. Lett.* **4** 3479–83
[39] Yang L, Chen Z, Zhang J and Wang C A 2019 $SrTiO_3/TiO_2$ heterostructure nanowires with enhanced electron–hole separation for efficient photocatalytic activity *Front. Mater. Sci.* **13** 342–51
[40] Bera S, Ghosh S and Basu R N 2018 Fabrication of Bi_2S_3/ZnO heterostructures: an excellent photocatalyst for visible-light-driven hydrogen generation and photoelectrochemical properties *New J. Chem.* **42** 541–54
[41] Hong W *et al* 2018 NiO quantum dot modified TiO_2 toward robust hydrogen production performance *ACS Sustain. Chem. Eng.* **6** 889–96
[42] Uddin M T *et al* 2015 New insights into the photocatalytic properties of RuO_2/TiO_2 mesoporous heterostructures for hydrogen production and organic pollutant photodecomposition *J. Phys. Chem.* C **119** 7006–15

[43] Gultom N S, Abdullah H and Kuo D H 2019 Effects of graphene oxide and sacrificial reagent for highly efficient hydrogen production with the costless Zn(O,S) photocatalyst *Int. J. Hydrogen Energy* **44** 29516–28
[44] Waterhouse G I N *et al* 2013 Hydrogen production by tuning the photonic band gap with the electronic band gap of TiO_2 *Sci. Rep.* **3** 1–5
[45] Enzweiler H *et al* 2020 Catalyst concentration, ethanol content and initial pH effects on hydrogen production by photocatalytic water splitting *J. Photochem. Photobiol.* A **388** 112051
[46] Saadetnejad D and Yıldırım R 2018 Photocatalytic hydrogen production by water splitting over Au/Al-$SrTiO_3$ *Int. J. Hydrogen Energy* **43** 1116–22
[47] Madhumitha A, Preethi V and Kanmani S 2018 Photocatalytic hydrogen production using TiO_2 coated iron-oxide core–shell particles *Int. J. Hydrogen Energy* **43** 3946–56
[48] Wang H, Chen F, Li W and Tian T 2015 Gold nanocluster-sensitized TiO_2 nanotubes to enhance the photocatalytic hydrogen generation under visible light *J. Power Sources* **287** 150–7
[49] López-Vásquez A, Suárez-Escobar A and López-Suárez F E 2020 Black sand-based photocatalyst for hydrogen production from EDTA solutions under UV–vis irradiation *Top. Catal.* **63** 1325–35
[50] O'Neill J S, Kearney L, Brandon M P and Pryce M T 2022 Design components of porphyrin-based photocatalytic hydrogen evolution systems: a review *Coord. Chem. Rev.* **467** 214599
[51] Guo F *et al* 2017 $Ni_3(C_3N_3S_3)_2$ coordination polymer as a novel broad spectrum-driven photocatalyst for water splitting into hydrogen *Appl. Catal.* B **210** 205–11
[52] Zhang S *et al* 2020 Strong-base-assisted synthesis of a crystalline covalent triazine framework with high hydrophilicity via benzylamine monomer for photocatalytic water splitting *Angew Chem.* **132** 6063–70
[53] Zhang K, Fujitsuka M, Du Y and Majima T 2018 2D/2D heterostructured CdS/WS_2 with efficient charge separation improving H_2 evolution under visible light irradiation *ACS Appl. Mater. Interfaces* **10** 20458–66
[54] Han B and Hu Y H 2016 MoS_2 as a co-catalyst for photocatalytic hydrogen production from water *Energy Sci. Eng.* **4** 285–304
[55] Yi X *et al* 2020 Boosting antiphotocorrosion and hydrogen-production activity of cadmium sulfide by cobalt lactate complex *Appl. Surf. Sci.* **512** 144786
[56] Liu H *et al* 2015 Facile preparation of cerium (Ce) and antimony (Sb) codoped SnO_2 for hydrogen production in lactic acid solution *Mater. Sci. Semicond. Process.* **40** 670–5
[57] Yuan H *et al* 2022 Enhanced photocatalytic hydrogen production based on laminated MoS_2/g-C_3N_4 photocatalysts *Colloids Surf.* A **641** 128575
[58] Hao X, Xiang D and Jin Z 2021 Zn-vacancy engineered S-scheme ZnCdS/ZnS photocatalyst for highly efficient photocatalytic H_2 evolution *Chem. Cat. Chem.* **13** 4738–50
[59] Dincer I 2000 Renewable energy and sustainable development: a crucial review *Renew. Sustain. Energy Rev.* **4** 157–75
[60] Su T-m *et al* 2015 Recent advances in the photocatalytic reduction of carbon dioxide *Environ. Chem. Lett.* **14** 99–112
[61] Su T *et al* 2018 Role of interfaces in two-dimensional photocatalyst for water splitting *ACS Catal.* **8** 2253–76

[62] Ismail A A and Bahnemann D W 2014 Photochemical splitting of water for hydrogen production by photocatalysis: a review *Sol. Energy Mater. Sol. Cells* **128** 85–101
[63] Ng K H *et al* 2021 Photocatalytic water splitting for solving energy crisis: myth, fact or busted? *Chem. Eng. J.* **417** 128847
[64] Abe R, Sayama K and Arakawa H 2003 Significant effect of iodide addition on water splitting into H_2 and O_2 over Pt-loaded TiO_2 photocatalyst: suppression of backward reaction *Chem. Phys. Lett.* **371** 360–4
[65] Tabata S, Nishida H, Masaki Y and Tabata K 1995 Stoichiometric photocatalytic decomposition of pure water in Pt/TiO_2 aqueous suspension system *Catal. Lett.* **34134** 245–9
[66] Galińska A and Walendziewski J 2005 Photocatalytic water splitting over Pt−TiO_2 in the presence of sacrificial reagents *Energy Fuels* **19** 1143–7
[67] Konta R, Ishii T, Kato H and Kudo A 2004 Photocatalytic activities of noble metal ion doped $SrTiO_3$ under visible light irradiation *J. Phys. Chem.* B **108** 8992–5
[68] Hu C C, Lee Y L and Teng H 2011 Influence of indium doping on the activity of gallium oxynitride for water splitting under visible light irradiation *J. Phys. Chem.* C **115** 2805–11
[69] Idris A M *et al* 2020 Sr_2CoTaO_6 double perovskite oxide as a novel visible-light-absorbing bifunctional photocatalyst for photocatalytic oxygen and hydrogen evolution reactions *ACS Sustain. Chem. Eng.* **8** 14190–7
[70] Zhang F *et al* 2012 Cobalt-modified porous single-crystalline $LaTiO_2N$ for highly efficient water oxidation under visible light *J. Am. Chem. Soc.* **134** 8348–51
[71] Qiao S *et al* 2020 CdS nanoparticles modified Ni@NiO spheres as photocatalyst for oxygen production in water oxidation system and hydrogen production in water reduction system *Chem. Eng. J.* **395** 125068
[72] Ye C *et al* 2020 The high photocatalytic efficiency and stability of $LaNiO_3$/g-C_3N_4 heterojunction nanocomposites for photocatalytic water splitting to hydrogen *BMC Chemistry* **14** 65
[73] Hao X *et al* 2018 Zn-vacancy mediated electron-hole separation in ZnS/g-C_3N_4 heterojunction for efficient visible-light photocatalytic hydrogen production *Appl. Catal.* B **229** 41–51

IOP Publishing

Materials Technology for the Energy and Environmental Nexus, Volume 2

R V Mangalaraja, Arunachalam Arulraj, Radhamanohar Aepuru and Chan Siew Hwa

Chapter 8

Recent advancements in tuning the electronic structures of transitional metal oxides for the electrocatalytic oxygen evolution reaction

Karthick Kannimuthu, Pawan Kumar, Md Abdullah Al Bari and Md Golam Kibria

The smooth transition from finite non-renewables to renewable energy conversion technologies will require efficient electrocatalysts which can harness intermittent energies to store in the form of chemical bonds. The oxygen evolution reaction (OER) impedes the widespread usage of water electrolyzers to convert H_2O into H_2 and persists as a bottleneck, including other energy conversion devices with sluggish four H^+/e^- kinetics. In this context, designing highly active and stable catalysts capable of driving a lower overpotential in the OER to produce continuous hydrogen (H_2) is a primary demanded. This chapter discussed the mechanism of the OER in conventional adsorbate oxygen and lattice oxygen participation in transition metal oxides (TMOs). Further, the influences of surface engineering, doping, and defects in the TMOs and understanding the electronic structure to screen electrodes towards the structure–activity relationship are highlighted. Specifically, the adsorption strength of O 2p is understood in detail as its binding ability over the surface of TMOs can be correlated directly to the OER activity. The iterative development of TMOs in terms of understanding electronic structural attributes is essential for the commercial deployment of energy conversion technologies. The comprehensive outlook of this chapter investigates thoroughly how TMOs can be used as significant materials for the OER in the near future.

8.1 Introduction

The dependence on fossil fuel reserves to meet the world's current energy requirements and the concomitant release of CO_2 into the environment has incited the

doi:10.1088/978-0-7503-5729-6ch8

movement toward other, renewable energy-based sources [1, 2]. Hydrogen (H_2), with its specific energy density of 120–142 MJ kg^{-1}, promises to be a versatile and clean energy carrier. After first being developed in the 1890s, water electrolyzers have manifested incredible advancements in later centuries and currently 4% of hydrogen is produced in this way [3, 4]. It is carbon-free and, except for water, no other by-products are formed. Most hydrogen is generated primarily from steam reforming and involves the high-temperature annealing of fossil fuels. The catastrophic climate change caused by CO_2 emissions has triggered researchers to focus on renewable energy resources for water electrolyzers to build a 'carbon neutral environment' [5]. Owing to the highly stable octet structure, the free energy change required to dissociate O_2 and H_2 is high, +237 KJ mol^{-1} which is equivalent to 1.23 V, a theoretical potential to split water molecules (H_2O). In the real scale, the requirements are elevated due to the complex mechanisms and kinetic parameters involved and are termed 'overpotentials'. The noble metal platinum (Pt), iridium oxide (IrO_2), and ruthenium oxide (RuO_2) are currently the most active and stable for use at the industrial scale, and developing non-noble, cost-effective electrodes with an improved industrial-scale performance is highly necessary [6, 7].

The earth-abundant 3d transition materials have shown substantial development in recent years and are being studied as oxides, hydroxides, chalcogens, and phosphides, mostly with a lower overpotential and higher stability in alkaline electrolytes. As the sluggish kinetics of the oxygen evolution reaction (OER) lacks the commerciality of water electrolyzer with four protons and four electrons combined process, more attention is being paid to the OER. In the case of the hydrogen evolution reaction (HER), it is comprised of a two-electron plus two-proton coupled process and involves just the hydrogen adsorption and desorption phenomena. However, the selection of materials, the approach to decorating materials over the external substrates, the method of development, and precise evaluation to meet industrial-scale applicability are the main aspects related to their performance [8, 9]. To improve the structure–activity relationship, enabling the most auspicious materials as the active surface, a prevalence of rich active sites, quick diffusion of ions to and from the electrode surface, a collective carrying rate-determining step with low free energy, and improving the kinetics using various metal heterostructure boundaries can be used, as depicted in figure 8.1. The bifunctionality of the electrodes provides many advantages, however, the most active material for OER in an alkali medium might have poor activity in the HER and could be better for HER in acidic media. Bifunctionality allows the effortless deployment of electrodes for industrial applications. However, the advancement of non-noble metal-based electrodes for the OER in an acidic electrolyte is still a bottleneck due to the poor stability under long-term processes. Amongst several earth-abundant elements, metal oxides such as cobalt (Co), manganese (Mn), and chromium (Cr) delivered improved results for the OER in acid with meaningful stability [10, 11].

However, the alkaline environment is promising with non-noble materials for the OER, HER, and total water splitting (TWS). Considering the growth of a material over an external substrate, the *in situ* growth of materials on three-dimensional (3D)

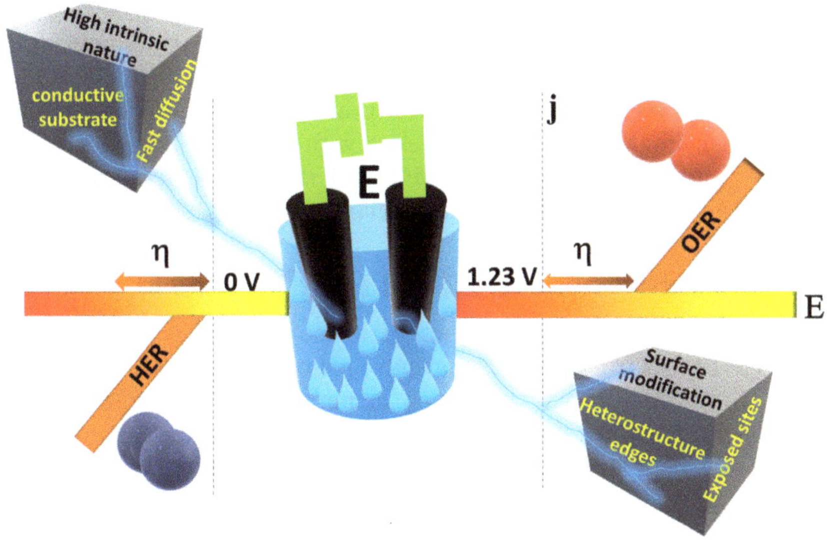

Figure 8.1. A schematic of electrocatalytic water splitting and the methods adopted to enhance the performance.

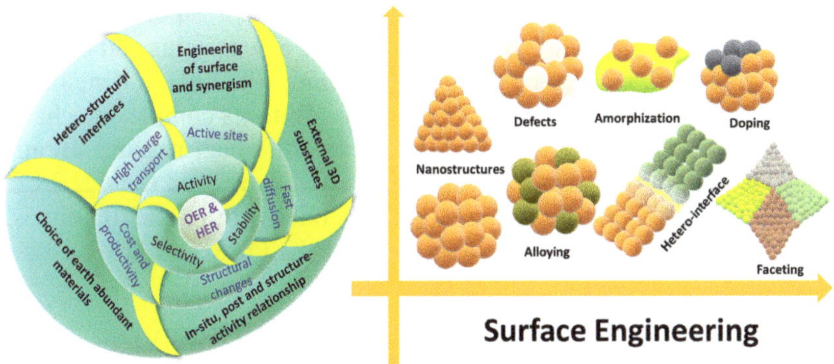

Figure 8.2. Methods for improving the activity of the OER and HER with structural modifications of electrocatalysts to allow the scalable production of hydrogen.

substrates possesses more advantages compared to the conventional coating. Further, in conventional coating, the use of commercial binders restricts the electrolyte contact through their polymeric nature. In the direct growth of external substrates, rich active sites, high conducting nature, and ease of mass transport assure reduced energy inputs. In summary, highly active materials grown over external substrates together alleviate the process of commerciality.

Numerous nano-architectures have been recognized as electrocatalysts in water splitting [12, 13] and countless efforts, namely structural variation, defects, amorphization, faceting, doping, hetero-interfaces, and alloying approaches, have been used to lay the way for the commercial scale, as pictured in figure 8.2. Considering the

OER, which mostly defies efficiency, recent progress in the metal oxides with successful modifications on the surface have ensured their potential advantages. By modifications on TMOs, such as surface engineering, doping, and defects, the OER activity can be improved. Further, the adsorption of the metal d-orbital with O 2p is explored for their binding nature to correlate with the OER activity. Overall, this chapter highlights the OER activity of TMOs with structure–activity relationships in water splitting for the concomitant commercial production of hydrogen [14].

8.2 The mechanisms of the OER and HER in different pH conditions

Electrocatalytic water splitting studies are carried out in strongly acidic (pH = 0) and alkaline (pH = 14) electrolytes, since they have plenty of protons and hydroxyl ions, respectively. Since conductivity is lacking in pure H_2O, and protons need to be extracted by bond breaking, this requires excess energy. In acidic electrolytes, for the HER, there is an adequate presence of H^+ ions, however, for the OER, H_2O bond breaking is required to produce the OH^- ions. Similarly, in alkaline electrolytes, OH^- ions are readily available and for the H^+ ions bond breaking of H_2O is required. Considering neutral electrolytes, H_2O bond breaking is required for the OH^- and H^+ ions to kickstart both the OER and HER and thus creates huge overpotentials [8]. Similarly, considering the mechanistic steps involved, information on the catalytic sites on the metal heterostructures is still a point of concern, and many *in situ* and post-structural studies demonstrate new insights and developments. Further, different chosen electrolytes, such as acid, alkaline, and neutral, follow various kinetic steps for both the OER and HER. With its complex mechanism, the OER is observed to have more kinetic steps, each of which requires an additional free energy change that requires excess overpotential compared to its HER counterpart. Because of such limitations, designing electrodes with bifunctionality is always challenging. In general, water splitting consists of the OER and HER, respectively, as follows:

$$4H^{+}+4e^{-}\rightarrow 2H_2\left(\text{HER—water reduction},\ E^0 = 0\text{V}\right) \tag{8.1}$$

$$2H_2O \rightarrow O_2 + 4H^{+} + 4e^{-}\left(\text{OER—water oxidation},\ E^0 = 1.23\ \text{V}\right). \tag{8.2}$$

8.2.1 The HER in acidic (pH = 0) and alkaline (pH = 14) electrolytes

In the HER the cathodic potential carries either protons or H_2O to the electrode surface and, eventually, after the mechanistic steps, such as hydrogen adsorption and desorption, H_2 molecules are evolved. The HER in acid is promoted by the reaction of H_3O^+ as follows:

$$S + H_3O^{+}\rightarrow S - H_{*} + H_2O\ (\text{Volmer}) \tag{8.3}$$

$$H^{+} + e^{-} + S - H_{*} \rightarrow H_2 + S\ (\text{Volmer–Heyrovsky}) \tag{8.4}$$

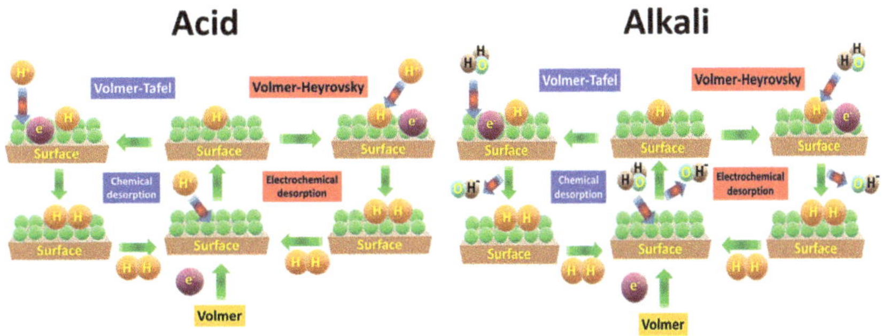

Figure 8.3. The mechanistic steps of the HER in acidic and alkaline environments, consisting of the Volmer, Heyrovsky, and Tafel routes.

$$2S - H_* \rightarrow H_2 + 2S \text{ (Volmer–Tafel)}. \quad (8.5)$$

During the Volmer step, an electron from the electrode surface reacts with the H_3O^+ for the generation of adsorbed hydrogen. The diffusion of another hydrogen and subsequent secondary electron from the electrode creates electrochemically adsorbed hydrogen leaving from the surface and is called the Volmer–Heyrovsky step. If the chemically adsorbed H* combine and leave the surface, it is called the Volmer–Tafel mechanism. A similar trend is followed in alkaline and neutral environments except for the reactant H_2O molecules. The mechanistic steps of the HER in acidic and alkaline electrolytes are given in figure 8.3. Further, if the adsorbed hydrogen on the surface of an electrocatalyst is loosely bonded, then the poor adsorption results in inferior activity. Likewise, strong adsorption affects the catalytic sites as the desorption process becomes difficult and the incoming protons are further hindered due to 'catalytic poisoning'. Thus it should have an optimum binding capability and for this a generalized scaling up of the activities of catalysts can be a better choice. The nature of the intrinsic activities of the electrocatalysts in the HER can be elucidated well with the volcano plot and from density functional theory (DFT) studies for the free energy changes and d-band structures [8, 15].

8.2.2 The role of the volcano plot in the HER

To have an accurate picture of the metal active sites with adsorbed hydrogen, the relationship between the hydrogen bonding energy and adsorption-free energy (ΔG_{H*}) depicts a 'volcano type' curve that determines the activity of hydrogen (figure 8.4(a)). According to the Sabatier principle, the material with zero ΔG_{H*} and high exchange current density (j_0) will have the highest HER activity. The Pt catalyst with near zero ΔG_{H*} sits at the top of the volcano plot (figure 8.4(b)).

The intrinsic activity of an electrocatalyst in the HER can be observed precisely with measured j_0 from the given volcano plot (figure 8.4(a)), and it is observed that a catalyst with positive ΔG_{H*} will have weak adsorption of H* and retards the Volmer step. Further, the exchange current density increases exponentially with an increase in the ΔG_{H*} positive direction. The catalysts remain in the positive direction and

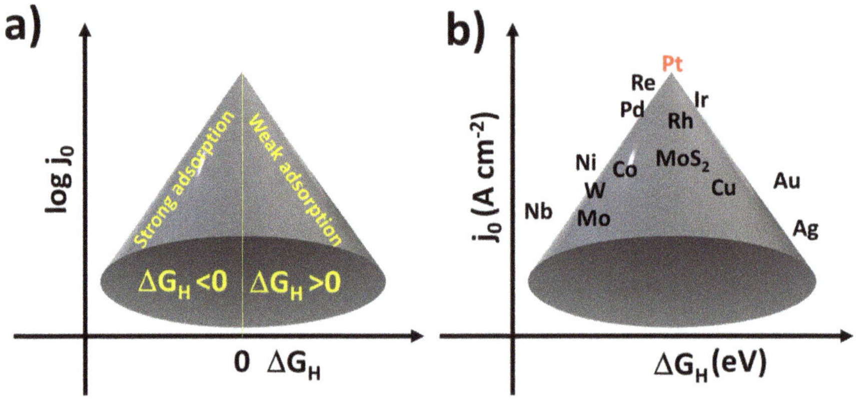

Figure 8.4. (a) Volcano plot model of log j_0 versus free energy of hydrogen adsorption and (b) volcano plot displaying the experimental j_0 versus ΔG_{H*}.

thus have low initiation on H* adsorption and H_2 evolution. Similarly, if the ΔG_{H*} values move in the negative direction, the H* adsorption is too strong, meaning that the Volmer step is rapid, however, due to the sticking of H with a strong M–H interaction, further desorption or Volmer–Tafel and Volmer–Heyrovsky reactions are affected. Figure 8.4(b) shows that the metal Pt is the best compared to the other metal active sites. This is attributed to its equilibrium potential that allows hydrogen evolution at zero overpotential. The development of other metals near Pt can be accomplished by combining other metals that are close to the volcano curve via pioneering approaches. Further, the DFT approach can be studied in the volcano plots to study the near Pt catalysts and recently molybdenum disulfide (MoS_2) electrocatalysts have shown improvement comparable to Pt as far as non-noble elements are concerned [16, 17].

8.2.3 The OER in acidic (pH = 0) and alkaline (pH = 14) electrolytes

The mechanism of the OER adheres to distinct pathways and comprises a complex four-electron oxidation process. The origin of the active sites, variance in the requirements of free energy during the rate-controlling step, the redox features of the active sites and the reactive species, and electrolyte pH affect the expected overpotential. For the OER in an acid, the first step is the interaction of H_2O with the surface of active sites. The mechanistic pathways of the OER in acid (S–metal surface) are as follows:

$$S + H_2O \rightarrow S - OH + H^+ + e^-(\text{dissociation of } H_2O) \quad (8.6)$$

$$S - OH \rightarrow S - O + H^+ + e^-(\text{dissociation of OH}) \quad (8.7)$$

$$S - O + S - O \rightarrow 2S + O^2(\text{direct combination}) \quad (8.8)$$

$$S - O + H_2O \rightarrow S - OOH + H^+ + e^-(\text{oxyhydroxide formation}) \quad (8.9)$$

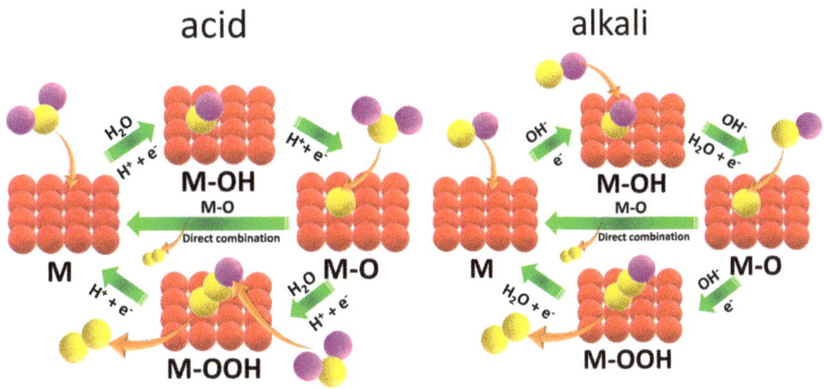

Figure 8.5. Electrocatalytic OER mechanism for various pH conditions with different kinetic steps in acidic, alkaline, and neutral environments.

$$S - OOH + H_2O \rightarrow S + O_2 + H^+ + e^-(O_2\text{formation}). \tag{8.10}$$

From these equations, even though many mechanisms are proposed, the common steps are followed as M–O and M–OH and only the production of O_2 from the surface is varied. The OER in an alkaline electrolyte is much favored by the direct presence of OH^- ions interacting with the catalytic surface. Like the OER in the acid, apart from the direct attack of OH^- ions, the remaining steps follow either oxide formation or oxyhydroxide formation to release O_2 from the surface. From all the mechanisms of the OER in acid and alkaline conditions, the most common steps followed are M–OH, M–O, and M–OOH, for which the change in free energies varies and the slowest one is the kinetically rate-controlling step. Upon the formation of O_2, M–OH to M–O and M–O to M–OOH cumulatively correspond to 3.2 eV. If the conversion of M–OH to M–O requires more energy with a high energy barrier, then this becomes the rate-controlling step and the conversion of M–O to M–OOH is too facile [18].

Based on this, the kinetic steps involved can be adjusted, mainly with the engineering of the electrode surfaces. The compiled electrocatalytic OER mechanisms in acidic and alkaline environments are provided in figure 8.5. As the mechanism is varied in different pH conditions, the choice of metal ions, with their difference in active sites and structural changes during the OER, cumulatively alter the energy required in the rate-controlling step. Further investigations on the relationship between the free energies with the kinetic steps of the OER and volcano plot are explored to relate the structure–activity relationships in the OER.

8.2.4 Free energy relationships and volcano plots for the OER

In the given electrocatalytic mechanistic steps of the OER, with four protons and four electrons, the formation of OH*, O*, and OOH* are the important steps in acidic and alkaline media. In the ideal situation, the summation of free energy changes of each step, namely, ΔG_1, ΔG_2, ΔG_3, and ΔG_4, is equal to 4.92 eV [18].

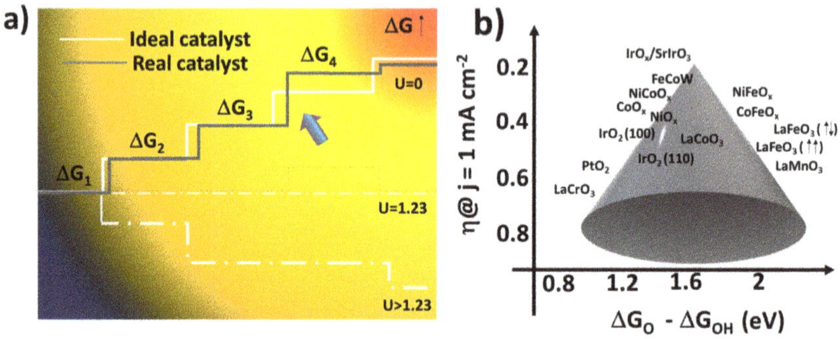

Figure 8.6. (a) Gibbs free energy change of the OER with four kinetic steps and (b) the volcano plot of current density of 1 mA cm^{-2} versus $\Delta G_{O*}-\Delta G_{OH*}$.

Dividing these by a factor of four produces 1.23 eV and hence the water oxidation takes place thermodynamically at 1.23 V. In a real case, for metal oxide ΔG_{OOH*} and ΔG_{OH*} is the rate-determining step with 1.6 ± 0.1 eV and is related to the process of the conversion of OH* to O* or OH* to OOH* (figure 8.6(a)). Further, with the uphill energy demand of 1. 6 eV, it means that the overpotential generally lies around 200–400 mV. The $\Delta G_{O*} - \Delta G_{OH*}$ of various metal oxide catalysts were studied with the overpotential demand for 1 mA cm^{-2}. This has developed as a volcano type curve as displayed in figure 8.6(b). The increased activity of 3d transition metal-based oxides was observed and these can further be modified as other counterparts with improved results. The modification of the valence states of active sites can downhill the free energy required and thus the overpotential [18]. Similarly, the electrochemical oxidation of the active sites during the OER can lead to oxide over the surface and the inner core will act as a highly conducting phase to accelerate the reaction. The state-of-the-art electrocatalyst for the OER in acid, iridium, as observed from the volcano plot, can be modified with other highly active materials from the volcano plot to expand the activity.

8.3 Transition metal oxides (TMOs) in the OER

The TMOs show variable oxidation states and are composed of different metal-to-oxygen ratios. Among them, TMOs with octahedral units such as spinel TMOs and perovskites are highly advantageous in the electrocatalytic OER applications. Based on the orbital, spin, and charge, the d-orbitals in the TMO play a major role in determining the electronic structural features for electrocatalytic OER studies [19, 20]. TMOs are studied widely for the OER due to their varying oxidation states that affect the reaction kinetics. The splitting of the d-orbital mainly relies on the transition metal oxidation state and the coordinating ligand. Before the interaction, 3d-orbitals lie in a degenerate level, and depending upon the ligand and coordination environment, the splitting is varied, such as the tetrahedral (T_d) and octahedral (O_h) arrangements. In T_d, in-between axis ligands (d_{xy}, d_{yz}, and d_{xz}) are the high-lying orbitals while the along the axis orbitals ($d_{x^2-y^2}$ and d_{z^2}) are the low-lying orbitals (figure 8.7). In the O_h arrangement, the opposite trend is observed. The effective charge on the specific metal

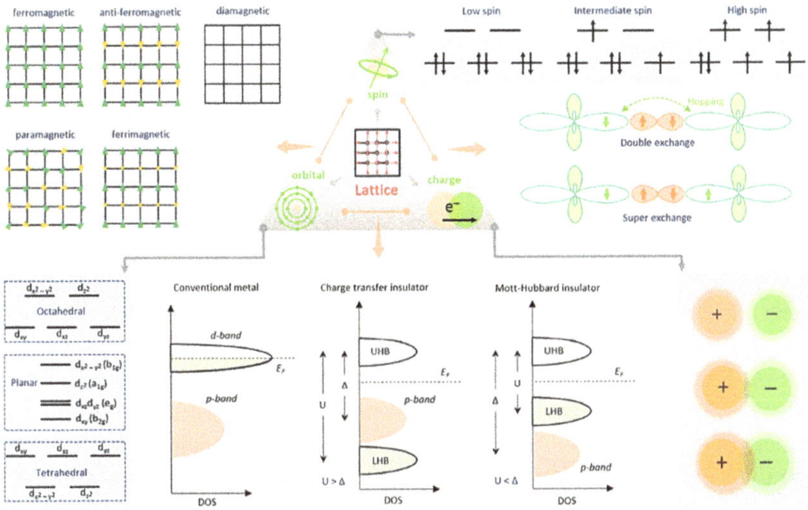

Figure 8.7. TMOs and their contributing factors for electrocatalysis, mainly the OER, such as orbital, charge, and spin. (Reproduced with permission from [21]. Copyright 2021 American Chemical Society.)

oxide can be fine-tuned via structural modifications such as metal replacement [21]. This particular replacement of the metal can induce charge transfer from the oxygen to further enhance the kinetics. Based on the oxidation state, the spin states are also varied in both the O_h and T_d arrangements and tend to improve the catalytic activity. This is mainly related to the spin-selected charge transport phenomenon which improves the conductivity of the TMOs during the OER. Other than this, the magnetism and bandgap further affect the nature of TMOs towards the OER, as shown in figure 8.7.

In the metal, the interactions between the orbital and electron show that the density of states (DOS) with the d-band is narrow near the Fermi level. In the case of TMOs, due to the strong Coulomb repulsion (U) and charge transfer energy (Δ) between the d and p orbitals, they mostly behave as insulators. Depending upon both U and Δ, the catalytic activity of the metal sites in the corresponding TMOs are affected [21]. The resulting band structure is mainly related to the charge transport factor that decides the kinetics of the reaction, however, it is still not clear to what extent it can contribute to the overall activity. In addition to the bandgap analysis, the magnetic ordering from the spin states also influences the catalytic activity, and among the ferromagnetic, antiferromagnetic, and paramagnetic nature of metal oxides, TMOs with aligned magnetic fields improved the trend of the OER activity. Recently, many reports have highlighted the influence of an applied external magnetic field to improve the activity trend. Mostly, ferromagnetic behavior aligns well with the external field whereby electron transfer is made facile, however, many effects still come into play, such as enriched mass transport and a decreased resistivity factor. Therefore, by altering the metal-to-oxygen ratio, cation or anion displacement, vacancies, and through the appliance of an external magnetic field, the OER activities can be improved further and, recently, many reports showed the industrial relevance of the TMOs with high-scale OER performance [21]. In this chapter we highlight the important modifications of TMOs with different methods of approach for improved OER activity.

In an interesting work by Yong-Chao *et al* the slow OER kinetics of spinel Co_3O_4 was improved by making it a defective $Co_{3-x}O_4$ which acted as an electron acceptor, where NiO has transferred electrons to the defective sites. Further, the modified structure ensured induced electron transfer with Ni^{3+} as the active sites. Here, the $NiO/Co_{3-x}O_4$ was prepared via a two-step synthesis approach. As shown in figure 8.8(a), the defective $Co_{3-x}O_4$ required annealing of the cobalt glycerolates at a higher temperature, and then underwent a hydrothermal route to prepare the $NiO/Co_{3-x}O_4$ electrodes [22]. Among the prepared Co_3O_4 and $Co_{3-x}O_4$, Co_3O_4 displayed increased particle size which is related to the Ostwald ripening under high-temperature annealing, even though both structures showed similar morphology. The Co metal defect is further confirmed by Raman spectra, where the Raman peak is shifted to the left, which arises due to the increased short-range disorder of $Co_{3-x}O_4$ structures. Further, an increased oxidation state is in high demand to abstract the

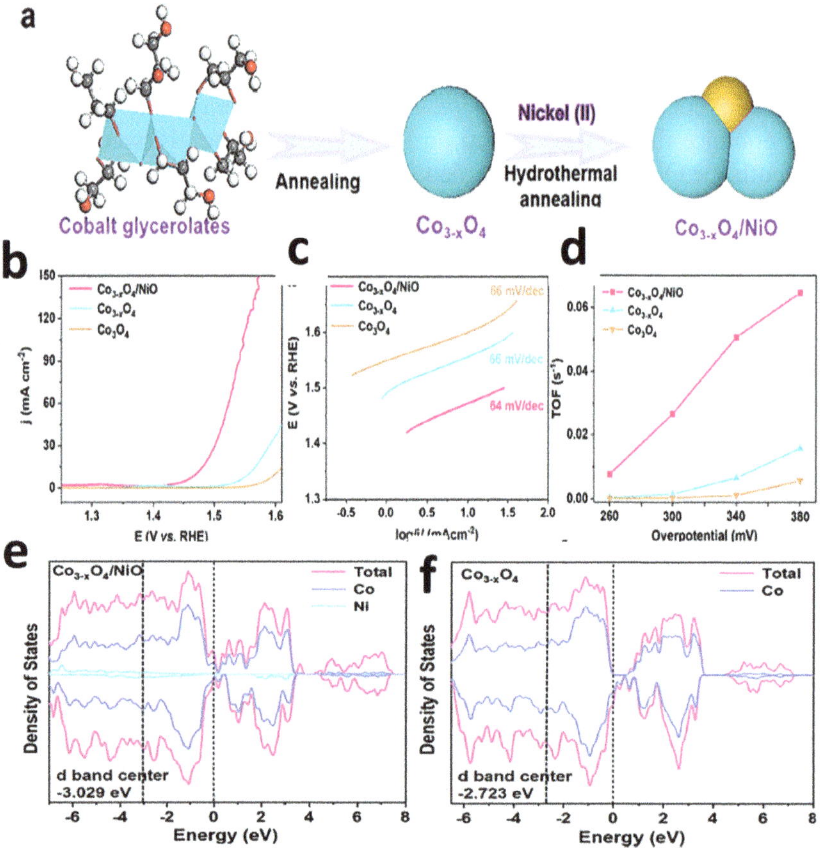

Figure 8.8. (a) Method of synthesis of defective $NiO/Co_{3-x}O_4$. (b)–(d) Electrocatalytic OER LSV, Tafel, and TOF plots. (e)–(f) DOS of oxides. (Reproduced with permission from [22]. Copyright 2022 Elsevier.)

electroactive species, particularly in the OER. The $NiO/Co_{3-x}O_4$ in electron energy loss spectra (EELS) shows that the L_3/L_2 ratio of the Ni L-edge is 2.27, which is smaller than for NiO (2.92), which further confirms the higher oxidation state of Ni. After this, the influence of TMOs was studied for the OER in a three-electrode cell with O_2 saturated 1 M KOH electrolyte.

As expected, linear sweep voltammetry (LSV) showed that the pristine spinel Co_3O_4 required a large overpotential of 370 mV at 10 mA cm^{-2} while $Co_{3-x}O_4$ displayed a low overpotential of 320 mV, which is 50 mV less than for normal spinel oxide [22]. When the $NiO/Co_{3-x}O_4$ electrode is tested, it required even lower overpotentials of 240 and 320 mV at 10 and 100 mA cm^{-2} current densities, respectively (figure 8.8(b)). From the compared specific activities, the $NiO/Co_{3-x}O_4$ showed high intrinsic activity, as discussed earlier, and the charge transfer from metal displacement ensured rapid kinetics. The charge transfer kinetics is further verified from the observed iR corrected Tafel slope, where $NiO/Co_{3-x}O_4$ shows a lower value of 64 mV dec^{-1} compared to $Co_{3-x}O_4$ (66 mV dec^{-1}) and Co_3O_4 (66 mV dec^{-1}) (figure 8.8(c)). Further, the turnover frequency (TOF) as an intrinsic activity parameter ensures a high O_2 turnover rate for $NiO/Co_{3-x}O_4$ (figure 8.8(d)). To further ascertain the nature of the band structure, a DOS study was performed for both $NiO/Co_{3-x}O_4$ and $Co_{3-x}O_4$ (figure 8.8(e)–(f)). From figure 8.8, when $NiO/Co_{3-x}O_4$ is present, due to its multi-metallic characteristics, the electron density is more populated near the Fermi level ensuring high electrical conductivity. From the overall OER results, it was concluded that the interfacial electronic structure provided the optimum binding strength for the adsorption of the oxygen intermediates to improve the kinetics of the OER [22].

Entropy engineering over the oxide structure to improve the activity and stability is another approach in using TMOs for the OER. Yue *et al* highlighted the role of entropy engineering in the $CoFe_2O_4$ with $(Co_{0.2}Mn_{0.2}Ni_{0.2}Fe_{0.2}Zn_{0.2})Fe_2O_4$, which exhibited high-rate active oxygen evolution during the OER [23]. Here, the presence of disordered multivalent cations resulted in lattice distortion. Because of this structural modification over the entropy-rich oxide, the oxygen vacancy is exposed to the surface leading to improved catalytically active sites. Here, the ball milling method was used to prepare various types of entropy-stabilized TMOs and studied for the OER. From x-ray diffraction (XRD) studies, the peaks are almost identical to the single-phase spinel oxide, however, a high angle shift was observed in the (220) and (400) planes.

After this, the OER activity was assessed in 1 M KOH for various spinel oxides and compared with a state-of-the-art RuO_2 catalyst for the overpotential at 10 mA cm^{-2} (E_{10}) and onset potential (E_{onset}), as shown in figure 8.9(a–b). Both E_{10} and E_{onset} were 326 and 230 mV for $S_{0.2}$, while RuO_2 showed 363 and 259 mV, respectively. As shown in figure 8.9(c), the entropy-rich multivalent metal sites in the O_h arrangement show different bonding energies towards the O atom and because of this the charge density is distributed around it and forms a distorted lattice structure [23]. The distorted metal cation shows entropy stabilization of a spinel oxide. The increased charge density around the unbalanced O atom is less prone to undergo separation and results in the formation of an oxygen vacancy. These vacant O sites are highly beneficial for

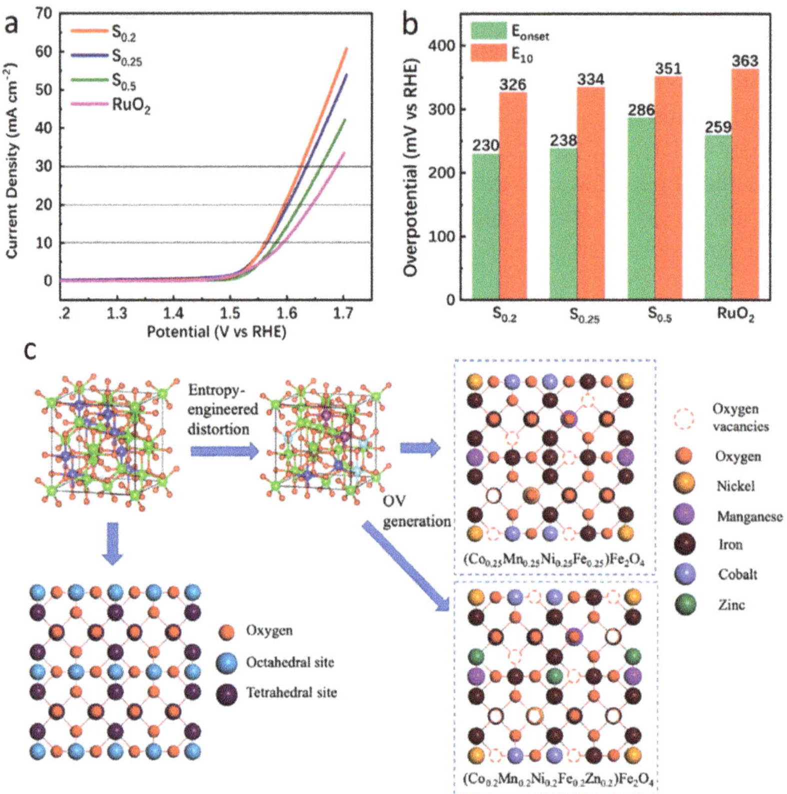

Figure 8.9. (a) OER LSV plot of entropy-stabilized oxide, (b) overpotential and onset potential difference of oxides, and (c) lattice distortion in the O-neighbored O_h and T_d lattices to form oxygen vacancies in the (400) surface of the TMO. (Reproduced with permission from [23]. Copyright 2020 American Chemical Society.)

electrolyte adsorption by increasing the surface energy states of the metal cations. This entropy-stabilizing strategy for TMOs is in high demand in terms of their distorted structure for increased activity and stability.

The surface modification with anions of the spinel oxides is another way to improving the performance in a rate-limiting OER. This approach can increase the electrochemically active surface area and thus can trigger high valence active sites in the OER. Surface tunability with metal undercoordination is required for the peroxidation of high valence metal sites and using amorphous borate ions over spinel oxides for enriching the OER was proposed by Junfeng *et al* via the molten-salt-protected pyrolysis technique [24]. As pictured in figure 8.10(a), initially the CoFe Prussian blue analog (CoFePBA) nanocubes were prepared and were annealed with borate at 400 °C to produce surface modified CoFe spinel oxides. Compared to pristine spinel oxides, in this case the molten salt pyrolysis ensured a high Co^{2+}/Co^{3+} ratio which is essential for rapid peroxidation of high valence Co sites and thus ensured the importance of highly exposed sites during the OER. X-ray photoelectron spectroscopy (XPS) studies further revealed the quantitative

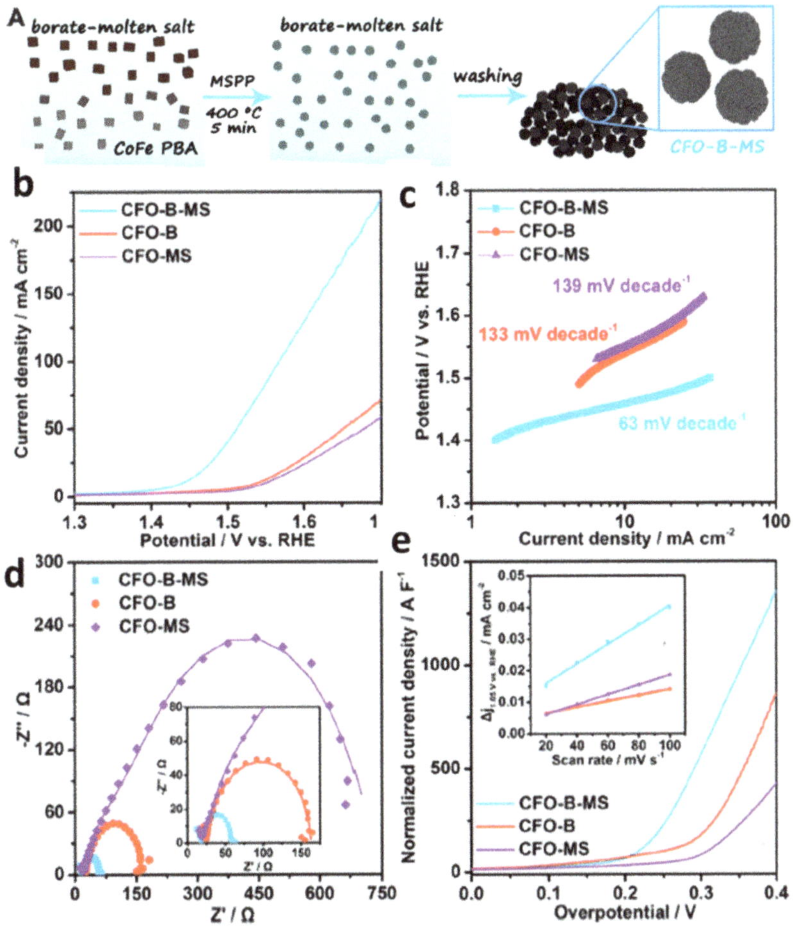

Figure 8.10. (a) Schematic showing the molten salt electrolysis of borate modified spinel oxide, (b) LSV, (c) Tafel slope, (d) Nyquist plot, and (e) ECSA normalized LSV. (Reproduced with permission from [24]. Copyright 2021 American Chemical Society.)

enrichment of Fe^{3+} sites. Further, due to the inclusion of borate species over the surface, the oxygen vacancies along with B^{3+} ions bonded to oxygen further demonstrate the surface modification over the CoFe spinel oxide with the borate species for the peroxidation of high valent Fe and Co metal centers [24]. When evaluated for the OER in 1 M KOH, to obtain a current density of 10 mA cm^{-2}, the CFO-B-MS required a lower overpotential of 208 mV compared to the CFO-B and CFO-MS, which directly conveys the importance of surface modification with molten salt pyrolysis compared to other methods (figure 8.10(b)).

Similarly, the charge transfer kinetics from the Tafel slope show that CFO-B-MS shows a lower value (63 mV dec^{-1}) compared to CFO-B (133 mV dec^{-1}) and CFO-MS (139 mV dec^{-1}). Electrochemical impedance spectroscopy (EIS) is an important technique to determine the charge transfer resistance (R_{ct}) in the faradaic region. As expected, in figure 8.10(c) the CFO-B-MS shows the lowest R_{ct} signifying the role of

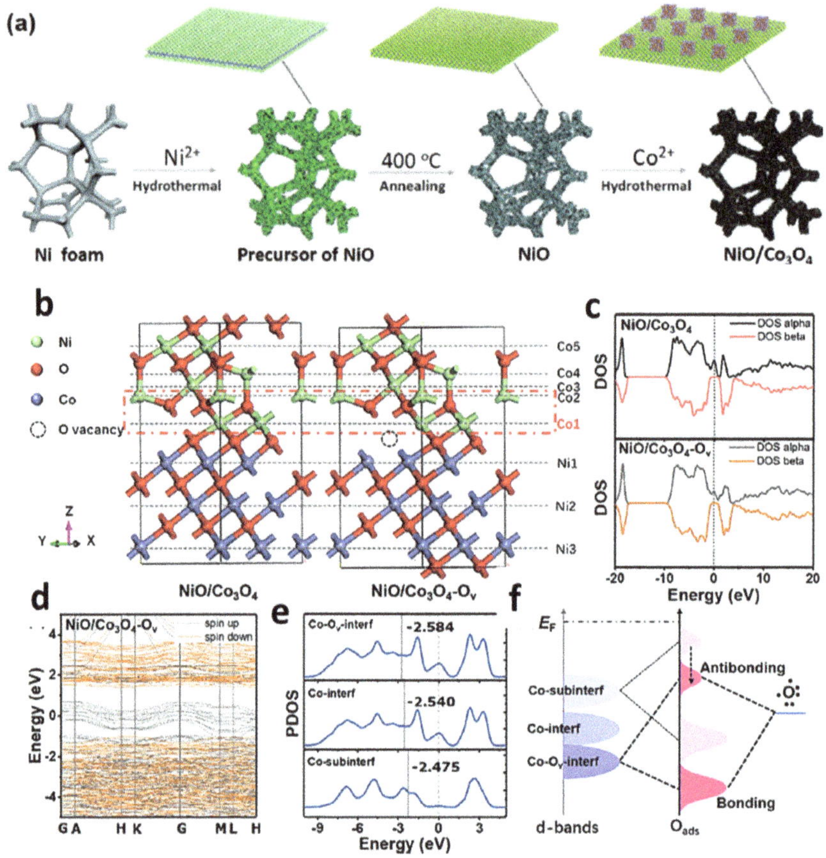

Figure 8.11. (a) Synthesis scheme for NiO/Co_3O_4, (b) DFT showing the heterostructure, (b) LSV, (c) DOS, (d) band structure, (e) PDOS, and (f) oxygen shifting d-band centers of Co. (Reproduced with permission from [25]. Copyright 2020 American Chemical Society.)

surface modification. In this case, the increased electropositivity of undercoordinated sites contributes to the enrichment of the Co/Fe ratio, thereby enhancing the OER [24].

The low coordination atoms (LCAs) in the heterointerface structures of oxides can further induce the kinetic ability for active species adsorption in the OER with prolonged stability. Jingyan *et al* showed the development of heterostructural NiO/Co_3O_4 electrodes with an oxygen vacancy and LCA sites, which together ensured higher OER activity [25]. These LCA sites could show an intermediate spin state of e_g^1 which is advantageous for the intermediate adsorption of oxygen-containing intermediates. Initially, as shown in figure 8.11(a), the NiO was developed on Ni foam and Co_3O_4 spinel oxide was decorated over NiO through a hydrothermal process. X-ray absorption near-edge structure (XANES) and extended x-ray absorption fine structure (EXAFS) studies revealed that the Ni K-edge is highly oxidized in NiO/Co_3O_4 compared to only NiO. In the case of the Co K-edge, the NiO/Co_3O_4 showed lower energy compared to Co_3O_4, which showed that electron density is migrated from Ni to Co. The same is witnessed in both XPS and EELS studies. By following the synthesis strategy of heterostructural oxides, in the presence of Ni^{3+} and CO^{2+} sites e_g unity is achieved. When measured for OER activity in 1 M KOH, the NiO/Co_3O_4 only required a 262 mV overpotential at 10 mA cm^{-2} compared to NiO (456 mV) and Co_3O_4 (490 mV). Figure 8.11(b) shows the developed structural model of NiO/Co_3O_4 heterostructure with oxygen vacancies and LCAs. As can be seen in figures 8.11(c)–(d), the bandgap was first analyzed for NiO and Co_3O_4 which were found to be semiconducting with 3.179 and 1.312 eV, respectively, while the NiO/Co_3O_4 showed a metallic nature [25]. Even after the introduction of the oxygen vacancy, the metallic nature persists, which can facilitate enhanced charge transfer during the OER. Further, the d-band center in figure 8.11(e) shows that the center of Co atoms is positioned far away from the Fermi level with a value of −2.540 eV compared to Co sub-interfaces. In figure 8.11(f) it can be seen that when the Co at the sub-interfaces lies far below the Fermi level, the bonding and anti-bonding states are shifted from their position where the strong bonding with the absorption is neglected. In other words, the combination of Ni and Co with O-vacancies ensured the optimum absorption and desorption of the electroactive species [25].

The morphology of the oxides being mesoporous in nature shows exposed surface-active sites. However, preparing such oxides requires micelle templates to prepare the intraparticle voids. In a noticeable work by Wenqiao *et al* a wet chemical method was followed to prepare the Ni/Mn-promoted cobalt oxides for the OER. Here, the mesoporous structure with Co-Ni(Mn)-O ensured high valent cobalt sites with Co^{3+} to Co^{4+} conversion along with surface vacancies [26]. Here, when compared to different molar concentrations of dopants, 5% Ni delivered superior OER performance by requiring just 381 mV at 10 mA cm^{-2} current density. Here, the evolution of Co^{4+} species could be related to the fast evolution of O_2 from M–OOH intermediate due to the high electronegativity. Further, the electronegativity of a metal in Co_3O_4 is crucial to develop the Co^{4+} species. Considering both Ni (1.91) and Mn (1.55), with higher electronegativity, Ni promotes the successful conversion of Co^{4+} sites to enhance the OER activity. The increased bond length of Co^{3+}–OH^- when Ni^{2+} ions replace the cobalt ions tends to adsorb OH^- ions effectively and leads to the fast conversion of Co^{4+} species [26]. Developing hetero-interfaces of oxides over layered double hydroxides (LDHs) over the conducting substrates can show ampere-level current densities under applied cell voltages. Anil *et al* prepared $NiFe_2O_4$/NiFe LDHs over stainless steel mesh through a one-step wet chemical synthesis [27]. Due to the advantageous nature of the heterostructure and stainless steel substrate, the overpotentials required at 10 and 100 mA cm^{-2} current densities were 150 and 190 mV, respectively. Ali *et al* highlighted the combination of phosphides to the oxide system, where mesoporous $CoP/CoCr_2O_4$ was developed using a KIT-6 hard silica template for the OER [28]. This heterojunction between CoP and $CoCr_2O_4$ ensured high OER activity and to reach 10 mA cm^{-2} in 1 M KOH it required only 290 mV as the overpotential. In addition to these modifications, the poor electrical conductivity of spinel oxides can be nullified by incorporating conducting metal nanoparticles (NPs) such as gold (Au) and silver (Ag). Also,

Ali *et al* showed the formation of Ag NPs over borophene that facilitated higher electrical conductivity [29]. Further, the high electrophilic character of Ag induces the conversion of Co^{3+} sites to the active phase of Co^{4+} sites for accelerating the OER. The B-sites on the borophene strongly bind to the Co and O atoms from Co_3O_4 and thus ensure high-scale stability. The catalyst required an overpotential of 270 mV at 10 mA cm^{-2} and a low Tafel slope of 62 mV dec^{-1}. Also, the Ag NPs 6–8 nm in size were well confined over the borophene structure and provided a maximum exposure of Co_3O_4 catalytic sites for the OER. From these observations, there is certainly still much room available to improve the performance of TMOs in the OER with extended activity and stability, mainly due to their tunable physicochemical properties and surface modifications with heterostructure features [30, 31]. Also, recently perovskites received attention with their improved catalytic performance in the OER and are discussed in the following.

Perovskites are ABO_3 type oxides with an A-sites corresponding to alkaline and alkaline-earth metals with the 12 coordination of oxygen. The B-site corresponds to the metal element with 6 coordination of oxygen atoms containing metals with 3d, 4d, and 5d metal elements. The e_g unity of perovskites is superior for the OER compared to most other catalysts. This is related to the optimum bonding strength of the e_g orbital towards the O 2p orbital for easy absorption/desorption kinetics. Since the e_g orbitals overlap effectively with the O_2 molecule compared to the t_{2g} orbitals in perovskite structures, it can easily be tuned in the A- and B-sites to bring the e_g unity to accelerate the OER kinetics [32]. Because of these fascinating properties, perovskites are widely studied in energy conversion processes such as the OER and oxygen reduction reaction (ORR). Further, the surface engineering with cation defects in the A-site showed accelerated OER activities and, further, the change in B-sites with conductive elements such as B or P affected the stability of the perovskites with higher electrical conductivities. The phosphorous-doped $LaFeO_{3-\delta}$ was reported by Zhishan *et al* for the OER under different temperature conditions [33]. Here, the presence of P-doping in the B-sites of the perovskite doubled the mass activity and specific activity of the pristine material for the OER. Further, the presence of Fe^{4+} sites during the OER improved the kinetics with the e_g unity, as discussed previously. Figure 8.12(a) shows the structure of the perovskite before and

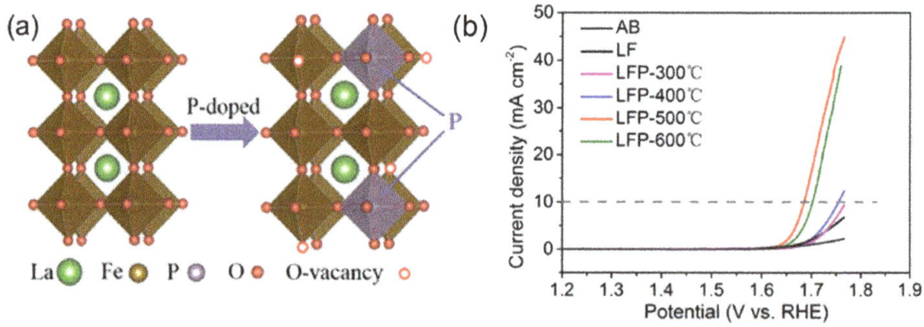

Figure 8.12. (a) Structure of the $LaFeO_{3-\delta}$ perovskite before and after P-doping. (b) LSV performance for the OER in 1 M KOH. (Reproduced with permission from [33]. Copyright 2018 Elsevier.)

after the P-doping that possesses oxygen vacancies for the exposed catalytic sites. DFT studies revealed that the addition of P to the $LaFeO_{3-\delta}$ increased the valence states of the Fe ions. Among the various catalysts developed, LFP-5 displayed higher OER activity with an onset potential of 1.623 V versus reversible hydrogen electrode (RHE) compared to others, signifying the role of P-doping here. Also, compared to the pristine LF, the LFP-5 also delivered a lower overpotential of 0.465 V at 10 mA cm^{-2} current density with a lower Tafel slope (50 mV dec^{-1}). Mossbauer spectroscopy revealed the formation of trace amounts of Fe^{4+} sites with high spin characteristics (t_{2g}^3 e_g^1). As the e_g unity is achieved here, the adsorption of OH* is made more facile along with the co-existence of the Fe^{2+}/Fe^{3+} sites. As the OH* adsorption is effective, the required overpotential is decreased after the doping of P instead of B-sites in the LFP perovskites. This study revealed that non-metal elemental doping is more advantageous in terms of the valence state and stability during the OER [33]. The role of A- and B-sites in the perovskites was further explored by Jing *et al* for the OER in 0.1 M KOH and revealed the critical role of cation deficiency and oxygen vacancy in determining the resulting catalytic activities [34]. In perovskites, two mechanisms are widely accepted, which are the lattice-oxygen-mediated mechanism (LOM) and the adsorbate evolution mechanism (AEM). In the LOM, the lattice oxygen promotes the evolution of O_2 from the lattice and the metal–oxygen covalency is modified based on the oxygen vacancy formed, which plays a major role in adjusting the band position of the metal d-orbital and O 2p orbitals.

In the AEM, the B-sites of perovskites act as real catalytic centers and the surface engineering of B-sites such as e_g filling, M–OH bond strength, d/p-center, and metal–oxygen hybridization together justifies the OER activity [35, 36]. Based on this discussion, in this work $Bi_{0.15}Sr_{0.85}Co_{1-x}Fe_xO_{3-\delta}$ (x = 0.2, 0.4, 0.6, 0.8, 1) was studied and the OER activity was decreased with an increased concentration of x. Here, the replacement of Bi^{3+} by Sr^{2+} at the A-site takes place because of the ionic compensation mechanism. The oxygen vacancy is created by the partial reduction of Co^{3+} and Fe^{2+} ions which facilitates the LOM pathway during the OER. This phenomenon was further confirmed by varying the pH conditions at 1.58 V, as shown in figure 8.13(a), where the activity is increased with increased pH, suggesting

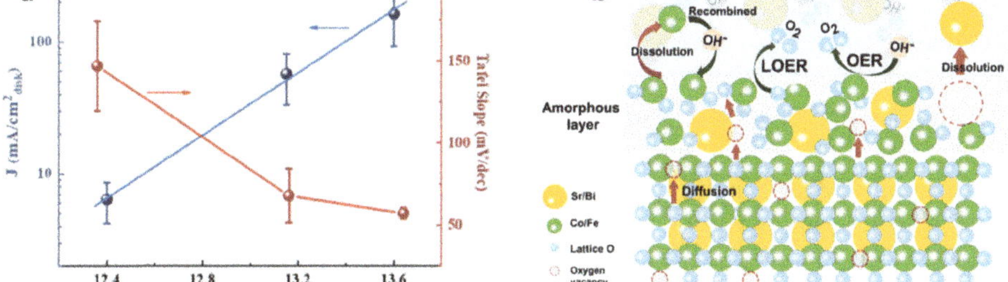

Figure 8.13. (a) Change in overpotential and Tafel slope as a function of pH in $BiSC_{0.8}F_{0.2}$. (b) Schematics of the OER process. (Reproduced with permission from [34]. Copyright 2023 Elsevier.)

the proton transference as the rate-determining step. Figure 8.13(b) highlights the overall mechanism in the OER where the evolution occurs at the interface of the amorphous layer and the KOH electrolyte [34].

Here, the first step is the oxidation of the adsorbed OH^- ions occurring at the B-site cations for the LOM mechanism and Co/Fe-O-OH (oxyhydroxide phase) is formed. The second step is the leaching of Sr A-sites which induced the exposure of B-sites for the OER, and the final step is the redeposition of Fe sites from the electrolyte for fresh active sites [34].

Also, the surface reconstruction of the perovskites showed tremendous improvement in the activity. For example, a work by Chunhua *et al* highlighted the use of $NaBH_4$ to produce surface amorphized structures of $La_{0.8}Sr_{0.2}Co_{0.8}Fe_{0.2}O_{3-\delta}$ (LSCF) which delivered high OER activity [37]. Upon $NaBH_4$ immersion, the surface was transformed into an amorphous structure with nanosized particles of around 20 nm. Further, the conversion of Co^{3+} to Co^{2+} sites enabled the oxygen vacancy in the perovskite structure. After this, the OER activity was measured in 1 M KOH, and to achieve a considerable current density of 10 mA cm^{-2}, LSCF-2 required 248 mV overpotential with 51 mV dec^{-1} when analyzed for the Tafel slope. Due to the stable structural aspect, the catalyst provided a negligible degradation in the OER even up to 20 h of reaction time. The nature of perovskites towards the OER can be accelerated further with double perovskites ($A_2BB'O_6$). Maria *et al* highlighted the role of $(La_{1.5}Sr_{0.5})_A(Ni_{0.5}Mn_{0.5})_B(Ni_{0.5}Ru_{0.5})_{B'}O_6$ in outstanding OER activity [38]. Here, the stability of the perovskite was increased by introducing two cations in the A-site. Also, for the catalytic activity, accommodating Ru ions in the B-sites ensured rapid kinetics towards the OER. This work provides further insights into the successful accommodation of active ions in the A- and B-sites resulting in improved OER activity and prolonged stability. From the discussion of perovskites, it is generally accepted that it follows both the AEM and LOM and the AEM volcano plot suggests that it stays uniform for all the perovskites, whereas for the LOM it varies with respect to the identity of the A-sites in the ABO_3 structure [39]. Therefore, by pioneering structural variations in perovskites, benchmark catalysts can be developed in the future [40].

8.4 Conclusion

Thus, developing earth-abundant electrodes for an effective OER is vital to improve the efficiency of energy conversion and storage applications such as electrolyzers, fuel cells, and metal–air batteries. In this chapter, engineering the structure of TMOs mainly in the spinel oxides and perovskites is discussed in terms of their role in orbital, spin, magnetism, and electronic structure for OER studies. From the discussion on TMOs for the OER, the activity and stability can be improved based on the M–O covalency, metal and non-metal doping, oxygen vacancy, surface reconstruction, heterointerface, band structure, and DFT for O-containing intermediate adsorption. However, the complex mechanism occurring over the surface is still not completely understood and a lot of questions need to be addressed. Despite the tuning of metal oxides for OER, more improvements are required for large-scale applications. Further, for industrial-scale applications, long-term stability is a matter of concern and is still elusive when considering the TMOs. As discussed briefly, the direct growth of the TMOs with highly active phases, such as hydroxides and LDHs, provided considerable activity with a minimum degradation in activity. Moreover, the surface engineering of TMOs can also be studied in other electrocatalytic applications in the future. Overall, this chapter demonstrate the applicability of TMOs and how the tuning of their surface via various approaches successfully improves activity and stability of the OER.

References

[1] Kolar J L 2000 Alternative energy technologies *Environ. Qual. Manag.* **10** 45–54

[2] Tee S Y, Win K Y, Teo W S, Koh L D, Liu S and Teng C P *et al* 2017 Recent Progress in Energy-Driven Water Splitting *Adv Sci.* **4** 1600337

[3] Khan M A *et al* 2018 *Recent Progresses in Electrocatalysts for Water Electrolysis* (Singapore: Springer)vol 1

[4] Dau H, Limberg C, Reier T, Risch M, Roggan S and Strasser P 2010 The Mechanism of Water Oxidation: From Electrolysis via Homogeneous to Biological Catalysis *Chem. Cat. Chem.* **2** 724–61

[5] Khezri B, Fisher A C and Pumera M 2017 CO_2 reduction: the quest for electrocatalytic materials *J. Mater. Chem.* A **5** 8230–46

[6] Reier T, Nong H N, Teschner D, Schlögl R and Strasser P 2017 Electrocatalytic oxygen evolution reaction in acidic environments—reaction mechanisms and catalysts *Adv. Energy Mater.* **7** 1601275

[7] Jin M *et al* 2022 Strategies for designing high-performance hydrogen evolution reaction electrocatalysts at large current densities above 1000 mA cm^{-2} *ACS Nano* **16** 11577–97

[8] Safizadeh F, Ghali E and Houlachi G 2015 Electrocatalysis developments for hydrogen evolution reaction in alkaline solutions—a review *Int. J. Hydrogen Energy* **40** 256–74

[9] Burke M S, Zou S, Enman L J, Kellon J E, Gabor C A and Pledger E *et al* 2015 Revised Oxygen Evolution Reaction Activity Trends for First-Row Transition-Metal (Oxy)hydroxides in Alkaline Media *J. Phys. Chem. Lett.* **6** 3737–42

[10] Xi P *et al* 2020 Iridium single atoms coupling with oxygen vacancies boosts oxygen evolution reaction in acid media *J. Am. Chem. Soc.* **142** 18378–86

[11] Lee Y, Suntivich J, May K J, Perry E E and Shao-Horn Y 2012 Synthesis and activities of rutile IrO_2 and RuO_2 nanoparticles for oxygen evolution in acid and alkaline solutions *J. Phys. Chem. Lett.* **3** 399–404

[12] Yan Y, Xia B Y, Zhao B and Wang X 2016 A review on noble-metal-free bifunctional heterogeneous catalysts for overall electrochemical water splitting *J. Mater. Chem.* A **4** 17587–603

[13] Anantharaj S, Karthick K and Kundu S 2017 Evolution of layered double hydroxides (LDH) as high performance water oxidation electrocatalysts: a review with insights on structure, activity and mechanism *Mater. Today Energy* **6** 1–26

[14] Gong M and Dai H 2014 A mini review of NiFe-based materials as highly active oxygen evolution reaction electrocatalysts *Nano Res.* **8** 23–39

[15] Xiao P, Chen W and Wang X 2015 A review of phosphide-based materials for electrocatalytic hydrogen evolution *Adv. Energy Mater.* **5** 1500985iR

[16] Du P *et al* 2016 Metallic 1T phase MoS_2 nanosheets as a highly efficient co-catalyst for the photocatalytic hydrogen evolution of CdS nanorods *RSC Adv.* **6** 74394–9
[17] Reier T, Oezaslan M and Strasser P 2012 Electrocatalytic oxygen evolution reaction (OER) on Ru, Ir, and Pt catalysts: a comparative study of nanoparticles and bulk materials *ACS Catal.* **2** 1765–72
[18] Hunter B M, Gray H B and Müller A M 2016 Earth-abundant heterogeneous water oxidation catalysts *Chem. Rev.* **116** 14120–36
[19] Xie X *et al* 2022 Oxygen evolution reaction in alkaline environment: material challenges and solutions *Adv. Funct. Mater.* **32** 2110036
[20] Giordano L, Akkiraju K, Jacobs R, Vivona D, Morgan D and Shao-Horn Y 2022 Electronic structure-based descriptors for oxide properties and functions *ACC. Chem. Res.* **55** 298–308
[21] Sun Y, Chen G, Xi S and Xu Z J 2021 Catalytically influential features in transition metal oxides *ACS Catal.* **11** 13947–54
[22] Zhang Y C, Han C, Gao J, Wu J, Zhu X D and Zou J J 2022 $Co_{3-x}O_4$/NiO with abundant Ni^{3+} active sites for boosting oxygen evolution reaction *Chem. Eng. J.* **446** 137036
[23] Zhang Y, Lu T, Ye Y, Dai W, Zhu Y and Pan Y 2020 stabilizing oxygen vacancy in entropy-engineered $CoFe_2O_4$-type catalysts for co-prosperity of efficiency and stability in an oxygen evolution reaction *ACS Appl. Mater. Interfaces* **12** 32548–55
[24] Xie J *et al* 2021 Molten-salt-protected pyrolytic approach for fabricating borate-modified cobalt-iron spinel oxide with robust oxygen-evolving performance *ACS Sustain. Chem. Eng.* **9** 14596–604
[25] Zhang J, Qian J, Ran J, Xi P, Yang L and Gao D 2020 Engineering lower coordination atoms onto NiO/Co_3O_4 heterointerfaces for boosting oxygen evolution reactions *ACS Catal.* **10** 12376–84
[26] Song W *et al* 2016 Ni- and Mn-promoted mesoporous Co_3O_4: a stable bifunctional catalyst with surface-structure-dependent activity for oxygen reduction reaction and oxygen evolution reaction *ACS Appl. Mater. Interfaces* **8** 20802–13
[27] Kashale A A, Yi C H, Cheng K Y, Guo J S, Pan Y H and Chen I W P 2020 Binder-free heterostructured $NiFe_2O_4$/NiFe LDH nanosheet composite electrocatalysts for oxygen evolution reactions *ACS Appl. Energy Mater.* **3** 10831–40
[28] Saad A *et al* 2020 Three-dimensional mesoporous phosphide-spinel oxide heterojunctions with dual function as catalysts for overall water splitting *ACS Appl. Energy Mater.* **3** 1684–93
[29] Saad A *et al* 2021 Ag nanoparticles modified crumpled borophene supported Co_3O_4 catalyst showing superior oxygen evolution reaction (OER) performance *Appl. Catal.* B **298** 120529
[30] Han X, Wu G, Du J, Pi J, Yan M and Hong X 2022 Metal and metal oxide amorphous nanomaterials towards electrochemical applications *Chem. Commun.* **58** 223–37
[31] Song F *et al* 2018 Transition metal oxides as electrocatalysts for the oxygen evolution reaction in alkaline solutions: an application-inspired renaissance *J. Am. Chem. Soc.* **140** 7748–59
[32] Huang K, Sun Y, Zhang Y, Wang X, Zhang W and Feng S 2019 Hollow-structured metal oxides as oxygen-related catalysts *Adv. Mater.* **31** 1–12
[33] Li Z *et al* 2018 Engineering phosphorus-doped $LaFeO_{3-\delta}$ perovskite oxide as robust bifunctional oxygen electrocatalysts in alkaline solutions *Nano Energy* **47** 199–209

[34] Li J *et al* 2023 The critical role of A, B-site cations and oxygen vacancies on the OER electrocatalytic performances of $Bi_{0.15}Sr_{0.85}Co_{1-x}FexO_{3-\delta}$ ($0.2 \leqslant x \leqslant 1$) perovskites in alkaline media. *Chem. Eng. J.* **451** 138646
[35] Zhang K and Zou R 2021 Advanced transition metal-based oer electrocatalysts: current status, opportunities, and challenges *Small* **17** 1–40
[36] Tokura Y and Nagaosa N 2000 Orbital physics in transition-metal oxides *Science* **288** 462–8
[37] Zhao C *et al* 2019 Surface reconstruction of $La_{0.8}Sr_{0.2}Co_{0.8}Fe_{0.2}O_{3-\delta}$ for superimposed OER performance *ACS Appl. Mater. Interfaces* **11** 47858–67
[38] Retuerto M *et al* 2019 $La_{1.5}Sr_{0.5}NiMn_{0.5}Ru_{0.5}O_6$ double perovskite with enhanced ORR/OER bifunctional catalytic activity *ACS Appl. Mater. Interfaces* **11** 21454–64
[39] Yoo J S, Rong X, Liu Y and Kolpak A M 2018 Role of lattice oxygen participation in understanding trends in the oxygen evolution reaction on perovskites *ACS Catal.* **8** 4628–36
[40] Hong W T, Risch M, Stoerzinger K A, Grimaud A, Suntivich J and Shao-Horn Y 2015 Toward the rational design of non-precious transition metal oxides for oxygen electrocatalysis *Energy Environ. Sci.* **8** 1404–27

IOP Publishing

Materials Technology for the Energy and Environmental Nexus, Volume 2

R V Mangalaraja, Arunachalam Arulraj, Radhamanohar Aepuru and Chan Siew Hwa

Chapter 9

Recent advances of nanomaterials in environmental applications

Govindan Rajivgandhi and Franck Quero

Two-dimensional (2D) nanomaterials are frequently used in all research fields, including environmental and biomedical applications, due to their unique shapes, surface, and anisotropic physical and chemical properties. In addition, they are very attractive because this new class of materials has a sheet-like morphology, transverse dimensions of 100 nm, and are thin (5 mm). In recent years, a greater number of 2D materials have emerged based on unique characteristic properties, notably double layered hydroxides, transition metal oxides, graphene, carbon nitride, hexagonal boron nitride, boron nanosheets, black phosphorus, and single, few, and multi-layers of transition metal carbides, nitrides, and carbonitrides. Among the various 2D nanomaterials, graphene and MXene are the top 2D nanomaterials, which have been reported on frequently worldwide in recent decades. Both have better efficiency of transparency, and electrical and thermal conductivities. They both have a large surface area and extraordinary mechanical properties. Based on the above advantages, the various 2D nanomaterials and their environmental effectiveness are being studies for removing environmental pollutants of heavy metal, dye degradation, various metal ions, gas molecules, radionuclides, and other environmental contaminants. This chapter concentrates on the effect of various 2D nanomaterials in the remediation of environmental pollution and the future challenges to understand their full potential effects in the environment and ecosystems.

9.1 Introduction

In recent years, nanotechnology has become a ubiquitous multidisciplinary science and technology that is applied in various fields of research and has acquired great recognition in different fields of rational knowledge [1]. In most developing

doi:10.1088/978-0-7503-5729-6ch9

countries, nanotechnology-based research is heightened in all fields, including soil crop production, soil improvement, agriculture sensing, food safety, medical diagnosis, biopharmaceutical and environmental monitoring, environmental treatment, removal of chemicals, wastewater treatment, energy conversion, etc. The word ‘nano’ refers simply to the dimension of the materials which are under 10^{-9} m in size in at least one direction. To-date, different definitions for ‘nano’ have been provided by various researchers, but is generally agreed to be 100 nm (100 times a millionth of a millimeter) in size of any structure [2]. The knowledge in the field of nanotechnology is classified based on the usage in the field, such as the formation of chemicals, colloidal, physical, materials, devices, sensors, polymers, vectors, and other formats. Different forms of the various developed nanomaterials are applied in several fields at the nanoscale level [3].

In 2002 the USA approved ‘The National Nanotechnology Initiative’ mission, which aimed to boost the development of nanotechnology in various fields within their country. Following this initiative, several developed countries are also concentrating on nanotechnology research for industrial level applications [4]. For example, Japan and the European Union have developed their own nanotechnology departments with highly equipped laboratories in several universities to develop nanotechnology use in various applications. In particular, agriculture, medicine, and the electronics and pharmaceutical industries have given a lot of interest and attention to using nanoscience and nanotechnology [5]. Furthermore, the nanotechnology is being used more to help produce larger amounts of nanomaterials to be applied in various types of industry, which is leading to new scenarios in the environmental and biomedical fields, which needs to be urgently evaluated. In general, nanotechnology is a diversified field of science that studies nanometric scale objects based on their size and characteristic nature [6].

9.1.1 The importance of nanotechnology

For the past few decades, nanotechnology has emerged as a tool in various fields and is considered as a primary tool for applied technology, demonstrating a way to work at the atomic scale. It is used in numerous fields such as soil crop production, soil improvement, agriculture sensing, food safety, medical diagnosis, biopharmaceutical and environmental monitoring, environmental [8]. The scientific community around the world is motivated to develop novel innovative nanosized materials with unique properties and shapes to address problems associated with end-user applications [5]. Usually, nanomaterials are divided into three basic classes based on their size and shape: one-dimensional (1D), two-dimensional (2D), and three-dimensional (3D). In addition, all the classes can be categorized by form as particles, sheets, rods, wires, and various other shapes based on their dimensionally confined electronic properties. The preparation and fabrication of all the 1D, 2D, and 3D nanomaterials can be the same or different depending on their surface sites and band structure [10]. The synthesized nanoparticles and its importance in various field was indicated in figure 9.1.

Figure 9.1. Overview of the importance of nanoparticles in various fields.

For example, the increase of antimicrobial therapy is creating further problems worldwide due to the indiscriminate usage of antibiotics creating a selective pressure on bacteria. Several attempts have been made to rectify such issues, such as the multiple experiments evaluated to increase the spectrum of action of antibiotics. Increasing drug efficiency with novel mechanisms and its action is an important factor in preventing bacterial resistance, thereby ensuring the health of people. Previously, antibiotics and drugs have been the main weapons to eradicate bacterial infections [23]. In addition, antibiotics are also used in the field of soil crop production, soil improvement, agriculture sensing, food safety, medical diagnosis, biopharmaceuticals, environmental treatment, the removal of chemicals, and wastewater treatment [12].

It is well evidenced that nanoparticles are the best suitable solution for eradicating all the associated problems in end-user applications due to its nanoscale level of intervention. Among the several nanoparticles, silver (Ag), zinc (ZnO), nickel (Ni), and copper (Cu) are notable and the most widely used materials in various applications, including environmental monitoring, environmental treatment, the removal of chemicals, and wastewater treatment [13]. These metallic, metal oxide, and plasmonic nanoparticles offer the benefits of enhancing the efficiency of the desired applications, which includes environmental treatment. Specifically, pollutant removal, wastewater treatment, toxicity removal, heavy metal removal, dye degradation, etc, are subjects of focus of the worldwide scientific community [2].

9.1.1.1 Nanoparticles in agriculture

In recent decades, the deficiencies of agriculture production have been global issues. Hence, improvements in agricultural production and the measurement of agronomic products are very important to global food security. Continuing the traditional approach of agricultural improvement by farmers and doing research in the same way is not sufficient to produce better outputs in the field of agriculture. Since the development of agriculture is hindered by different aspects, which include the scarcity of sufficient natural fertilizers, which hinders growth and creates environmental pollution due to the use of artificial fertilizers. Diseases caused by microbes including bacteria, viruses, fungi, yeasts, and pests reduce the productivity and yield of agriculture [14]. In addition to this, different varieties of abiotic stressors and environmental conditions stimulated by anthropogenic and natural conditions prevent the growth of plants. Thus there is an emerging need for alternative methods to avoid these limitations, with increased efficiency and productivity, and decreased pollution, to address this dilemma of agriculture [15].

In this context, nanotechnology has emerged to help in the field of agriculture by means of manipulating and monitoring substances. Chemical, physical, and biological mediated nanoparticles are also used for the development of agriculture [16]. Nanoparticles have unique properties including chemical, physical, mechanical, optical, electrical, and magnetic characteristics, with <100 nm size and a large surface area. As mentioned, environmental stressors and behavioral changes in the field of agriculture have affected food security [17]. Thus, the use of nanotechnology can improve the yield and quality of agricultural products and other produce, and also decrease the effect of fertilizers and pesticides in the agriculture field due to environmental changes [18].

9.1.1.2 Nanoparticles in the food industry

The urbanization of human populations from rural areas and modern lifestyles cause an excess amount of foodwaste without livestock farming. Also, the larger volume of pre- and post-consumption foodwaste affects the environment. In manpower-based industries, the preparation of food for workers is vital, in which there is an unavoidable occurrence of foodwaste, producing a large volume of environmental pollution. Based on statistical data, approximately only one third of food production is used for human consumption worldwide [19]. Also, approximately ~1.3 billion tons of waste is produced by the food industry. Of this, household activities produced ~42% of foodwaste, with 39% from food production industries, hotels, restaurants, and catering, and 15% from other sources [20].

Among this waste, green vegetables, fruits, plants, and roots have the highest wastage rates [21]. The tons of foodwaste around the world cause a major threat to human health, ecosystems, and livestock, create eutrophication, and allow the multiplication of pathogens and diseases. Importantly, kitchen-generated waste materials (organic, inorganic), including seeds, peels, fruits, and vegetables, are the reason to produce more wastage (such as phenolic compounds) by exhibiting the rate of 15% [22]. Thus, an effective way of utilizing this waste to make valuable products is of prime importance. In this respect, almost all of the foodwaste can be used for the synthesis of various nanoparticles and it can become a substantial phase in active waste consumption. The nanomaterials synthesized using materials such as foodwaste are often referred as being created using 'green synthesis'. The advantages of nanoparticles synthesized using waste products are being applied in different fields of application, including biosensors, energy production, environmental remediation, environmental monitoring, etc. In this way, the foodwaste is considered as a potential source to produce greener nanoparticles.

In addition, nanoparticles are applied in the food industry for the effective packaging and preservation of foods to enhance their shelf life. Generally, plenty of methods exist to preserve foods, vegetables, fruits, meats, dairy products, baked items, and juices. However, the use of nano-based products as the preserving agents, where the nanoparticles can be mixed directly with food products to aid in long shelf life and are less toxic.

9.2 The effects of environmental pollution

Environmental pollution due to agricultural discharges, industrial waste, and infectious pathogens is another emerging problem, and the respective remediation or treatment approaches are being explored worldwide [29]. Environmental degradation based on photocatalysis systems involves different kinds of homogeneous and multiphase semiconducting photocatalytic nanomaterials. The most widely used semiconducting photocatalytic nanomaterials are TiO_2, CuO, MoS_2, and polymer carbon nitrate. Among these nanomaterials, polymer carbon nitrate is an important nanomaterial which is used frequently in environmental remediation. It has the qualities of excellent thermal stability, decreased toxicity, and unique electronic structure [30]. In addition, it is used for the degradation of organic pollutants, the conversion of solar energy, and greenhouse gas capture. The band gap between the molecules of polymer carbon nitrate is 2.7 eV and edge adsorption at 450 nm aids in utilization of the visible wavelength [24].

Hazardous and non-hazardous dyes are mostly used in the textile, handloom, food, pharmaceutical, and analytical industries in larger quantities. The sludge of dyes from these industries are mixed with the water nearby, thus contaminating the water and the environment, causing a threat to aquatic and marine ecosystems [25]. The mixed dyes in the environment (water) cause carcinogenic and mutagenic effects in human and their removal is necessary. The wastewater from industrial discharges is the prime cause of reducing light diffusion into water resources, causing oxygen deficiency [23]. Most marine organisms are affected by these dyes due to the mixing of industrial discharges with sea water [23]. Complete analysis or detection methods need to be developed for the identification of these trace elements. The existing technologies to remove or control such pollutant contamination include membrane filtration, flocculation using biological methods, photocatalysis, and sludge treatment reverse osmosis, however, most of these applied process are temporary solutions to treat the contamination [28]. In recent decades different structured nanomaterials as photocatalysts, which can be used for the removal of pollutants in the waste water through various purification processes. The photocatalytic nanomaterials help in the removal of and/or treating toxic dyes in wastewater, thereby ensuring a sustainable supply of clean drinking water.

Additionally, photocatalysis is another important method to remove the complete mineralization of textile dyes, which is very efficient and inexpensive, and the by-products are reported to be clean and eco-friendly. Among the various dyes, methylene blue is a thiazine dye used in various industries for coloring paper and clothing, and as a photosensitizer and toxic indicator. Nanomaterials are effectively applied in dye degradation using wide band gap materials such as TiO_2, ZnO, etc. It is also cost effective, unique, accurate, and with low toxicity and high stability [31].

In the last ten years, organic waste, organic pollutants, heavy metals, dyes, and other industrial wastes have damaged the environment heavily and are threatening the global environment and human health, and causing other issues [32]. The increasing rate of pollution of water bodies is a major alarming factor in the development of industries. The most common heavy metal present in industrial wastewater is chromium. It is present at very high levels and is highly polluting in the environment. It is a mutagenic agent, very carcinogenic, and highly toxic to humans. Toxic levels above 0.05 ppm in drinking water cause dangerous infections in humans. The reduction of hexavalent chromium ions, which is a heavy metal, should be done and it should be removed from industrial wastewater. Apart from chromium, methylene blue is also an organic pollutant which is present in water, creating a high level of environmental issues [33]. It is an organic dye, and a very well-known type of non-degradable and harmful pollutant in drinking water and wastewater and is a great threat to human health. Thus the removal of heavy metals, dyes, and other compounds that are major threats in industrial wastewater, and the removal of pollutant hydrocarbons, is very important. The current removal processes include precipitation, membrane separation, adsorption, and photocatalytic oxidation. In various industries, the removal of hydrocarbons using a photocatalysis method is most advisable [34].

Based on their ecological risk, toxicity, biomagnification, bioaccumulation, and non-degradability, heavy metal pollution is a major threat to every group of living organisms [35]. Considerable or plentiful amounts of heavy metals or elements are present everywhere as particulate matter or bound to the surface layers of soil. The continuous presence of excess levels of metals greatly affects the physiological activation and metabolic activities in both humans and other organisms. Also, they affect and damage the entire body of living beings strongly, both physically and internally. In addition, one of the most important emerging threats is industrial effluents, which continuously affect water quality due to the frequent discharge of wastewater coagulation into running water without any restrictions [36, 37]. Metals and metalloids are present at higher levels in the soil and are also present in some other lithogenic sources, including volcanic ash and water drainage. From one hundred years ago, more anthropogenic sources have been reported. All these sources are accumulated more directly into the aquatic environment. Notably, heavy metals, such as Pb, Cr, Zn, Ni, Hg, As, Cu, and Cd, are present in industrial wastewater, industrial effluents, and other soil contaminated sites. The removal of heavy metals is an important process as they contribute 80%–90% of environmental pollution and human health issues through the consumption of heavy metal ion containing water. There are various experiments and techniques used for the removal of heavy metals, but only a few methods are frequently applicable for effective removal. Electrocoagulation is one of the important methods used for the removal of heavy metals present in industrial wastewater. Adsorption using natural ingredients or artificial adsorbents, advanced oxidation processes, and specific membrane separation methods are more efficient. In the adsorption process, the metal ions and adsorbents will bond to each other on the surface and this is called competitive binding. Various time intervals, temperature conditions, adsorption doses, initial metal concentrations, pH levels, and other physiological factors contribute to the removal of heavy metals. Cost-effective, efficient, and low operational cost methods are highly preferable in most industries. For the adsorption process, mostly metal and/or metal oxide-based nanoparticles can be used as adsorbents, including magnetic nanoparticles, carbon-based nanoparticles, synthetic nanoparticles, and polymer-based nanoparticles. Specifically, the toxic contamination of water is creating very dangerous diseases through the water column, including in humans, animals and the majority of aquatic flora and fauna [38].

9.3 The role of nanoparticles in environmental applications

Nanoparticles are emerging tools in environmental applications including dye degradation, wastewater treatment, heavy metal pollution, remediation, sensors, etc [39]. Nanoparticles (1D, 2D, and 3D) provided plenty of benefits using advanced methods in environmental applications. Recently, different types of metal and metal oxide nanoparticles have been used in various environmental applications, and include silver, gold, copper oxide, zinc oxide, tin oxide, graphene oxide, MXene, boron oxide, etc [40]. The most notable dyes, such as methylene blue, alizarin red, methyl orange, Congo red, and gentian violet, can be treated effectively using various structured nanoparticles. Previously, many researchers reported that catalysts suffer many drawbacks due to the factors of UV, temperature, NaCl, H_2O_2, and concentration of dyes. Also, the catalyst processes are slow in the degradation of dyes in industrial waste [41].

Nanoparticles are of great interest in environmental applications due to their unique chemical, optical, electrical, and magnetic properties, and their small size and high surface area [42]. The synthesized nanoparticles are applied in various environmental applications and they are indicated clearly in figure 9.2. Nanoparticles are being applied in the remediation of environmental pollution that causes very severe harm to the eco-system and humans. Various nanomaterials such as tin oxide, cerium dioxide, nickel oxide, manganese dioxide, and vanadium pentoxide are notable nanomaterials involved in the photocatalysis process, with their energy dissipation, light resistance, and non-toxicity [43]. In addition, tungsten trioxide, molybdenum disulfide, zirconium oxide, cadmium sulfide, and iron oxide are also used as photocatalysts applied in the decomposition of organic and inorganic pollutants (Table 9.1).

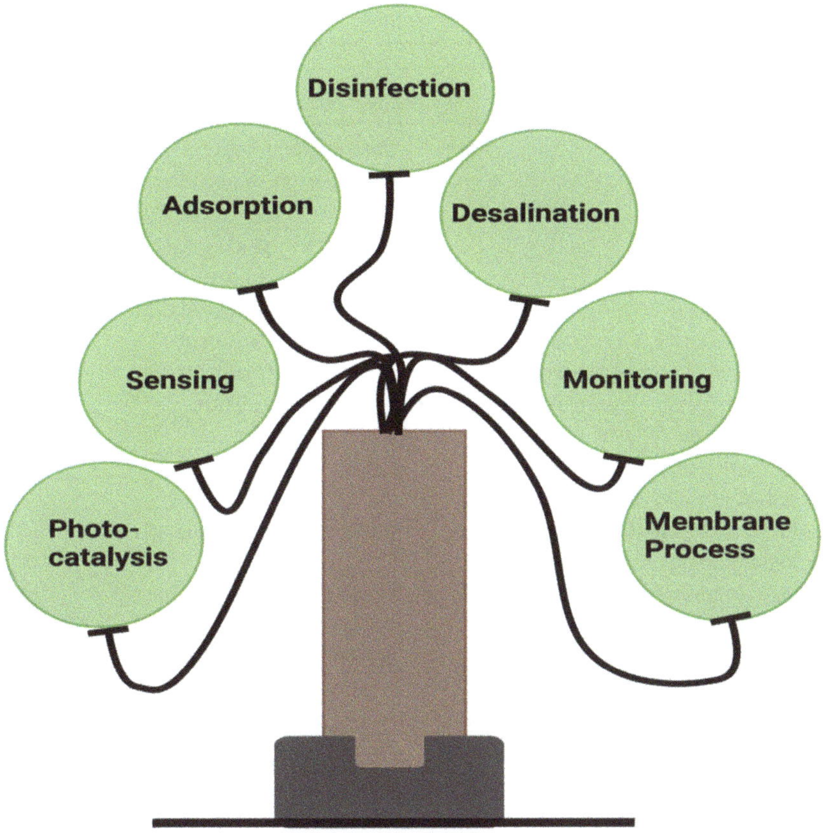

Figure 9.2. Nanoparticles in environmental applications.

Titanium dioxide (TiO_2) is one of the benchmarking semiconducting photocatalysts. Mechanistically, the metals with a continuous electron mode, semiconductors, have an energy free region. The covalent band lies between the atoms of the valance band and the other energy levels located at a higher location, called the conduction band [44].

In recent years, metal nanoparticles have been used extensively in the field of catalytic degradation, heavy metal removal, wastewater treatment, and industrial effluent degradation. Among the various nanoparticles, silver nanoparticles are more attractive for environmental applications, in particular dye degradation, effluent treatment, and catalytic degradation. It has spherical morphology, higher surface area, and more free surface energy than other nanoparticles.

Silver (Ag) nanoparticles (NPs) can be produced in different forms such as nanoparticles, nanocubes, nanowires, nanodisks, nanoprisms, etc. In addition, AgNPs are cheaper than other metallic particles, such as gold and palladium, in treating environmental applications. Recently, Bonnia *et al* [36] reported that biological mediated silver nanoparticles effectively degraded methylene blue dye.

Table 9.1. Applications of various nanoparticles in environmental fields.

S. No.	Nanoparticles	Environmental applications	References
1	Silver nanoparticles	Catalytic degradation	[29]
2	Silver nanoparticles	Degradation of accumulated drug waste	[31]
3	Silver nanoparticles	Congo red dye degradation	[32]
4	Silver nanoparticles	Degradation of azo dyes	
5	rGO supported silver	Photocatalytic degradation	[34]
6	Graphene material	Wastewater treatment	
7	Graphene/Fe_3O_4/agro-based waste material	Wastewater treatment	[37]
8	Graphene oxides magnetized with $NiFe_2O_4$ nanoparticles	Biodiesel production	
9	Ni nanoparticles@graphene oxide-crosslinked poly(vinyl alcohol)	Catalytic degradation	[39]
10	Graphene capsuled sulfur nanoparticles	Wastewater treatment	[40]
11	Zinc oxide nanoparticles	Catalytic activity	[41]
12	Zinc oxide nanoparticles	Photocatalytic and biotic degradation	[42]
13	Zinc oxide nanoparticles	Wastewater treatment	[43]
14	Zinc oxide nanoparticles	Wastewater treatment	
15	Graphene-ZnO nanoparticles	Plasma catalytic degradation	[45]
16	Laccase immobilized on Fe_3O_4@SiO_2 nanoparticles	Photothermal catalytic degradation of textile dye	[46]
17	Gold nanoparticles	Catalytic degradation	[47]
18	Cobalt ferrite nanoparticles	Catalytic degradation of refractory phenol sulfonic acid	[48]
19	Copper oxide nanoparticles	Methylene blue via photocatalysis	[49]
20	Zn substituted $CuFe_2O_4$ nanoparticles	Catalysis degradation	[50]
21	NaY zeolite films	Catalysis degradation	[51]

Ramesh and Mala [38] also reported that AgNP and ZIF-8 composites contributed to the photocatalytic dye degradation of methylene blue and Congo red. Similarly, the authors of [29, 31] reported industrial waste and its accumulating drug degradation using AgNP mediated catalytic degradation.

9.3.1 Applications of 2D nanomaterials in the environmental field

2D nanoparticles have emerged recently as promising candidates to be used in various applications at high rates [52]. The most well-known 2D material is graphene, which was discovered in 2004 by the exfoliation of graphite materials and is fundamentally made up of a complete 2D carbon layer. In addition, the synthesis of graphene from graphite was a revolution in the era of 2D materials. The production of graphene from graphite is strengthening research, and it covered by one 0r multi thick layer of atoms performed very accurate result in most of the fields. After the introduction of 2D nanomaterials in the research field, different kinds of 2D nanomaterials have been produced [53]. Next to graphene, a decade ago, a new class of 2D materials called 'MXene' was derived from MAX phased materials. The materials were initially identified by a research group at Drexel University in 2011 [54]. The initial MXene material discovered was titanium carbide (Ti_3C_2) from the etching and exfoliation of titanium aluminum carbide (Ti_3Al_2C). Followed by this discovery of new materials, currently more than 30 types of 2D nanomaterials have been derived from the MXene family. These 2D nanomaterials possess a large diameter–length ratio and exhibit exceptional properties which are unique compared to their bulk materials. In addition to this, graphene is an inferior nanomaterial in semiconductive devices which have high electrical conductivity behavior [55].

The most promising 2D nanomaterial in recent years has been MXene (consisting of either transition metal carbides, or nitrides or carbonitrides), which is covered by a multi-layered nanostructure with large class layered sheets. Based on previous research [56], MXenes are excellent materials for energy production and environmental applications. Generally, MXene is the most widely used in energy storage devices such as supercapacitors, electromagnetic interference shielding, photocatalysis, and electrocatalysis. The large amount of metal atoms in MXene nanosheets are terminated with other functional groups such as –OH, –F, and –O [57]. It has outstanding characteristics including larger surface area, and excellent hydrophilic, biocompatibility and conductivity properties. This creates more reactive sites and a low charge transfer path in the photocatalytic reaction process [58]. MXene is also used extensively in the process of photohydrogen production and photocatalytic oxidation–reduction. Also, it is used easily and efficiently in the recycling process and is used as an external magnet (table 9.2).

Graphene is a single layer of carbon atoms arranged in a honeycomb lattice pattern with a unique arrangement of sp^2 hybridized bonds. It is the basic building

Table 9.2. Advancement of 2D nanomaterials in environmental applications.

S. No.	Nanomaterials	Applications	References
1	Graphene oxide	Removal of dye	[63]
2	Graphene	Adsorption	[62]
3	Graphene	Gas sensing	
4	Graphene oxide	Heavy metal absorbent	[61]
5	$MXene/TiO_2/Bi_2S_3$	Photocatalytic sterilization	[67]
6	Ti_3C_2 $MXene/TiO_2$	Methyl orange degradation	[68]
7	ZnO/MXene	Photocatalytic transformation	[66]
8	MXene	Removal of hazardous waste	
9	$Ti_3C_2T_X$ MXene-PAN membrane	Wastewater treatment	[71]
10	Magnetic-MXene nanocomposite	Wastewater treatment	[72]

block of graphite, which is composed of multiple layers of graphene stacked on top of each other. The interlayer spacing between each graphene sheet in graphite is approximately 3.34 Å, and it is held together by weak van der Waals forces. Initially, a single layer of graphene was produced through a process known as micro-mechanical exfoliation, where graphite is successively peeled apart using adhesive tape. This process allows the separation of individual graphene sheets from the bulk graphite structure, revealing their unique properties. Graphene has unique physical, chemical, and mechanical properties that make it an attractive material for a wide range of applications in various fields [59]. Graphene in the nanoscale range is known for its incredibly high specific surface area. This is attributed to the fact that every atom of a single-layer graphene sheet is exposed from both sides to its environment, making it the most extreme case of high surface area materials. With a theoretical value of 2630 $m^2\ g^{-1}$, graphene has the highest specific surface area of all materials, making it an ideal candidate for processes that involve adsorption or surface reactions. Additionally, because of its high surface area, graphene can act as an excellent support for anchoring chemical functionalities or nanomaterials. As a result, research into graphene-based nanocomposites has become a growing field in the search for novel materials [60].

Several research works have highlighted the utilization of graphene-based materials as effective adsorbents for the elimination of inorganic contaminants present in water. Among them, graphene oxide (GO) has gained widespread attention due to its high content of oxygen-containing functional groups, which play a crucial role in the adsorption of metal ions. The superiority of GO over pristine graphene as an adsorbent is attributed to the presence of these oxygen-containing functional groups, which provide many sites for metal ion adsorption. This has been confirmed by several studies, which have compared the adsorption performance of pristine graphene and graphene oxide for the removal of Pb (II) ions from water. These studies have shown that GO displays significantly higher adsorption efficiency compared to pristine graphene [61].

The utilization of GO as an adsorbent for the removal of inorganic contaminants from water has been studied extensively, and its effectiveness has been attributed to the presence of oxygen-containing functional groups. These groups provide a large number of sites for the adsorption of metal ions, leading to the enhanced performance of GO compared to pristine graphene [62].

The handling and elimination of organic dyes produced during dye manufacturing and textile finishing is one of another significant issue. Many dyes exist either in the form of cationic or anionic ions in a dissolved form, with only a limited number being dispersive. Although it is a relatively new technology, 3D graphene based membrane (GBM) has been applied to treat a range of different dyes, including methyl orange (MO), malachite green (MG), methylene blue (MB), acridine orange (AO), calcian (CA), Bordeaux red (BR), brilliant green (BG), and methyl violet (MV). For instance, the GO and DNA hydrogels are types of composite materials that have been used in the field of materials science for various applications, including the loading of dyes. The GO/DNA hydrogel has been evaluated against the ability to load safranin O, a commonly used dye in biological research. The results show that the hydrogel had a high dye-loading capacity of 960 mg · g^{-1}, which is comparable to that of other carbon nanomaterials. The combination of GO and DNA in a hydrogel can result in unique properties that make it useful with a high level of loading the dye. The high surface area and electrical conductivity of GO, combined with the biocompatibility and biodegradability of DNA, can make this composite hydrogel an attractive material for applications such as drug delivery, tissue engineering, and water treatment. It is important to note that the report demonstrated the potential of GO/DNA hydrogels for dye-loading applications. Kanani *et al* focused on exploring the use of magnetic manganese ferrite-graphene oxide ($MnFe_2O_4$-GO) as an adsorbent for the removal of rhodamine 6G dye from aqueous solutions. The results showed that $MnFe_2O_4$-GO has a higher absorption capacity compared to unmodified graphene oxide (GO), which is a commonly used adsorbent material. The absorption capacity of $MnFe_2O_4$-GO was found to be 24.96 mg g^{-1}, while that of unmodified GO was 14.82 mg g^{-1}.

Graphene-based environmental sensors utilize the unique properties of graphene, such as its high electrical conductivity, large surface area, and chemical stability, to detect various environmental parameters, such as temperature, humidity, and pollutants. These sensors are highly promising due to their high sensitivity, fast response times, and low cost compared to traditional sensors. Graphene-based sensors can be used in a variety of applications, including air quality monitoring, water quality monitoring, and soil analysis. For example, graphene-based gas sensors can detect various gases, such as carbon monoxide, nitrogen oxides, and volatile organic compounds, which are important indicators of air pollution. Graphene-based humidity sensors can be used to measure the relative humidity in the environment, while graphene-based temperature sensors can provide highly accurate and rapid temperature readings. Graphene-based environmental sensors have the potential to revolutionize environmental monitoring by providing cost-effective, sensitive, and fast-response solutions for monitoring various environmental parameters [64].

Experimental studies have shown that pristine graphene has the potential to be used as a gas sensor for toxic gases. These studies have aimed to understand the factors that affect the selectivity and limit of detection of the sensor, including the chemical nature of the gas molecules, the physicochemical characteristics of graphene, and the experimental conditions. For example, graphene films prepared through chemical vapor deposition (CVD) and supported on SiO_2/Si substrates have shown high sensitivity towards ammonia (NH_3), methane (CH_4), and hydrogen (H_2). When tested with a mixture of NH_3 and CH_4, the graphene sensor was able to display a clear response for NH_3 even in the presence of an excess of CH_4. This result indicates that NH_3 has a greater ability to alter the conductivity of graphene compared to CH_4, which makes the sensor more selective and sensitive to the presence of NH_3 [55].

Similarly, reduced graphene oxide (rGO) has also been used widely for gas sensing due to its improved electrical conductivity and charge carrier transport capacity compared to pristine graphene. The conversion of GO to rGO is typically performed through chemical reduction using hydrazine vapor or solutions. The reduction process results in a material with fewer structural defects, which leads to improved sensing performance. For example, rGO sensors obtained through the reduction of GO with liquid hydrazine have been used for the detection of gases such as ammonia (NH_3) and nitrogen dioxide (NO_2). Additionally, these sensors have also been shown to be sensitive to the presence of 2,4-dinitrotoluene (DNT) with a detection limit of 28 parts per billion [65].

In recent years, the MXenes, advanced 2D nanomaterials comprising carbides or nitrides and/or carbonitrides, have been receiving increasing attention after the discovery of Ti_3C_2. Commonly, MXenes are derived from etching elements, simply the A layers of the MAX materials, where M, A, and X denotes transition metals, IIIA/IVA group elements, and carbon and nitrogen. MXenes are used widely in the field of energy storage and conversion, dye degradation, catalysis, heavy metal removal, wastewater treatment, and adsorption. Soyoung *et al* [69] reported that MXene has excellent electrical conductivity behavior affected by its metallic nature and it can be applied as a co-catalyst in the photocatalytic process which aids in the transfer and separation of photo-induced electron–hole pairs.

Weibing *et al* [66] studied the ZnO/MXene combination which was used effectively to transfer the electron in the process of photocatalytic activity. Also, Haimeng *et al* [67] reported MXene/TiO_2/BiS_3 for UV-light photocatalytic sterilization. Huapeng *et al* [68] supported the previous report, with the MXene/TiO_2 acting as a co-catalyzer and boosting the photocatalytic activity at a high rate and degrading methyl orange and hydrogen production. In the preparation of MXene and its composites, the simple hydrothermal method is used by most researchers [66, 67]. Among these, the electrostatic self-assembled experiment can inhibit the self-stacking in TiO_2 and Ti_3C_2 vigorously [68]. Also, the larger surface area of this composite enhances the photocatalytic activity efficiently. Also, this kind of experiment is very simple, effective, and low cost, and it can be controlled by the density level in TiO_2 on the Ti_3C_2 nanosheet.

Nadeem *et al* [70] reported MXene based 2D nanomaterials used as adsorbents for the removal of hazardous pollutants. In addition, the use of MXenes in environmental applications does not release any weak van der Waals effect, thereby they possess excellent removal efficiency of heavy metals, organic dyes, and radionuclides due to the presence of more active sites on their surface. Importantly, MXene is resistant against powerful radiation, and it has excellent thermal and chemical stability. Thus, the adsorbent rate is very high in radioactive elements from nuclear waste. In pollution removal, MXene is an important 2D material due to its interlayer space within its stack of layers. In addition, the interlayer space of MXene can be heightened by a variety of intercalants in the processes of intercalation and exfoliation. MXene eliminates more kinds of radioactive particles, such as coordination, reduction sequestration, and ion exchange. In addition, MXene has a more efficient characteristic nature of permeability, hydrophilicity, and catalytic deficiency.

Compared with other nanomaterials, 2D MXene is more efficient, with surface complexation and electrostatic interaction with ions of phosphate and nitrates, thereby helping to remove toxic ions. Compared to other 2D nanomaterials, the MXene

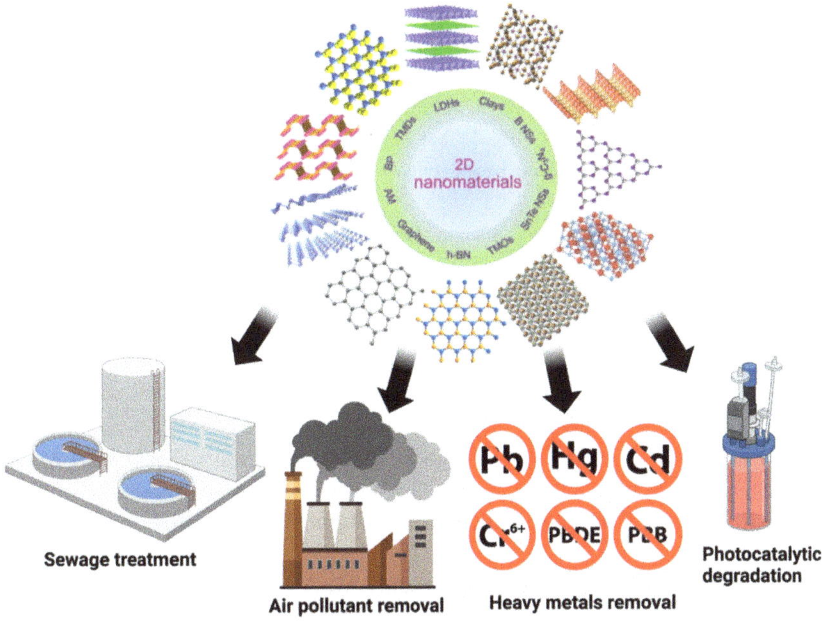

Figure 9.3. Applications of 2D nanomaterials in environmental applications.

nanomaterials possess a very efficient characteristic nature of their association, with excellent thermal, chemical, and mechanical stability. The O, –OH, and F functional groups present in the MXenes have the potential property to absorb the maximum number of selected contaminants. Compared with other 2D nanomaterials, the adsorption behavior is very high and excellent [73]. Based on previous experiments, the effect of 2D nanomaterials in environmental applications is very high (figure 9.3).

9.4 Conclusion

In recent years, the 2D nanomaterials have been emerging in various fields including environmental remediation. Researchers are intensifying their research in the field of 2D nanomaterials due to their advantages of high efficiency, low cost, and accurate results. All the 2D materials have excellent performance in energy storage and conversion, dye degradation, catalysis, heavy metal removal, wastewater treatment, and adsorption. Due to these advantages, 2D nanomaterials are used widely in the environmental field with high success rates.

References

[1] Arulraj A, Senguttuvan G, Veeramani S, Sivakumar V and Subramanian B 2019 Photovoltaic performance of natural metal free photo-sensitizer for TiO2 based dyesensitized solar cells *Optik (Stuttg)* **181** 619–2

[2] Arulraj A, Murugesan P K and Zamorano R C A T and Mangalaraja R V 2023 Nanoarchitectonics of layered metal chalcogenides-based ternary electrocatalyst for water splitting *Energies* **16** 1669

[3] Rajivgandhi G N, Ramachandran G, Kanisha C C, Alharbi N S, Kadaikunnan S, Khaled J M, Alanzi K F and Li W-J 2021 Effect of Ti and Cu doping on the structural, optical, morphological and anti-bacterial properties of nickel ferrite nanoparticles *Results Phys.* **23** 104065

[4] Dhanker R, Rawat S, Chandna V, Deepa, Das S, Sharma A and Kumar V 2022 Recovery of silver nanoparticles and management of food wastes: obstacles and opportunities *Environ. Adv.* **9** 100303

[5] Maruthupandy M, Rajivgandhi G, Muneeswaran T, Anand M and Quero F 2021 Highly efficient antibacterial activity of graphene/chitosan/magnetite nanocomposites against ESBL-producing *Pseudomonas aeruginosa* and *Klebsiella pneumonia Colloids Surf.* B **202** 111690

[6] Kalssom U, Ahsan Z, Bhatti H N, Amin F, Nadeem R, Aftab K and Bilal M 2022 Iron oxide nanoparticles immobilized *Aspergillus flavus* manganese peroxidase with improved biocatalytic, kinetic, thermodynamic, and dye degradation potentialities *Process Biochem.* **117** 117–33

[7] Rajivgandhi G, Vimala R T V, Maruthupandy M, Alharbi N S, Kadaikunnan S, Khaled J M, Manoharan N and Li W-J 2021 Enlightening the characteristics of bioflocculant of endophytic actinomycetes from marine algae and its biosorption of heavy metal removal *Environ. Res.* **200** 111708

[8] Heyse A, Kraume M and Drews A 2020 The impact of lipases on the rheological behavior of colloidal silica nanoparticle stabilized pickering emulsions for biocatalytical applications *Colloids Surf.* B **185** 110580

[9] Maruthupandy M, Riquelme D, Rajivgandhi G, Muneeswaran T, Cho W-S, Anand M, Manoharan N and Quero F 2021 Dual-role of graphene/bacterial cellulose/magnetite nanocomposites as highly effective antibacterial agent and visible-light-driven photocatalyst *J. Environ. Chem. Eng.* **9** 106014

[10] Ramanaviciene A, Voronovic J, Popov A, Drevinskas R, Kausaite-Minkstimiene A and Ramanavicius A 2016 Investigation of biocatalytic enlargement of gold nanoparticles using dynamic light scattering and atomic force microscopy *Colloids Surf.* A **510** 183–9

[11] Rajivgandhi G, Bhavya K, Vidhya B, Alharbi N S, Kadaikunnan S, Khaled J M, Alanzi K F and Nandhakumar R 2022 Fabrication of graphene oxide-p-phenylenediamine nanocomposites as fluorescent chemosensors for detection of metal ions *Environ. Res.* **204** 111914

[12] Gao B-H, Ding S-N, Kargbo O, Wang Y-H, Sun Y-M and Cosnier S 2013 Enhanced electrochemiluminescence of peroxydisulfate by electrodeposited Au nanoparticles and its biosensing application via integrating biocatalytic precipitation using self-assembly bi-enzymes *J. Electroanal. Chem.* **703** 9–13

[13] Sripriya N, Vasantharaj S, Mani U, Shanmugavel M, Jayasree R and Gnanamani A 2019 Encapsulated enhanced silver nanoparticles biosynthesis by modified new route for nano-biocatalytic activity *Biocatal. Agric. Biotechnol.* **18** 101045

[14] Preethi S, Arulraj A, Mangalaraja R V, Ravichandran V and Subramanian N 2022 Biomass-derived carbon-based materials for microbicidal applications *Biomass-Derived Carbon Materials* (New York: Wiley) pp 63–92

[15] Sonawane H *et al* 2022 Fungi-derived agriculturally important nanoparticles and their application in crop stress management—prospects and environmental risks *Environ. Res.* **212** 113543

[16] Aghashiri A, Hashemian S and Fotooh F K 2022 Mesoporous $LaFeO_3$ nanoparticles decorated onto activated carbon from agricultural paste (synthesis, characterization and adsorption properties) *J. Mol. Struct.* **1260** 132822

[17] Salama D M, Abd El-Aziz M E, El-Naggar M E, Shaaban E A and Abd El-Wahed M S 2021 Synthesis of an eco-friendly nanocomposite fertilizer for common bean based on carbon nanoparticles from agricultural waste biochar *Pedosphere* **31** 923–33

[18] Narasaiah B P, Koppala S, Kar P, Lokesh B and Mandal B K 2022 Photocatalytic and antioxidant studies of bioinspired ZrO_2 nanoparticles using agriculture waste durva grass aqueous extracts *J. Hazard. Mater. Adv.* **7** 100112

[19] Rai A, Sharma V K, Jain A, Sharma M, Pandey A, Singh H B, Gupta V K and Singh B N 2022 Microbe-fabricated nanoparticles as potent biomaterials for efficient food preservation *Int. J. Food Microbiol.* **379** 109833

[20] Al-hakimi A *et al* 2022 Enhanced structural, optical, electrical properties and antibacterial activity of selenium nanoparticles loaded PVA/CMC blend for electrochemical batteries and food packaging applications *Polym. Test.* **116** 107794

[21] Malaguti Figueiredo C C *et al* 2023 Biosynthesis of silver nanoparticles of *Tribulus terrestris* food supplement and evaluated antioxidant activity and collagenase, elastase and tyrosinase enzyme inhibition: *in vitro* and *in silico* approaches *Food Bioprod. Process.* **138** 150–61

[22] Lin L, Mao X, Sun Y, Rajivgandhi G and Cui H 2019 Antibacterial properties of nanofibers containing chrysanthemum essential oil and their application as beef packaging *Int. J. Food Microbiol.* **292** 21–30

[23] Rajivgandhi G, Gnanamangai B M, Ramachandran G, Chackaravarthy G, Chelliah C K, Maruthupandy M, Alharbi N S, Kadaikunnan S and Li W-J 2022 Effective removal of heavy metals in industrial wastewater with novel bioactive catalyst enabling hybrid approach *Environ. Res.* **204** 112337

[24] Khaled J M, Alyahya S A, Govindan R, Chelliah C K, Maruthupandy M, Alharbi N S, Kadaikunnan S, Issac R, Murugan S and Li W-J 2022 Laccase producing bacteria influenced the high decolorization of textile azo dyes with advanced study *Environ. Res.* **207** 112211

[25] Mohanraj R, Gnanamangai B M, Rajivgandhi G N, Li W-J, Vijayalakshmi G, Ponmurugan P, Alharbi N S, Kadaikunnan S, Khaled J M and Alanzi K F 2021 Monitoring the decolourisation efficacy of advanced membrane fabricated phytosilica nanoparticles in textile effluent water treatment *Chemosphere* **273** 129681

[26] Rajivgandhi G *et al* 2022 Metal tolerance and biosorption of Pb ions by *Bacillus cereus* RMN 1 (MK521259) isolated from metal contaminated sites *Chemosphere* **308** 136270

[27] Ramachandran G, Chackaravarthi G, Rajivgandhi G N, Quero F, Maruthupandy M, Alharbi N S, Kadaikunnan S, Khaled J M and Li W-J 2022 Biosorption and adsorption isotherm of chromium (VI) ions in aqueous solution using soil bacteria *Bacillus amyloliquefaciens Environ. Res.* **212** 113310

[28] Liu B-B *et al* 2022 Halophilic archaea and their extracellular polymeric compounds in the treatment of high salt wastewater containing phenol *Chemosphere* **294** 133732

[29] Hassan S S, Carlson K, Mohanty S K, Sirajuddin and Canlier A 2018 Ultra-rapid catalytic degradation of 4-nitrophenol with ionic liquid recoverable and reusable ibuprofen derived silver nanoparticles *Environ. Pollut.* **237** 731–9

[30] Pandey S, Do J Y, Kim J and Kang M 2020 Fast and highly efficient catalytic degradation of dyes using κ-carrageenan stabilized silver nanoparticles nanocatalyst *Carbohydr. Polym.* **230** 115597

[31] Hameed R S A, Obeidat S, Qureshi M T, Al-Mhyawi S R, Aljuhani E H and Abdallah M 2022 Silver nanoparticles—expired medicinal drugs waste accumulated at hail city for the local manufacturing of green corrosion inhibitor system for steel in acidic environment *J. Mater. Res. Technol.* **21** 2743–56

[32] Albeladi S S R, Malik M A and Al-thabaiti S A 2020 Facile biofabrication of silver nanoparticles using *Salvia officinalis* leaf extract and its catalytic activity towards congo red dye degradation *J. Mater. Res. Technol.* **9** 10031–44

[33] Zhang Z, Zhang N, Liu Y, Fang Q, Xi J, Xiao Y, Zhou P and Xu L 2022 Efficient degradation of organic dyes and reduced Cr(VI) in environmental water purification by *in-situ* deposition of silver nanoparticles on polydopamine-modified M-ATP/PCN *Catal. Commun.* **172** 106528

[34] Abo-Dief H M, El-Bahy S M, Hussein O K, El-Bahy Z M, Shahid M and Shakir I 2022 Synthesis and characterization of rGO supported silver doped bimetallic $ZnCo_2O_4$ spinel oxides for enhanced photocatalytic degradation of industrial effluents *J. Alloys Compd.* **913** 165164

[35] Kumari P and Samadder S R 2022 Valorization of carbonaceous waste into graphene materials and their potential application in water and wastewater treatment: a review *Mater. Today Chem.* **26** 101192

[36] Bonnia N N, Kamaruddin M S, Nawawi M H, Ratim S, Azlina H N and Ali E S 2016 Green biosynthesis of silver nanoparticles using 'Polygonum *Hydropiper*' and study its catalytic degradation of methylene blue *Procedia Chem.* **19** 594–602

[37] Nasiri R, Zarei M, Arsalani N, Pezhhanfar S, Someh A A and Panahian Y 2023 One-pot synthesis of novel 3D graphene/Fe_3O_4/agro-based waste material (*Sesamum indicum*) nanocomposite for wastewater treatment and artificial neural network modeling *Chem. Eng. Res. Des.* **190** 451–63

[38] Chandra R and Nath M 2020 Controlled synthesis of AgNPs@ZIF-8 composite: efficient heterogeneous photocatalyst for degradation of methylene blue and congo red *J. Water Process Eng.* **36** 101266

[39] Metwalli O I, Eisa W, Anis B, Salem A N M, Khalil A S G and Khalil M M H 2022 High performance Ni nanoparticles@graphene oxide-crosslinked poly(vinyl alcohol) free-standing film for catalytic applications: reductive and oxidative reactions *J. Mater. Res. Technol.* **20** 1142–56

[40] Liu P, Xing Y, Liu Y, Si Z, Xie L, Weng D and Kang F 2020 Facile method of synthesizing multilayer graphene capsuled sulfur nanoparticles for water treatment *Appl. Surf. Sci.* **502** 144194

[41] Bhuyan B, Paul B, Purkayastha D D, Dhar S S and Behera S 2016 Facile synthesis and characterization of zinc oxide nanoparticles and studies of their catalytic activity towards ultrasound-assisted degradation of metronidazole *Mater. Lett.* **168** 158–62

[42] Fakhri L A, Ghanbarzadeh B, Dehghannya J, Hosseini M and Dadashi S 2021 Photocatalytic and biotic degradation of polystyrene packaging film: effect of zinc oxide photocatalyst nanoparticles and nanoclay *Chemosphere* **283** 130972

[43] Liu C, Usman M, Ji M, Sha J, Zhou L and Yan B 2023 Response mechanisms of anaerobic fermentative sludge to zinc oxide nanoparticles during medium-chain carboxylates production from waste activated sludge *Chemosphere* **317** 137879

[44] Zhang Y-T, Wei W and Ni B-J 2021 Revealing the mechanism of zinc oxide nanoparticles facilitating hydrogen production in alkaline anaerobic fermentation of waste activated sludge *J. Clean. Prod.* **328** 129580

[45] Guo H, Yang H, Huang J, Tong J, Liu X, Wang Y, Qioa W and Han J 2022 Theoretical and experimental insight into plasma-catalytic degradation of aqueous p-nitrophenol with graphene-ZnO nanoparticles *Sep. Purif. Technol.* **295** 121362

[46] Zhu Q, Song J, Liu Z, Wu K, Li X, Chen Z and Pang H 2022 Photothermal catalytic degradation of textile dyes by laccase immobilized on Fe_3O_4@SiO_2 nanoparticles *J. Colloid Interface Sci.* **623** 992–1001

[47] Punnoose M S, Bijimol D and Mathew B 2021 Microwave assisted green synthesis of gold nanoparticles for catalytic degradation of environmental pollutants *Environ. Nanotechnol., Monitor. Manag.* **16** 100525

[48] Shi S, Wang Y, Ahmed I and Zhang B 2022 Catalytic degradation of refractory phenol sulfonic acid by facile, calcination-free cobalt ferrite nanoparticles *J. Environ. Chem. Eng.* **10** 107616

[49] Ahmad A, Khan M, Khan S, Luque R, Abualnaja K M, Alduaij O K and Yousef T A 2022 Bio-construction of CuO nanoparticles using Texas sage plant extract for catalytical degradation of methylene blue via photocatalysis *J. Mol. Struct.* **1256** 132522

[50] Sarkar B J, Kundu M, Mondal B, Mukherjee S, Bandyopadhyay A and Roy U K 2023 Microstructural investigation of sonochemically synthesized Zn substituted $CuFe_2O_4$ nanoparticles for heterogeneous green catalytic click chemistry and dye degradation *J. Mol. Struct.* **1274** 134493

[51] Riaz M, Khan N, Khan S A, Saeeduddin, Ahmad Z, Khan M A, Iqbal M, Hemeg H A, Bakhsh E M and Khan S B 2022 Enhanced catalytic reduction/degradation of organic pollutants and antimicrobial activity with metallic nanoparticles immobilized on copolymer modified with NaY zeolite films *J. Mol. Liq.* **359** 119246

[52] Berry V 2013 Impermeability of graphene and its applications *Carbon* **62** 1–10

[53] Cohen-Tanugi D *et al* 2014 Quantifying the potential of ultra-permeable membranes for water desalination *Energy Environ. Sci.* **7** 1134–41

[54] Zhang J, Lu M, Li W, Xue H, Tian J and Jiang T 2023 0D/2D $CuFe_2O_4$/MXene Z-scheme heterojunction for improved photocatalytic selective oxidation reaction *Appl. Surf. Sci.* A **611** 155670

[55] Fowler J D *et al* 2009 Practical chemical sensors from chemically derived graphene *ACS Nano* **3** 301–6

[56] Zhang S, Cai M, Wu J, Wang Z, Lu X, Li K, Lee J-M and Min Y 2023 Photocatalytic degradation of TiO_2 via incorporating Ti_3C_2 MXene for methylene blue removal from water *Catal. Commun.* **174** 106594

[57] Pang X, Xue S, Zhou T, Xu Q and Lei W 2022 2D/2D nanohybrid of Ti_3C_2 MXene/WO_3 photocatalytic membranes for efficient water purification *Ceram. Int.* **48** 3659–68

[58] Tian Y, Wang C, Zhang Q, Cui H, Liu Y, Wang K, Wei J, Long L and Qian J 2023 Tailored Ti_3C_2 MXene/SnS_2 nanocomposites to realize both sensitive photoelectrochemical determination and efficient photocatalytic detoxification of Cr(VI) *Sensors Actuators* B **382** 133496

[59] Hontoria-Lucas C *et al* 1995 Study of oxygen-containing groups in a series of graphite oxides: physical and chemical characterization *Carbon* **33** 1585–92

[60] Marcano D C *et al* 2010 Improved synthesis of graphene oxide *ACS Nano* **4** 4806–14

[61] Sitko R, Zawisza B and Malicka E 2013 Graphene as a new sorbent in analytical chemistry *TrAC Trends Anal. Chem.* **51** 33–43

[62] Zhao J *et al* 2014 Graphene in the aquatic environment: adsorption, dispersion, toxicity and transformation *Environ. Sci. Technol.* **48** 9995–10009

[63] Kanani M *et al* 2022 Removal of Rhodamine 6G dye using one-pot synthesis of magnetic manganese graphene oxide: optimization by response surface methodology *Environ. Nanotechnol., Monitor. Manag.* **18** 100709

[64] Nagarajan L, Arulraj A, Senguttuvan G and Kumaraguru K 2020 Solar assisted reduced graphene oxide as adsorbent for carbon dioxide and its kinetic studies *Physica E: Low-dimensional systems and Nanostructures* **116** 113739

[65] Gautam M and Jayatissa A H 2011 Gas sensing properties of graphene synthesized by chemical vapor deposition *Mater. Sci. Eng.* C **31** 1405–11

[66] Zhou W, Yu B, Zhu J, Li K and Tian S 2022 Hierarchical ZnO/MXene (Nb_2C and V_2C) heterostructure with efficient electron transfer for enhanced photocatalytic activity *Appl. Surf. Sci.* **590** 153095

[67] Huang H, Zhang J, Tang C, Li A, Zhang T, Xue H and Zhang D 2022 Efficient visible-light-photocatalytic sterilization by novel Z-scheme MXene/TiO_2/Bi_2S_3 *J. Environ. Chem. Eng.* **10** 108654

[68] Li H, Sun B, Gao T, Li H, Ren Y and Zhou G 2022 Ti_3C_2 MXene co-catalyst assembled with mesoporous TiO_2 for boosting photocatalytic activity of methyl orange degradation and hydrogen production *Chin. J. Catal.* **43** 461–71

[69] Park S, Kim S, Yea Y, Saravanakumar K, Lee E, Yoon Y and Park C M 2023 Adsorptive and photocatalytic performance of cobalt-doped $ZnTiO_3/Ti_3C_2T_x$ MXene nanohybrids towards tetracycline: kinetics and mechanistic insight *J. Hazard. Mater.* **443** 130165

[70] Solangi N H, Karri R R, Mubarak N M, Mazari S A, Jatoi A S and Koduru J R 2023 Emerging 2D MXene-based adsorbents for hazardous pollutants removal *Desalination* **549** 116314

[71] Imsong R and Purkayastha D D 2023 Dual-functional superhydrophilic/underwater super-oleophobic 2D $Ti_3C_2T_X$ MXene-PAN membrane for efficient oil–water separation and adsorption of organic dyes in wastewater *Sep. Purif. Technol.* **306** 122636

[72] Hojjati-Najafabadi A, Mansoorianfar M, Liang T, Shahin K, Wen Y, Bahrami A, Karaman K, Zare N, Karimi-Maleh H and Vasseghian Y 2022 Magnetic-MXene-based nanocomposites for water and wastewater treatment: a review *J. Water Process Eng.* **47** 102696

[73] Arulraj A, Mangalaraja R V and Khalid M 2022 MXene: Pioneering 2D materials *Fundamental Aspects and Perspectives of MXenes* ed M Khalid, A N Grace, A Arulraj and A Numan (Frankfrut: Springer) pp 1–16

IOP Publishing

Materials Technology for the Energy and Environmental Nexus, Volume 2

R V Mangalaraja, Arunachalam Arulraj, Radhamanohar Aepuru and Chan Siew Hwa

Chapter 10

Recent advancements in the use of nanomaterials in electrochemical sensors for the detection of pesticides and organic pollutants

Padmanaban Annamalai, Manikandan Ramalingam, Muthusankar Eswaran, Bavatharani Chokkiah, Prabhu Saravanan, Seung-Cheol Chang, Hector Valdes and R V Mangalaraja

Pesticides are an effective way to increase crop yields and assure food security. Human health around the world is under threat due to food pollutants and discarded pharmaceutical intermediate compounds. Food pesticides and pharmaceutical contaminants in the environment and food pose several unresolved problems. For the detection of chemical and toxic pollutants, it is important to design sensitive, specific, and economical sensors. The use of electrochemical sensors is a reliable way to do research on the monitoring of environmental contamination. Ideally, the operating costs are low, the response is quick, and the sensor has an extremely high level of sensitivity and selectivity. Electrochemical sensors based on nanocomposites have an enhanced surface area, increased catalytic activity, and superior stability. Electrochemical sensors based on nanomaterials are becoming increasingly popular for applications in the testing of potentially hazardous pesticides in the environment. Researchers have concentrated their efforts primarily on metal chalcogenides, nanomaterials based on carbon and non-carbon, and the fabrication of efficient electrodes for electrochemical sensors to detect pesticides and organic pollutants in real sample analysis. This review places emphasis on the metal and non-metallic nanomaterial electrochemical sensors that were created for applications related to the environment. In addition, an review will be conducted into the advancements that have been made in this area.

doi:10.1088/978-0-7503-5729-6ch10

10.1 Introduction

Organic pollutants make up a significant portion of water toxins. Their existence in environmental water is harmful to marine creatures and also to human life on Earth via potable water. Organic dyes, drugs, mycotoxins, pesticides, and nitroaromatics are all examples of organic pollutants. Organic dyes are a diverse group of complex aromatic mixtures employed in textiles, paints, leather, plastics, and papers. Pesticides are utilized in agriculture to protect crops from weeds and pests [1, 2]. Organic dyes and pesticides have drawn significant global interest as environmental contaminants since they are enormously toxic and nonbiodegradable, and are capable of transforming into carcinogenic, even mutagenic, and teratogenic agents. Antibiotics are usually employed to treat bacterial infections in animals and can also have negative effects in the environment.

In recent years, the increase in global concern about environmental contamination and its worldwide health effects instigated by poisonous and toxic pollutants is related to rapid urbanization, heavy industrialization, etc. Furthermore, contaminated potable water has become a primary worldwide issue for nearly half of the world's population [3]. People are continuously endangered by numerous adverse health issues due to the harmfulness of these contaminants, chiefly by consumption of polluted potable water and products, and breathing in dirty air [4]. Toxic pollution can cause a variety of health problems, including physical and mental dysfunctions, neurological and organ disorders, a weakened immune system, decreased life expectancy, and cancers, as well as leading to death in many cases [5, 6]. The aforementioned concerns have highlighted the importance of the emerging reliable, highly selective and sensitive, easy, and affordable tools for measuring contaminants even at low concentration ranges. To detect environmentally hazardous pollutants, existing analytical approaches such as capillary electrophoresis [7], spectrofluorimetry [8], high-performance liquid chromatography (HPLC) [9], gas chromatography/mass spectrometry (GC/MS) [10], flow injection chemiluminescence [11], and atomic absorption spectroscopy (AAS) [12] have been employed. Such analytical processes have been laborious, highly expensive, require advanced assistance to function, and are difficult to organize owing to the large size of the instrumentation or devices.

On the other hand, electrochemical sensors, which employ an electrode as the transduction element, are an essential subclass of chemical sensors that are well-matched for meeting the cost, portability, power, and miniaturized size requirements for remote environmental monitoring. Improved sensitivity and selectivity, a broad concentration range, power requirements, negligible space, and economical instrumentation are all features of electrochemical sensing systems. These electrochemical sensing devices have initiated several applications in medical/healthcare [13], environmental [14, 15], agricultural [16], and industrial analyses [17]. For decades, electrochemical devices have been utilized in the field to screen various parameters to check the quality of water continuously (e.g. pH, dissolved O_2, conductivity, etc). A large range of environmental uses has resulted, such as the trace metals measurement in natural water samples, carcinogenic investigations (e.g. aromatic amines or

N-nitroso compounds), the fabrication and development of biosensors for organic pollutant detection (e.g. phenols, pesticides, etc) in water samples, as well in clean energy conversion [18] and environmental protection. Electrochemical sensors that can be remote and underwater have been developed in response to the growing number of security issues. These sensors are used to monitor chemical agents and explosives. These kinds of technologies allow direct and dependable monitoring by sending back analytical data in a manner that is not only timely but also secure and budgeted (including evaluating the target analytes' gradient and destiny).

Based on the various effective demonstrated research on the multiple applications of conducting polymer nanocomposites, including the fabrication and development of electrochemical energy conversion [19, 20] and storage devices [21], electrochemical healthcare devices [22], electrochemical environmental monitoring [23], etc. This chapter details the electrochemical detection of pesticides, nitroaromatics, phenolic compounds, and organic dyes using conducting polymers and advanced nanomaterials as nanocomposites/nanohybrids.

10.2 Detection of organic pollutants using electrochemical sensors

10.2.1 Organic pollutants: detection of pesticides using polymer nanocomposites

The organophosphate derivatives of pesticides, insecticides, antifungals, and weedkillers are widely utilized in farming contexts around the globe to manage insects in agricultural crops and increase agricultural yield, as evidenced by relevant research. However, as discussed in the preceding section, widespread use poses substantial hazards to protecting the well-being of humans and the natural world. Here, the detection of organophosphates pesticides derivatives such as imidacloprid (IMI), fenitrothion (FNT), methyl parathion (MP), paraquat (PQ), and simazine (SMZ) using versatile electrochemical techniques is discussed.

10.2.1.1 Organophosphates

The organophosphate compounds have a side chain of phosphate esters that is reactive to chemical reactions made up of core phosphorous molecules that are one-to-one bound to two aryl or alkyl groups, for instance, as two methoxy ($-OCH_3$) or ethoxy ($-OCH_3CH_3$) categories, and double links to atoms of oxygen or sulfur can be present. Pesticides based on organophosphates have been utilized widely as pest management chemicals in agricultural crops during the past few decades, yet they continue to pose a significant health risk to people. Fruit and vegetable agricultural goods with organophosphate residues need to be assessed and measured rapidly to establish a problem with consumer health and safety. Although heavily utilized in agricultural growth, organophosphates pose severe environmental and human health risks. One of the most important organophosphate pesticides, fenitrothion (dimethoxy-(3-methyl-4-nitrophenoxy)thioxophosphorane), is used to control pests on plants and fruits. Chemicals with an organophosphate base have carcinogenic effects on humans and other mammals. Furthermore, poisoning from pesticides and organochlorines affects nearly 3 million people and results in 20 000 fatalities annually. As a result, the US Environmental Protection Agency (EPA) has restricted the application of organophosphates in agriculture. The development of straightforward on-site technologies for detecting nerve drugs and toxins is a significant problem in protecting the environment and people due to their toxicity. Therefore, a quick and effective method for better organophosphate detection in fruits and vegetable samples is essential. Additionally, the identification of acetylcholinesterase inhibitor pesticides has been shown utilizing various analyses, including electrochemical sensors, high-performance liquid chromatography, mass-selective detection, photothermal biosensors, flame photometry, colourimetry, and photoluminescence detectors. The processes mentioned above are often carried out in sophisticated laboratories and take a while to obtain findings. More importantly, electrochemically based analytical methods provide a straightforward approach to obtaining extremely sensitive and focused data. Researchers are currently concentrating their efforts on the application of electrochemical sensing platforms based on organophosphate in several ways. Janjani *et al* presented a unique method for fabricating an electrochemical measuring device on a screen-printed electrode (SPE) for the direct measurement of organophosphate insecticide chlorpyrifos enhanced by Mn^{2+}/Fe^{3+} metal–organic frameworks (MnFe-MOFs) [24]. The electroanalysis was conducted using a square wave voltammetry (SWV) technique, and the limit of detection (LOD) was kept at 0.29 ppb or 0.85 nM, throughout a broad concentration range of 1.0×10^{-7} to 1.0×10^{-9} M. The detectors displayed outstanding reproducibility and interference-resistant abilities. Hetero-metal synergy was developed among the Mn^{2+}/Fe^{3+} ions; the MnFe-MOF showed strong sorbent and electrocatalytic characteristics to chlorpyrifos, which led to its use for non-enzymatic electro-identification of chlorpyrifos as the electrode material. PEDOT:PSS enhanced electrodes have been made via one-step electropolymerization and used as electrochemical sensors for the measurement of organophosphorus and nitrophenols chemicals [25], as shown by Hryniewicz *et al.* By using spectroscopic and electrocatalytic processes, the electrochemical technique performance of PEDOT towards the redox processes of nitrophenol isomers was investigated. The standard curve showed high linearity in the 0–80 μmol l^{-1} range, and the isomers might be distinguished by linear sweep voltammetry (LSV) utilizing the intermediates' oxidation peak. In a composite sample, the sensitivities for 4-NP, 3-NP, and 2-NP were determined to be 0.1292, 0.1902, and 0.1921 AL^{-1} mol^{-1}, respectively. With advantages linked to improved stability, selectivity, and sensitivity for the analysis of organophosphate pesticides, Wei *et al* produced a new type of low-cost polished α-Al_2O_3 nanomaterial electrode surface [26]. It has been established that polished α-Al_2O_3 nanomaterials are effective sorbents with a great affinity for the phosphoric group in organophosphorus insecticides. In laboratory environments, a 0.035 ng ml^{-1} detection limit is observed.

In their research, Gopi *et al* concentrated on creating a catalyst called bismuth vanadate that can detect the occurrence of toxic organophosphorus substances [27]. For electrochemical biosensors, the suggested electrocatalyst was effectively used as the working electrode sample to modify a screen-printed carbon electrode (SPCE). According to the research results, the electrode modified with bismuth vanadate showed better performance than a control electrode for the electrochemical sensing of ethyl paraoxon, with a sensitivity of 0.345 $\mu A\ \mu M^{-1}\ cm^2$ and an LOD of 0.03 μM and superior accuracy, stability, reproducibility, and consistency. A solid mounted resonator (SMR) adapted for self-assembled single layer (SAM) organophosphorus vapor sensing was designed by Wang *et al* [28]. The electrode surface was successfully fabricated to detect organophosphate ions using a Cu^{2+}/11-mercaptoundecanoic acid monolayer that assembles itself. The resonance frequency decreased correspondingly when the mass of the self-assembled monolayer evolved due to the adsorption of organophosphate pollutants. The findings of the studies demonstrate that the modified solid mounted resonator can provide a quick, reversible, sensitive, and repeatable reaction to the neurotoxin dimethyl methyl phosphonate vapor. A 3D absorbent bimetallic alloy framework with ultrathin nanowires (NWs) (PdNi, PdCo, and PdCu) and a MoS_2 nanosheet was designed by Song *et al.* The prepared bimetallic materials are used for electrochemical biosensors with the selective and sensitive determination of organophosphate compounds [29]. PdNi NWs, in particular, performed better than the other two bimetallic alloys. For the electroanalytical study of omethoate, the developed biosensor based on PdNi/m-MoS_2 exhibits excellent presentation under ideal circumstances, including a broad concentration range (10^{-13}–10^{-7} M), an LOD of 0.05 pM, sensitivity, and enduring rigidity. The quick field monitoring technologies developed by Wu *et al* are very suitable for the sensitive and selective determination of organophosphate insecticide traces in aquatic and farm products [30]. The Er-GRO-Nafion nanocomposite's conductive and 2D layered network makes it an effective material for electrode modification. Acetylcholinesterase (AChE) associated with the biodegradable substance is adsorbed with, dichlorvos utilizing the modified electrode, electrochemical detection was made. The suggested electrochemical biosensor demonstrated a detection limit of 2.0 ng ml^{-1}, heightened response, selectivity, excellent consistency, and a concentration range from 1.0–20 ng ml^{-1} and 5.0–100 ng ml^{-1}. This was accomplished by carefully influencing the variables for electrochemical reduction and adjusting the electrode fabrication criteria. The silver nanoparticle-N-F-MoS_2 nanosheet has excellent conductivity, electron movement, and a sizeable electrocatalytic surface, as reported by Song *et al* [31]. The generated silver nanoparticle-N-F-MoS_2 was used as a base for detecting enzymatic inhibition and organophosphate pesticide detection. The limit of detection was 1 pg ml^{-1} (3 pM) and 0.05 pg ml^{-1} (0.2 pM) and the ($CNTs$-NH_2)/Ag NPs-N-F-Mo had excellent repeatability, extended storage, and selectivity. Mukherjee *et al* established a device to detect organophosphate pesticides from vegetables with green leaves and fruits [32]. The constructed biosensor demonstrated a concentration range of 0.1–100 ng ml^{-1} insecticide made of organophosphate with a slope of 235.678 ng ml^{-1}. Ravi *et al* designed an uncomplicated MnO_2-based sensor to develop a non-toxic, specific, economical, and sensitive sensor to monitor nitro-group-containing pesticides in vegetable washing water [33]. To monitor 4-nitrophenyl phosphate (4-NPP), MnO_2 nanorods have been fabricated successfully on a platinum disk electrode (MnO_2/Pt). With a detection limit of 10 nM and a fast response of 11.68 $\mu A\ \mu M^{-1}$, the proposed detector showed a linear relationship from 100 to 900 nM. The sensor demonstrated strong selectivity against several inorganic ions, such as dimethoate and quinalphos, two of the more important organophosphates. However, the selectivity is minimal for other aromatic compounds containing the nitro-group. Zhao *et al* suggested using 3D holey Co-based bimetallic oxides constructed randomly using porous ultrathin nanosheets (NSs) [34]. The 3DH-CoCu-O /GCE, GDY/3DH-CoNi-O/GCE, 3DH-Co-O/GCE, 3DH-CoNi-O/GCE, and 3DH-CoFe-O/GCE electrodes were used for monitoring dipterex and omethoate with a electrochemical acetylcholinesterase (AChE) biosensor. The biosensor exhibits exceptional sensing capability, a broad linear range of 4.69×10^{-13}–4.69×10^{-7} M, and a detection limit of 0.033 pM for omethoate, and a linear range of 3.88×10^{-12}–3.88×10^{-6} M and 0.36 pM for diptetex. Using multilayer films containing an AChE liposome bioreactor and chitosan, Guan *et al* produced a novel acetylcholinesterase (AChE) biosensor [35]. The electrochemical sensor displays better AChE activity with an LOD of 0.86 ± 0.098 g l^{-1} and a dichlorvos concentration range of 0.25–1.50 and from 1.75–10.00 μM. Zhang *et al* successfully synthesized a Ag-rGO-NH_2 sample with a responsive amperometric AChE biosensor [36]. Initially, the surface of the electrode was electropolymerised using 4-di (furan-2-yl) benzo thiadiazole (FBThF). The interaction between Ag-rGO-NH_2 and poly(FBThF) created a wet surface for the binding of AChE, and the biosensor displayed good catalytic activity and better electronic conductivity. The malathion analyte has a regression coefficient of 0.9947 with a linear range of 0.099–9.9 μg l^{-1}. Trichlorfon has a regression rate of 0.9969 through a concentration range of 0.020–2.06 μg l^{-1}. The calculated LODs for malathion and trichlorfon are found to be 0.032 and 0.001 μg l^{-1}. Additionally, the biosensor demonstrated long-term consistency and repeatability for organophosphate pesticides. Novel acetylcholinesterase biosensors were constructed by Cancar *et al* [37], who also demonstrated that conjugated polymers make a suitable base for sensing organophosphate (OP) pesticides. The detection of OPs in 0.1 M KCl aqueous background solution necessitated the development of another device on the ZrO_2/AuNPs electrode. By employing SWV, the sensor demonstrated [38] a very low LOD of 1 ng ml^{-1}. To detect and monitor exposed OP pesticides, a super sensitive immune-tagged MWCNTs-Au nanocomposite coated with SPCE was immobilized with AChE [39]. Furthermore, a new technique for sensing applications was developed by employing an electrode modified with coomassie brilliant blue dye surface-confined carbon nano black (GCE/CBnano@CoomBB). Strong connections between the CoomBB and CBnano instigated the composite to have an agglomerated particle-like assembly. An LOD of 20 μL was also described for the effective electrochemical activity [40]. Recent research based on the electrochemical sensing of organophosphate derivates is shown in table 10.1.

10.2.1.2 Imidacloprid (IMI)

Lei *et al* developed imidacloprid (IMI) electrochemical sensing by employing poly (carbazole)/chemically rGO customized glassy carbon electrodes (GCEs) [56]. They created PCz/chemically rGO (CrGO) and polycarbazole (PCz) customized GCEs by means of a simple procedure and used them in customized electrodes for electrochemical examination of IMI insecticides. Furthermore, differential pulse voltammetry (DPV) and cyclic voltammetry (CV) were employed as the detection

Table 10.1. Comparison of organophosphate derivative detection.

Modified electrode	Analyte	Sensing technique	Linear range	LOD	References
(a) HPH-N-Cu NCs	Methamidophos	DPV	1.00×10^{-9}–1.00×10^{-3} g l^{-1}	2.58×10^{-11} g l^{-1}	[41]
(b) Au-His-GQD-G	Chlorpyrifos	DPV	1×10^{-16}–1.0×10^{-11} M	3.8×10^{-17}M	[42]
(c) AChE/Ce/UiO-66@ MWCNTs/GCE	Paraoxon	CA	0.01–150 nM	0.004 nM	[43]
(d) GC/NaCl•15-crown-5 ether	Profenofos	DPV	1.025–153.645 nM	0.341 nM	[44]
(e) AuNPs/neutral red-bio-functionalized graphene	Methyl parathion	DPV	0.02–0.153 μM and 0.153–1.36 μM	6 nM	[45]
(f) Immunoassay	Glyphosate and chlorpyrifos	EIS	0.3–243 ng ml^{-1} and 1–243 ng ml^{-1}	1 ng ml^{-1} and 1 ng ml^{-1}	[46]
(g) CS/PPy	Acetamiprid	CC CA	0.347 pg ml^{-1} and 0.065 fg ml^{-1}	1.0 pg ml^{-1} –0.1 mg ml^{-1} and 1.0 fg ml^{-1}–0.1 ng ml^{-1}	[47]
(h) AChE-Chit/MXene/ AuNPs/MnO_2/Mn_3O_4/GCE	Methamidophos	DPV	10^{-12}–10^{-6} M	1.34×10^{-13} M	[48]
(i) CCM/UiO-66/GCE	Methyl parathion	DPV	20–20 000 ng ml^{-1}	0.98 ng ml^{-1}	[49]

(*Continued*)

Table 10.1. (*Continued*)

Modified electrode	Analyte	Sensing technique	Linear range	LOD	References
(j) TiO_2@DA@S/H/E	Ethyl paraoxon, methyl parathion, methyl paraoxon	SWV	0.5–100 μM	0.2 μM	[50]
(k) LaPO4@Ag	Phoxim	DPV	0.10–600.0 μM	0.051 μM	[51]
(l) AChE@CHIT/TiO_2/MoS_2/SPE	Chlorpyrifos	DPV	10 nM–50 pM	50 pM	[52]
(m) MnO_2-GNPs/SPCE	Carbosulfan (CBS) fenobucarb (FNB), and carbaryl (CBR),	DPV	1–40 μM, 5–150 μM, and 50–600 μM	14.90 μM, 1.30 μM, and 0.30 μM	[53]
(n) HSA-Pd-CdTe/GCE	Chlorpyrifos (CPS)	EIS	0.5 pM–500 nM	0.16 pM	[54]
(o) AChE/GLD-CS/CdCoS2(2)@Ag/ITO	Chlorpyrifos	PCE	0.001–270 μg ml^{-1}	0.57 ng ml^{-1}	[55]

(a) Hierarchical porous hollow N-doped Cu-based nanocages. (b) Gold-histidine-functionalized graphene quantum dots–graphene. (c) AChE/cerium/UiO-66-template@multi-walled carbon nanotubes (MWCNTs). (d) Glassy carbon /NaCl•15-crown-5 ether. (g) Chondroitin sulfate/polypyrrole. (i) Metal-organic framework UiO-66(Zr) loaded curcumin (CCM). (j) TiO_2@dopamine@serine/histamine/glutamic acid. (m) Manganese dioxide nanoparticles–graphene nanoplatelets/screen-printed carbon electrode. (n) Human serum albumin-Pd-CdTe/GCE. (o) Acetylcholinesterase-glutaraldehyde-chitosan/$CdCoS_2$(2)@Ag/ITO.

measures. The two customized electrodes had satisfactory electrocatalytic performance for reducing IMI based on the investigations. Moreover, the existence of CrGO improved the customized GCE's signals electrochemically, enhancing the sensor's responsiveness. Moreover, the LOD and limit of quantitation (LOQ) values for CV were 0.22 and 0.74 μM, respectively, in contrast they were 1.52 and 0.44 M for DPV. Due to its high energy levels, a functionalised MWCNT/EDTA electrode was used for sensing IMI [57], as reported by Suresh *et al.* The sensor demonstrated increased sensitivity of 10.70 μA nM^{-1}, along with satisfactory interference resistance of 6.11% over a broad concentration range (0.04 nM to 0.001 mM). The prepared electrocatalytic detector exhibited distinctive electrochemical activity when there was water present in matrices collected from different sample locations. Based on a functionalized MWCNT/EDTA electrode composite, the sensor had an LOD of 3.1×10^{-3} pM and substantial concentration ranges of 0.001–0.004 mM, 0.001–0.04 μM, and 0.001–0.05 nM. In real time, the electrochemical sensors based on composites of MWCNT/EDTA showed good electrocatalytic activity towards IMI residues in environmental contaminants. Using electrochemical techniques, Ajermoun *et al* evaluated the phytotoxic effect of the IMI on the plant *Phaseolus vulgaris* L. Increased imidacloprid concentrations were applied to ordinary bean plants, and numerous analyses were performed on various plant tissues [58]. After the extraction of imidacloprid from the bean plant tissues, SWV was employed to find imidacloprid using a metallic silver electrode. In the bean plant, the impact of various imidacloprid doses (starting from 5.0×10^{-3}–5.0×10^{-2} mol l^{-1}) was assessed on the leaves, seeds, stems, and roots. According to Li *et al*, annealed carbon cloth (ACC) exhibits improved electrocatalytic activity for imidacloprid monitoring [59]. Higher imperfections, a hydrophilic surface, functional groups, an ion-accessible surface area, and rougher surfaces are all characteristics of ACC. Additionally, with a high sensitivity of 35.58 μA mM^{-1} cm^2, an LOD of 0.04 μM, and a concentration range of 5–100 μM, the ACC electrode exhibits improved electrocatalytic reduction activity toward imidacloprid. In addition, this sensor can detect imidacloprid in apples and grapes with a 96.8% to 104.1% recovery rate. Silver decorated on mesoporous carbon and derived from natural source iron ore were used to design a ternary nanocomposite (Ag@meso-C/hematite ore) in an innovative electrochemical sensor by Rashed *et al* [60]. The catalytic performance showed a $4e^-$ and $4H^+$ electroreduction process of irreversible diffusion-controlled kinetics. The amperometric technique was used to demonstrate an imidacloprid concentration range of 10.80–195.50 μM, an LOD of 0.257 μM, and a sensitivity of 0.8113 μA μM^{-1} cm^{-2}. Furthermore, the LSV method shows an LOD of 1.06 μM, a sensitivity of 0.1955 μA μM^{-1} cm^{-2}, and a wide linear range of 63–870 μM for the detection of imidacloprid. An excellent response was shown by the silver@meso-C/hematite ore sensing electrode in terms of reproducibility, repeatability, storage stability, sensitivity, and selectivity. To detect imidacloprid residues found in guava and thyme leaves, Elbaz *et al* created aluminum silicate and chitosan nanoparticles coated on the carbon paste electrode (CsNP/ALS/MCPE) [61]. When employing square wave voltammograms for the sensing of IMD at pH 8.0 and a scan rate of 100 mV s^{-1}, the (CsNP/ALS/MCPE) electrode was found to have a correlation coefficient of 0.999, a linear range of 2×10^{-8}–1×10^{-4} mol l^{-1}, a quantification limit of 4.5×10^{-9} mol l^{-1}, and an LOD of 1.5×10^{-9} mol l^{-1}. The linearity range, repeatability, sensitivity, and selectivity for the amount of imidacloprid residues found in thyme and guava leaves was measured and was improved by the CsNP/ALS/MCPE electrodes.

10.2.1.3 Fenitrothion (FNT)

Electrochemical research into FNT was described by Surucu *et al* using rGO/poly (E)-1-(4-((4-(phenylamino)-phenyl) diazenyl) phenyl) ethanone (DPA) on a pencil graphite electrode (PGE) film [62]. Analyses revealed that the customized PGE performed admirably in detecting FNT. The rGO/DPA/PGE SWV current response to FNT changed linearly with concentration variations ranging from 0.096 to 1.912 μmol l^{-1} and a detection limit of 3.480×10^{-9} M l^{-1}; the results suggest appropriate selectivity of rGO/DPA/PGE in anti-interference investigations and its application recognized FNT. Moreover, the process results in an accurate FNT recovery from spiked tomato samples ranging from 96.4% to 106.9%, demonstrating its effectiveness in FNT detection for industrial and agricultural uses in actual environmental samples. By thermally oxidizing a deposited metallic coating, Kalinic *et al* formed Co_3O_4 nano-petals on Si-based photoanodes [63]. They then used them for quick chemisorption and photoelectrochemical identification, followed by removal of the dangerous fenitrothion. The fenitrothion nitro-groups' direct interaction with the Co (II) centers, which prevent Co(II) oxidation, enables the nanostructures to function effectively for a fenitrothion sensitivity of 0.1 A (cm^{-2} M^{-1}). Additionally, Kalinic *et al* show that a small quantity of this nanostructured film, polarized at a modest bias (1.45 V versus RHE) and illuminated with visible light (100 mW cm^{-2}) without the requirement for catalyst remembrance, can remove pesticide from water successfully with a time constant of 1 h and remains stable, thus this technology is appropriate for pollution control applications. A synthetic method for rGO-tungsten carbide nano-polyhedron (rGO-WC NPH) composites was reported by Sakthivel *et al* [64]. The detection of fenitrothion in aquatic environments and rural soil samples was accomplished inexpensively by fabricating rGO-WC NPHs on a screen-printed carbon electrode. The results show a concentration range of 10 nM–471.2 μM and an LOD of 1.15 nM. The prepared rGO-WC NPH materials exhibit an excellent sensitivity of 5.3760 μA μM^{-1} cm^2 for the detection of fenitrothion. To verify the robustness of the developed sensor (rGO-WC NPHs/SPCE), the outcomes were also contrasted to those from the accepted HPLC method. Through a simple hydrothermal procedure, Rajaji and colleagues synthesized strontium hexaferrite decorated on ($SrFe_{12}O_{19}$/g-C_3N_4) material with outstanding electrochemical performance for the detection of fenitrothion [65]. The new biosensor has a sensitivity of 19.23 μA μM^{-1} cm^{-2}, an improved detection range of 0.005–378.15 μM, and a detection limit of 0.0014 μM. Additionally, the sensor has a high predictive range of 0.005–378.15 M. The profitable determination of fenitrothion in diverse fruit samples with good recoveries demonstrated the applicability of the proposed $SrFe_{12}O_{19}$/g-C_3N_4 modified detector. Kilele *et al* analysed a discerning and discriminating electrochemical sensor for FNT contamination in samples taken from fruit utilizing the nanocomposite IL@Co-Fe_2O_4NPs@MWCNTs [66]. The co-precipitation approach was used to synthesize 1-ethyl-3-methylimidazolium tetrafluoroborate ([Emim][BF_4]) ionic liquid (IL)@cobalt ferrite nanoparticle ($CoFe_2O_4$NP)@multi-walled carbon nanotube (MWCNT) nanocomposite. The obtained linear concentration FNT range was 0.02–160 μM. The actual LOD was 0.0253 μM and the estimated LOQ value was 0.0135 μM. The detection of FNT in grapefruit and orange extract was employed as a means of assessing the analytical capabilities of the IL@Co-Fe_2O_4NPs@ jMWCNTs@GCE sensor, with satisfactory findings.

10.2.1.4 Methyl parathion

Zeng *et al* created a electrochemical detector for measuring methyl parathion (MP) that used an MWCNT poly(acrylamide) nanocomposite (MWCNTs PAAM) NC film customized GCE [67]. Compared to several nitroaromatic and metal ion compositions in organic pollutants, the MWCNTs PAAM/GCE showed higher adsorption efficiency and greater affinity toward MP. Noticeably, the absorbed MP composition on the MWCNTsPAAM/GCE estimated the equilibrium value after 5 min of absorption. Following that, a calibration chart for MP was identified in the range of 5.0×10^{-9}–1.0×10^{-5} M, and the detection limit was 2.0×10^{-9} M. Subsequently, studies showed that MWCNTsPAAM/GCE is among the detecting tools for sensing MP in samples quickly and selectively with good sensitivity. For the electrochemical sensing of MP, Xu *et al* designed and crafted a new combination made of poly(malachite green)/graphene sheets (PMG/GrNSsNF) composite film customized GCEs [68]. As a result of the outstanding electrocatalytic performance of the PMG/GrNSs/NF composite film, p-nitrophenol, the alkaline hydrolysis byproduct of MP, was significantly identified at a considerably decreased potential with the usage of the proposed electrode. GrNSs improved the PMG durability and charge transfer rate in a composite film. The chronoamperometric (CA) response current was proportional towards the proportion of MP ranging from 0.02 to 1.5 μM with 2.0 nM as the LOD based on optimum circumstances. As a result, the researchers used their sensing device to determine MP in actual samples (tap and river water, and cabbage), yielding promising results and a recovery rate between 97.20% and 104.53%. To develop a nanocomposite consisting of hybrid NiCo-LDH and F-HNTs for the electrochemical sensing of parathion, Kokulnathan *et al* used a hydrothermal method [69]. The electrochemical sensor based on NiCo-LDH/F-HNTs provides exceptional performance with features such as a concentration range of 33.4–0.01 μM, an LOD of 0.002 μM, and sensitivity of 13.0 $\mu A\ \mu M^{-1}\ cm^{-2}$. The nanocomposite ternary hybrid composed of Fe_3O_4@MoC MFs/gCN is reported by Devi *et al* and is a driving force behind the sensing of the organophosphorus insecticide parathion (PAT) [70]. More effective interaction with the reactive catalytic sites will be made possible by forming a hierarchical nanostructure from the core level. This suggested sensor has a broad linear range (0.5–600 μM), excellent sensitivity, and an LOD of (7.8 nM) to determine parathion under ideal conditions. The Fe_3O_4@MoC MFs/gCN/GCE sensor can be a perfect candidate for real-time parathion detection, according to the positive test findings of the food samples. To determine the organophosphates in biological samples, Nagabooshanam *et al* developed an electrochemical sensor with high sensitivity linking with polyaniline and carboxyl fMWCNT [71]. A linear range of 120–10 ng l^{-1} was used to evaluate the sensor. It was determined that the lowest amount of chlorpyrifos that could be detected (the LOD) was 8.8 ng l^{-1}, and the sensitivity was 0.41 mA ng^{-1} l^{-1} cm^{-2}. The LOD for methyl parathion was 10.2 ng l^{-1}, and the sensitivity was 0.58 mA^{-1} ng^{-1} l^{-1} cm^{-2}. Furthermore, water-washed samples collected from cucumbers were also examined. It was observed that the average recovery rates for chlorpyrifos and methyl parathion were 98.05% and 96.63%, respectively. Gong *et al* developed the stripping voltammetric sensing of methyl parathion pesticides using Ni/Al-LDHs/GCEs with a solid-phase extraction method [72]. The nitroaromatic pesticides were examined by stripping analysis with an LOD of 0.6 ng ml^{-1} and linear ranges of 0.001–0.1 and 0.2–1.0 μg ml^{-1} MP. The square wave voltammetric results exhibit better performances due to MP intercalated into LDHs. Using multilayer films containing an AChE liposome bioreactor and chitosan, Guan *et al* produced a novel acetylcholinesterase (AChE) biosensor [35]. The electrochemical sensor displayed better AChE activity with an LOD of 0.86 ± 0.098 g l^{-1} and a dichlorvos concentration range of 0.25–1.50 μM and 1.75–10.00 μM.

10.2.1.5 Paraquat

Frequently employed as a bactericidal herbicide, the quick-acting, broad-spectrum paraquat (PQ), often referred to as methyl violet, is a highly poisonous pyridine compound. Within 2–3 h, green leaves exposed to PQ become yellow. PQ is frequently used in horticulture and agriculture because, although it becomes unresponsive when in contact with soil, it has no adverse effects on plant roots. Even at low concentrations, PQ is exceedingly poisonous to humans and hazardous to them (oral LD50 = 4 mg kg^{-1}) [73]. Severe organ damage brought on by acute PQ poisoning includes chronic lung damage brought on by pulmonary fibrosis that ultimately causes death [74, 75]. By electrochemical polymerization of o-phenylenediamine onto functionalized MWCNTs, Jeyapragasam and colleagues developed poly(o-phenylenediamine) (PoPD)-MWCNT NCs, which served as effective moderators for the sensing of paraquat electrochemically [76]. Among the GCE, MWCNT/GCE, and PoPD/GCE, the PoPD-MWCNT/GCE displayed the most considerable cathodic current for paraquat. SWV was employed to obtain the sensor calibration graph for paraquat in a linear concentration range from 8.8×10^{-7} to 2.5×10^{-8} M. The LOD value for the developed sensor was 1.4×10^{-9} M. The fabricated sensor was employed significantly to determine paraquat in cabbage samples. By using a one-pot synthesis method, Niu and colleagues created a para-sulfonylcalix [4]arene(pSC_4)-gold nanoparticles@zeolite imidazole skeleton-8 (ZIF-8) composite [77]. A glassy carbon electrode utilized for paraquat sensing was covered with an electrocatalyst of pSC_4-AuNPs@ZIF-8 material. The LOD for paraquat was found to be 0.49 pM. To spot trace pollutants in an aquatic environment, the approach of signal amplification based on pSC_4-AuNPs@ZIF-8 offers potential usefulness. Zhou *et al* used the hydrothermal process to synthesize an Fe_2O_3-coated MIL-100 (octahedral metal–organic framework) material. They identified paraquat at low concentration levels in milk and honey using sensitive and specific electrochemical methods. Having a detection threshold of 2.6 nM and a concentration range of 0.01–30 μM, the Fe_2O_3-MIL-100-based biosensor demonstrated unique sensing capability under ideal circumstances. Also, after 20 days of storage time, the electrode built on the Fe_2O_3-MIL-100 material produced an excellent current response to PQ. The Fe_2O_3-MIL-100-based biosensor, in particular, demonstrated excellent specificity and sensitivity for PQ identification in honey and milk specimens. Using three different MTsPc (NiTsPc, CuTsPc, or FeTsP), the SPCEs modified by Kavazoi *et al* were assessed for detecting paraquat electrochemically [78]. For PQ sensing in river water and ultrapure water, the FeTsPc/SPCE demonstrated a superior differential pulse response when used in the test method. In the standard sample, the DPV analysis revealed an LOD of 0.97 μmol l^{-1}, with a concentration range of 6.0–60 mol l^{-1} and a recovery of 103.3% in the riverbed specimen. A Fe_2O_3-BCN (boron/nitrogen-co-doped carbon) nanosheet modified on the GCE surface was formed by Li *et al.* Paraquat was detected using the SWV method in river water samples [79]. The detection limit for paraquat by electrochemical means was 2.74 nmol l^{-1}. According to the findings, the Fe_2O_3-BCN/GCE electrode performed well electrochemically against the hazardous pesticide paraquat.

10.2.1.6 Simazine

To precisely and selectively determine simazine (SMZ) in contaminated environmental matrices, Zhang *et al* reported a sensing system on the basis of molecularly imprinted polymer membranes (MIPM) of poly(o-aminothiophenol) with a AuNP customized Au electrode [80] (figure 10.1). By using electropositive ATP@AuNP in

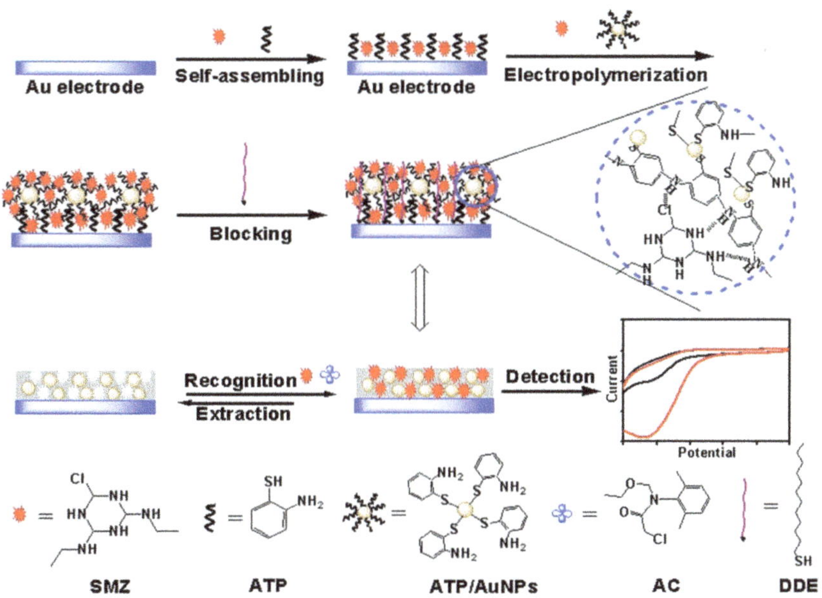

Figure 10.1. A diagrammatic representation of the construction process of an SMZ imprinted sensor. (Reproduced with permission from [80]. Copyright 2017 Elsevier.)

the presence of a SMZ pattern and self-assembling o-aminothiophenol (ATP), a nanostructured MIPM with enhanced electroactive surface area was obtained. The peak current density had a linear progression with SMZ with 0.013 μM LOD and a linear range of 0.03–140 μM; the SMZ imprinted sensor showed exceptional long-term durability and excellent reproducibility. Furthermore, the spiking recovery went from 91.4% to 96.8% when it was used effectively for SMZ assessment in several real samples, including river and tap water samples, as well as soil samples, demonstrating a promising possibility for applied applications. Zhang *et al* developed a sensor based on electrochemistry that combined a molecularly imprinted composite membrane of poly(o-aminothiophenol) with AuNPs as a sensitive method for determining the pesticide simazine in real samples [80].

A large specific surface area nanoscaled MIPM was made using self-assembly of o-aminothiophenol (ATP) and electrochemically modified ATP@AuNPs with a pattern is present in the simazine. Simazine was evaluated by voltammetry techniques with a negative potential and quick rebinding kinetics towards the reduction of simazine in a 0.01 M H_2SO_4 solution in pH 1.7 medium. From 0.03 to 140 μM, the peak current depended on linear factors of simazine concentration, with an LOD of 0.013 μM. The occurrence of AuNPs, which lowered the electron transfer impedance and enhanced imprinting sites, resulted in a special enhancement of the extremely low concentrations of simazine in a hydrophilic solution.

10.2.2 Organic pollutants: the detection of nitroaromatics and phenolic derivatives

As intermediates in the synthesis of complex nitroaromatic compounds, phenolic derivatives have been connected to contamination in the environment, including soil, food, medicine, and surface water, and has become a critical issue because of the relative toxicity of the substances and their environmental persistence, carcinogenic nature, harmfulness, therapeutic effects, and mutagenesis. The United States EPA has set a permissible level [81] of 1 mg l^{-1}. As a result, several researchers have focused on simplifying and effectively detecting nitroaromatic compounds and phenolic derivatives in the environment. It is vital to develop precise and practical analytical techniques for nitroaromatic and phenolic derivative substances. Here, some notable works related to nitroaromatic compounds and phenolic derivatives, such as as nitrobenzene (NBz) and 2,4,6-trinitrotoluene (TNT), benzenediol, nitrophenol, and aminophenol are discussed.

10.2.2.1 Nitrobenzene

Synthetic aromatic nitro compounds are used to make insecticides, anilines, lubricants, synthetic rubber, dyes, explosives, and as solvents in the flavoring, petroleum, and leather industries. Over 95% of nitrobenzene (NBz) is used to make food additives, acetates, polyurethanes, and cellulose ethers. It is released from effluents into industrial waterways. The United States EPA classifies NBz as a primary hazardous pollutant. According to the USA's Clean Water Act, 17 g l^{-1} is the absolute tolerable limit for NBz in drinking water. To detect nitrobenzene, an electrode made of immobilized silver NP (AgNP)-PMEMWCNT over a GCE substrate was developed by Ren *et al* to fabricate a highly sensitive electrochemical sensor [82]. The current density at its peak was enhanced significantly compared to the GCE or other customized electrodes owing to the customized electrode's outstanding capacity to reduce nitrobenzene and contribute electrocatalytically to this reduction. The sensor exhibited 0.55 μM as the LOD for NBz with a broader linear calibration band from 20 to 1000 μM between 1000 and 6000 μM. The electrocatalyst worked well as an NBz sensor in samples taken from local faucets and rivers, with a retrieval band of 97.2% to 104.6%. It also had excellent durability, reproducibility, and specificity. A portable electrochemical sensor for highly sensitive identification of gaseous phenol and nitroaromatics was presented by Andrews *et al*, and it is based on carbon cloth functionalized with manganese porphyrin. The LOD for aqueous phenol and nitrobenzene (NB), respectively, was found to be 4.0178×10^{-10} and 5.9268×10^{-10} M for the electrode that was developed. Furthermore, the proposed transportable electrochemical sensor exhibits high specificity and repeatability toward phenolic and nitroaromatic chemicals and can be used in actual environmental aquatic models. Due to the exceptional electrocatalytic properties fo the detector, ecological point of care (POC) testing is anticipated to be made possible [34, 83]. A CPE-customized nanocomposite (NC) made using MWCNTs and the conducting polymer PEDOT was employed by Xu *et al* to construct an extremely sensitive and electrochemically selective sensor [84]. This NC demonstrated outstanding electrocatalytic performance toward the electrochemical reduction of NBz, and the NBz reduction potential decreased significantly with an increase in the reduction current. With 83 nM as the LOD, the customized CPE could recognize NBz in a range of concentrations from 0.25 to 43 μM. This very reliable sensor, provided with functional capability, might be used to determine NBz in effluents without being interfered with by other commonly encountered nitro compounds.

According to Yi *et al* electrochemical techniques can simultaneously determine N-(4-hydroxyphenyl)acetamide (ACT) and 1-chloro-4-nitrobenzene (CNB) at low concentrations [85]. For usage as an electrocatalyst in the oxidation of CNB and ACT, a nitrogen-doped carbon black (NCB) nanocomposite was utilized. The limit of detection for ACT and CNB was found to be 0.008 M and 0.007 μM, respectively. The concentration ranges for the ACT and CNB analytes were 0.05–50.0 μM. Kokulnathan *et al* proposed GO/MgO nanocomposite for nitrobenzene (NBZ) sensing using DPV and CV for water samples [86]. Compared to other electrodes, the GO/MgO nanocomposite exhibited remarkable nitrobenzene sensing capabilities. The GO/MgO-NC/GCE electrode had an LOD value of 0.01 μM and a concentration range of 58.5–2333.5 and 0.1–38.9 μM. The GO/MgO-NC/GCE electrode was extended to analyze nitrobenzene in groundwater sources with 99% recovery, giving an efficient method for nitrobenzene in real sample analysis. To modify the GCE surface for the sensitive detection of nitrobenzene, a zinc stannate-graphitic carbon nitride (ZSO-gCN) sample was synthesized *in situ* through pyrolysis [87]. To analyze nitrobenzene electrochemically, the ZSO-gCN/GCE has a charge transfer coefficient of $\alpha = 1.171$ and an electron transfer rate constant of $ks = 0.746\ s^{-1}$. With an increasing nitrobenzene concentration range of 100–30 μM, the LSV approach shows a rise in cathodic peak current that is linear in nature. It was found to be the case that the sensitivity and LOD were 0.058 57 $\mu A\ \mu M^{-1}$ cm and 22.2 μM, respectively.

10.2.2.2 2,4,6-Trinitrotoluene

2,4,6-trinitrotoluene (TNT) is a well-known dynamite chemical that is also utilized in the production of fireworks and matches, and in aviation, forensics, and military applications. TNT can easily build-up in soil and groundwater through dumping, storage, and long-term use. TNT is a poisonous and mutagenic substance that is dangerous to humans, wildlife, and the environment, requiring rapid, sensitive, and practical detection methods. To more accurately and sensitively measure TNT residues in an aqueous background solution, Mahmoud *et al* constructed a versatile electrochemical sensor using poly(styrene-co-acrylic acid) PSA/SiO_2/Fe_3O_4/AuNPs/lignin customized GCEs [88]. The LOD was 35 pM, which matched the TNT's LOD as recorded by an amperometric *I–t* graph. Significantly, the tightly packed Fe_3O_4/AuNPs and lignin layer on the electrode surface can be attributed to the customized electrode's increased sensitivity and excellent electrocatalytic performance. Additionally, it was discovered that 98% and 96% of TNT was recovered completely from sea and tap water samples, respectively. Additionally, after five iterations of desorption/absorption, the film electrode was highly stable, and the newly developed electrochemical detection offered high precision, flexibility, and selective assay for *in situ* TNT measurement in a challenging water medium. A rutile TiO_2 coating was used by Sanghyeon *et al* for an electrochemical method to detect TNT. The DPV analysis showed that the LOQ and LOD for 2,4,6-TNT sensing employing the calcined TiO_2 layer, which served as the working electrode, were 36.0 and 10.8 ppm, respectively, with a linear concentration range from 10 to 100 ppm. This appealing method that can be used to determine other dangerous nitro-organic substances such as pentaerythritol tetranitrate and nitroglycerin allows the active measurement of TNT utilizing a crystalline metal for transitional support [89].

To determine nitroaromatic detonation chemicals on a gold-customized GCE covered with poly(o-phenylenediamine-aniline) film, Saglam *et al* constructed a sensing system using a GCE/P(O-PDA-CO-ANI)-Au nanoelectrode [90]. The current intensity versus concentration calibration plot showed a linear band of 2.5–40 mg l^{-1} for TNT with 2.1 mg l^{-1} as the LOD, 5–100 mg l^{-1} for tetryl, and 2–40 mg l^{-1} for 2,4-dinitrotoluene (DNT). The inter-assay and intra-assay measures' variation coefficients were 1.2% and 0.6%, respectively ($N = 5$), indicating more excellent assay repeatability. The method was used successfully to examine the nitroaromatics found in comp B, octol, and tetrytol, three types of army explosives. A straightforward technique for the detection of 2,4,6-trinitrotoluene using α-Al_2O_3-supported GCE. Lima *et al* described the low-potential electrocatalytic reduction of TNT by Al_2O_3 on GCEs [91]. Particularly in an acidic medium, the adsorption of TNT on α-Al_2O_3-supported GCEs increased the transmission of electrons between TNT and the glassy carbon electrode. Other nitroaromatic substances, including DNT, nitrotoluene, and nitrobenzene, exhibited a similar tendency, although they did not affect TNT's first reduction peak. The presence of α-Al_2O_3

significantly improved the electrochemical reduction of the dissolved oxygen, but it did not interfere with the ability to detect TNT. The alumina-polished GCE offered a wide concentration range (0.5–1000 μmol l^{-1}) and an LOD value of 0.83 nmol l^{-1} and did not suffer from interference from any other nitroaromatic explosives. Zhang *et al* produced novel metal-free base washing-nitrogen, sulfur-codoped rGO nanoribbons (BW-NS-rGONRs) [92]. Carbonaceous oxidative waste in graphene oxide nanoribbons (GONRs) formed by unzipping carbon nanotubes affects their electrocatalytic performance. After base washing to remove adsorbed oxidative waste, the BW-NS-rGONRs showed strong electrical conductivity in addition to an abundance of defect active sites, enhancing the electrochemical TNT performance. The BW-NS-rGONR enhanced catalytic sensor demonstrated a particular and sensitive outcome for TNT with a 0.1 ppb detection limit and a concentration ranging from 5.1 to 0.008 ppm. The BW-NS-rGONR-based TNT platform for detecting threats was reliable and repeatable. The BW-NS-rGONR-based sensing technology can detect TNT in tap and lake water. The excellent sensing capability is due to excellent active sites from dual doping of S and N atoms, improved electrical conductivity, and full exposure of active sites in BW-NS-rGONRs following base washing. Salaria *et al* describe a microwave-assisted approach using PEG-coated cadmium sulfide (P-CdS) nanoparticles to monitor the explosive (TNT) [93]. For electrocatalytic sensing, the CV and CA techniques were used. The results reveal that TNT can be detected selectively using amperometry, with a sensitivity of 47.8 μA cm^{-2} M^{-1} and an LOD value of 1.045 nM.

10.2.2.3 Benzenediols

Phenolic substances and variants are used to make industrial intermediates. Catechol, hydroquinone, and resorcinol are the isomers of dihydroxybenzene, used to make herbicides, insecticides, secondary coloring materials, medicines, dyes, medications, antioxidants, cosmetics, stabilizers, and preservatives. The hazardous phenolic substances severely affect humans, in particular those that cause headaches, impaired liver function, and lethargy. Researchers have found solutions using electrochemical approaches for the sensing of benzenediols in the environment. Using ZnO/MnO_2 hybrid 1D nanostructures on working electrodes that are based on printed contacts, Dubey *et al* created a nano-enabled electrocatalytic detector. The sensing results show good sensitivity (0.45 k nM^{-1} mm^{-2}) and a low LOD of 2 nM. As a means of identifying chlorinated phenolic contaminants in water at ambient settings, the developed sensor offers a unique, quick, and affordable method. This is advantageous for environmental and human health [94]. A GCE-altered composite film made of polydiphenylamine (PDPA) and MWCNTs was reported by Yang *et al* for the electrocatalytic oxidation of phenol [95]. The authors polymerized the MWCNT-customized GCE with PDPA utilizing methods of cyclic voltammetry in H_2SO_4 (0.1 M) electrolyte medium. In a phosphate buffer solution of pH 7, the PDPA-MWCNT composite film demonstrated appropriate electrochemical performance for phenol oxidation, displaying a visible anodic peak at 0.65 V (compared to the reference electrode). The introduction of MWCNT in the PDPA-MWCNT composite film resulted in a peak current densities of 2.36 and 8.48, respectively. As the concentration of phenol rose, the peak current climbed linearly. Additionally, a concentration between 80 and 9.8 μM for the CA carbolic acid determination at the composite film customized electrode was observed. To eliminate environmental toxins, Annamalai *et al* proposed using Ce and Sn doped ZnO nanoparticles for electrochemical sensing of catechol (CC) and the degradation of hazardous dyes [96]. The electrochemical determination of catechol revealed that the sensitivity and detection limit for 5% Ce doped ZnO were 0.494 μM μA and 0.21 μM, respectively. Several approaches, including cyclic voltammetry and photodegradation, were used to detect and effectively destroy environmental contaminants such as catechol, rhodamine B dye, and crystal violet. Du *et al* developed the nanocomposites based on ERGO/CNTs/AuNPs@CeMOFs. Through the use of a signal synergistic amplification method, the electrochemical sensing platform was used to concurrently monitor hydroquinone (HQ) and CC [97]. The synergy of the signal depended on the regulated pore nano-restriction effect of MOFs, the excellent conductivity with a great degree of consistency of the AuNPs, and the extremely rapid rate at which the electrons are moving, created by ERGO/CNTs. Utilizing a network structure that is 3D improved the performance of the electrocatalytic detector greatly. The sensor has an LOD for HQ and CC of 0.44 and 0.73 μM, respectively, with concentration ranges of 3–200 and 3–190 μM. It has been used to sense phenolic isomers in actual models, with retrieval rates ranging from 93.5% to 1040% for HQ and 100.1% to 104.5% for CC. Tian *et al* fabricated 1D PEDOT-Gr composites by electropolymerizing 3,4-ethylene dioxythiophene *in situ* on Gr [98]. The 1D PEDOT-Gr composite demonstrated exceptional electrochemical performance for NO_2^- ions and dihydroxybenzene isomers. HQ, resorcinol (RS), nitrite, and CC were all detected simultaneously by DPV employing a PEDOT-Gr/Ta electrode, with current separations of 392 mV between RS and CC, 108 mV between CC and HQ, and 188 mV between nitrite and RS. The LOD for each sample was, respectively, 0.08, 0.06, 7, and 0.16 mM. Additionally, the 1D PEDOT-Gr composite demonstrated excellent electrochemical performance, fast response times, and good selectivity. For electrochemical detection as a high-performance platform for the sensitive identification of dihydroxybenzene isomers in domestic wastewater, Kugalur *et al* described hybridized one-dimensional–two-dimensional $MnMoO_4$-MXene nanomaterials. The one-dimensional $MnMoO_4$ and 2D MXene nanoarchitectures abundant enabled deficient edges showed significant electrochemical performance toward catechol and hydroquinone oxidation and enhanced the operation of combinatorial signal amplification. For HQ and 1,2-dihydroxybenzene, $MnMoO_4$-MXene-glasssy carbon electrodes demonstrated respective oxidizing potentials of 0.203 and 0.102 V. It had a significant anodic peak current and a distinct and simultaneous sensing range of 0.101 V. It is worth noting that the suggested 1D–2D hybridized $MnMoO_4$-MXene-GCE electrochemical sensor showed a broad linear relation for catechol and hydroquinone from 5 to 65 nM. Additionally, CC and HQ, with strong durability, demonstrated a corresponding low LOD of 0.30 and 0.26 nM. Using the DPV approach with recovering values, the practical 1D–2D $MnMoO_4$-MXene NC-based biosensor efficiently identified catechol and hydroquinone in hazardous water contaminants [99].

By electropolymerizing PEDOT on GCE in sequence and electrochemical deposition of Au dendritic nanostructures (DGNs) on top, Cheng *et al* created a sensor based on electrochemical reactions. With an LOD of 1.3×10^{-6} M for RS, 1.6×10^{-6} M for CC, and 9.5×10^{-7} M for HQ, the fabricated sensor with an Au dendritic assembly, large accessible active surface area, and strong electrical conductivity demonstrated remarkable sensing performance toward dihydroxybenzene isomer detection [100]. In the end, the sensor gave appropriate outcomes for effective implementation to detect dihydroxybenzene in cosmetic samples.

10.2.2.4 Nitrophenol

In terms of environmental applications and national security, explosives detection has attracted much attention. Compared to primary explosives, secondary explosives are more common at a military sites. 4- hexahydro-1,3,5 trinitroazine (RDX), nitrophenol (NP), 2,4-dinitrophenol (DNP), 2,4,6-trinitrophenol (TNP), dinitrotoluene (DNT), and trinitrotoluene (TNT) are examples of secondary explosives. These pollutants contain the nitro (NO_2) group, which interacts with biota and is extremely dangerous to the environment. The accurate and quick detection of these contaminants is therefore crucial. $ZnFe_2O_4$/PANI@RGO aerogel was synthesized by Wei and colleagues as an easy-to-use electrochemical sensor for p-NP detection [101]. The prepared $ZnFe_2O_4$/PANI@ RGO aerogel demonstrated effective electrocatalytic reduction and excellent sensing towards p-NP. Over a calibration range between 100 and 1 μM, an LOD of 0.083 μM with superior sensitivity of 36.898 mA mM^{-1} cm^{-2} for p-NP was also shown by the $ZnFe_2O_4$/PANI@RGO customized GCE. The $ZnFe_2O_4$/PANI@RGO/GCE provided good anti-interference capabilities and extended durability for the sensing of p-NP. A recovery value of 101.16% was maintained when the sensing efficiency of tap water was measured as well. To simultaneously sense 4-NP and 2-NP, Tang *et al* acquired Gr-CHIT composite film customized GCEs (Gr-CHIT/GCEs) [102] and examined the electrochemical performance of 2-NP and 4-NP by CV. The customized GCE demonstrated good electrochemical performance to reduce o-NP and p-NP. Dhanasekaran and colleagues reported that AgNP coated Co-Al LDHs/PoPD (Ag@Co-Al LDHs/PoPD) material was synthesized using a straightforward chemical process [103]. The Ag@Co-Al LDH/PoPD materials were drop-cast onto the GCE surface and utilized for the sensing of both 2,4-DNP and 4-nitrophenol 4-NP. With an LOD of 50 nM for 2,4-DNP and 63 nM for 4-NP, the Ag@Co-Al LDH/PoPD/GCE demonstrated improved behavior related to electrochemical sensing. Padmanaban *et al* developed a facile solvothermal decomposition method which was used to construct a single-phase zinc bismuthate ($ZnBi_{38}O_{60}$) binary composite with nanobundle shaped morphology [104]. Using DPV, a GCE reformed with $ZnBi_{38}O_{60}$ nanobundles as the electrocatalyst was utilized to investigate the quantitative and qualitative properties of 4-nitrophenol (4-NP) (DPV). The LOD value of 35×10^{-9} M and the modified electrode's sensitivity (1.024 mA mM^{-1} cm^{-2}) were enhanced. Accordingly, the $ZnBi_{38}O_{60}$ nanobundles have the potential to be an additional choice of electrocatalyst for the determination of 4-NP, and their applicability can be broadened to a variety of nitro-containing samples.

10.2.2.5 Aminophenol

For aminophenol (AP) voltammetric detection, Fan *et al* established electrocatalytic sensors constructed on Gr-PANI NCs [105]. The CV response at the Gr-PANI-customized GCE via p-aminophenol was shown to be enhanced by analysis of the electrochemical responses of the 4-AP at the Gr-PANI customized GCE employing CV. With an LOD of 6.5×10^{-8} M and a greater sensitivity of 604.2 μA mM^{-1}, it demonstrated a respectable analytical function for 4-AP detection. Additionally, in the wider dynamic region, it might be able to concurrently recognize paracetamol and 4-AP without any perturbation from them. Furthermore, the Gr-PANI NC may offer a fresh platform for developing electrochemical processes based on graphene biosensors and sensors. Praveen Kumar *et al* designed a sensitive and selective electrochemical approach to simultaneously measure paracetamol (PR) and 4-AP [106]. Using the differential pulse voltammetry method, a poly(chromium Schiff base complex) was fabricated onto a GCE. Simultaneous determinations of paracetamol and 4-AP were then carried out. The peak separation between 4-AP and paracetamol was observed to be 300 mV using differential pulse voltammetry. Both analytes' electrochemical responses can be placed in the category of a concentration range of 8–133 nM and 8–125 nM, respectively, with LODs of 5.6 and 6.8 nM for 4-AP and paracetamol, respectively. For the detection and quantification of PR and 4-AP in the pharmaceutical industry at nanomolar level dosage forms, the electropolymerized chromium Schiff base complex was straightforward, quick, economical, and accessible.

10.2.3 Organic pollutants: the detection of organic dyes

A brand-new ratiometric electrochemical sensing platform was created on the foundation of MXene. We are unaware of any reports on the combination of MXene with CuNPs or the application of such as composite for ratiometric electrochemical detection (figure 10.2).

A reference signal that was consistent was present and obtained using a quick, one-step adsorption method based on the robust MB adsorption by MXene. This reduces significantly the difficult process of changing a reference substance in an earlier set-up for a ratiometric sensor. The system has a remarkable sensitivity that may be due to the special structure of MXene that is layered, which not only acts as an highly accessible surface area composite with the ability to combine with other nanomaterials, but also offers valuable conductive channels to speed up electron transmission. CuNPs are employed in this article as a model, and it is anticipated that the system will be adaptable and able to be paired with NPs made of other elements, such as Mn, Fe, and Ni, among others. The CuNPs' insertion increases the sensor's performance in electrocatalysis. When combined with a ratiometric detecting approach, the suggested detector displays a broad linear range, a low LOD, repeatability, good sensitivity, durability, and stability. Easily-made electrode samples of human serum and piroxicam in tablets were analyzed using this method, with excellent recovery, demonstrating the platform's applicability and dependability. As a result, it is possible to predict that the constructed device will function as a useful diagnostic device in various domains [107].

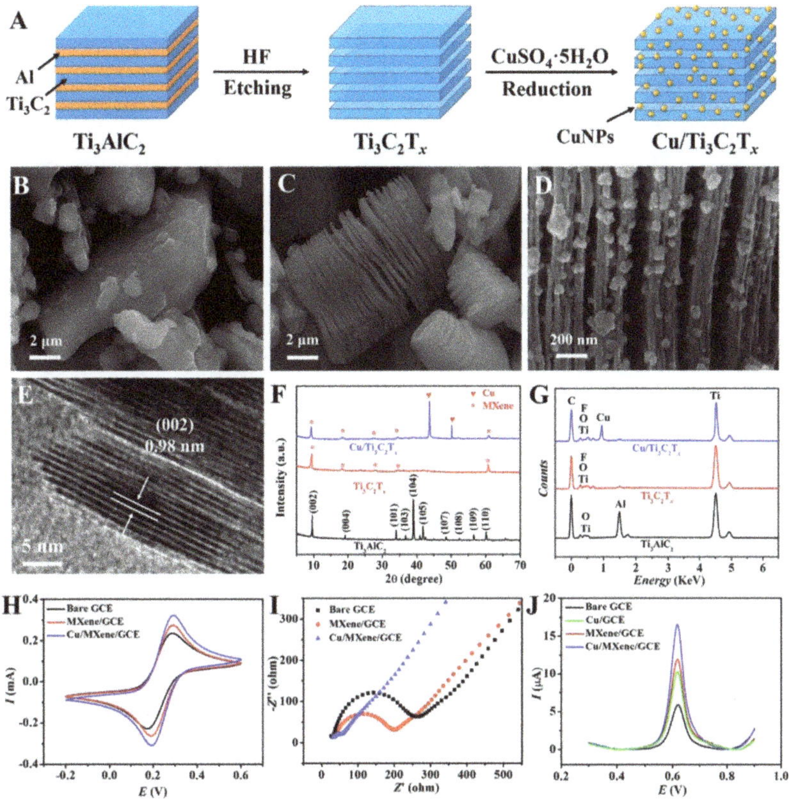

Figure 10.2. (A) The reaction strategy for composite Cu/Ti_3CT_x. SEM images of composites made of (B) Ti_3AlC_2, (C) $Ti_3C_2T_x$, and (D) Cu/ $Ti_3C_2T_x$. (E) HRTEM image of $Ti_3C_2T_x$. (F) XRD patterns and (G) EDS spectra of composites made of Ti_3AlC_2, $Ti_3C_2T_x$, and $Cu/Ti_3C_2T_x$. (H) Cyclic voltammograms of EIS on bare GCE, MXene/GCE, and Cu/MXene/GCE in 5 mM $[Fe(CN)6]^{3-/4-}$ containing 0.1 M KCl. (I) Nyquist diagrams of EIS. The CV scan rate was set to 100 mV s^{-1} and the EIS frequency was in the range of 0.01 Hz to 100 kHz. (J) SWV graphs of the response to 50 M piroxicam in 0.1 M phosphate buffer for bare GCE, Cu/GCE, MXene/GCE, and Cu/MXene/GCE (pH 2.0). (Reproduced with permission from [107]. Copyright 2019 American Chemical Society.)

To quickly create NiNP-decorated carbon porous (Ni/ap) materials, Veerakumar *et al* [108] devised an evaporation-induced self-assembly (EISA) process using microwave heating (figures 10.3 and 10.4). The mesoporosity and high surface areas of these innovative Ni/CPM materials make them ideal for dispersing NiNPs, making them reusable heterogeneous electrocatalysts for the degradation of organic dyes and the electrocatalytic location of HMIs. It was discovered that the magnetically separated Ni/CPM catalysts could reduce organic dyes such as RhB and MB with remarkable reactivity when $NaBH_4$ was present as the reduction agent. Usually, even with an electrocatalyst loading of only 2 mg, complete reduction of RhB and MB can be accomplished in less than 1013 min at a rate constant of 0.6 min^{-1}. The Ni/CPM-customized GCEs were also discovered to have outstanding sensitivity, selectivity, low detection, and electrocatalytic activity, and a low LOD

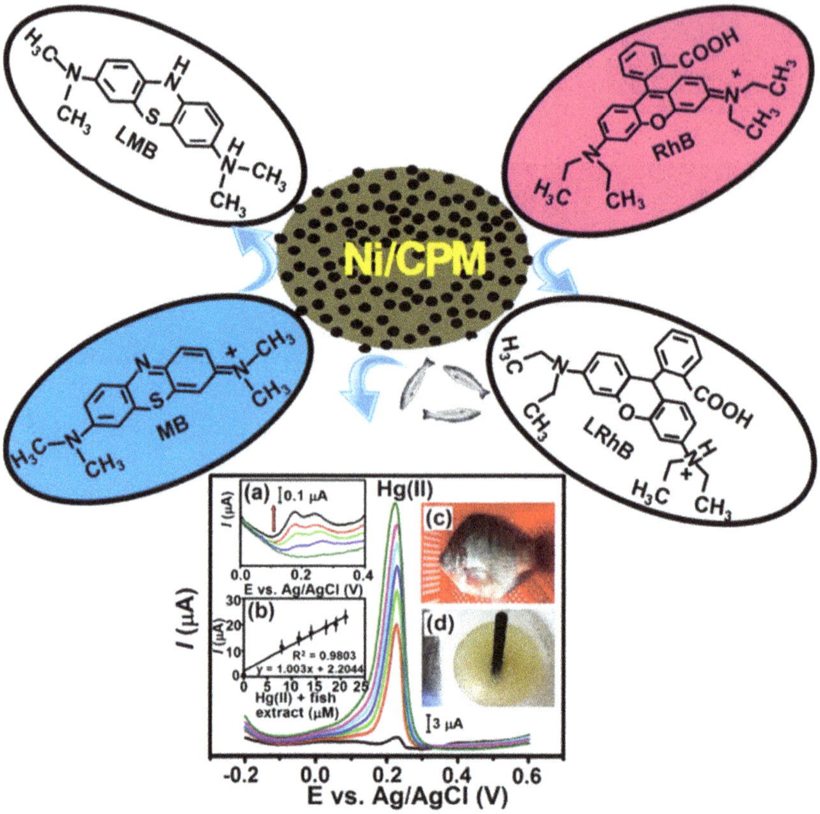

Figure 10.3. A diagramtic representation of Ni/CPM materials used for organic dye reduction and heavy metal detection. (Reproduced with permission from [108]. Copyright 2015 American Chemical Society.)

for Hg(II) ion detection, even in real models, and the concurrent presence of other heavy metal ions. These Ni/CPM materials could therefore have the capability to be used as effective nano-electrocatalysts for organic dyes reduction in wastewater management in addition to electrochemical sensing for the precise identification and evaluation of heavy metal ions (HMIs) in realistic models.

10.3 Conclusion

Metal oxides and nanocomposites based on conducting polymers have unique properties like high thermal and mechanical strength as well as electrical conductivity and chemical stability. Moreover the tuneable features of metal based nanocomposites, in terms of the their construction, light weight, streamlined procurement, chemical structure, high electroactive spots, and robust adhesion to the electrodes surface, make them ideal candidates for the formation of practical electrochemical sensors. This chapter outlined recent advancements in the electrochemical detection of various pesticides. The creation of electrochemical sensors for the detection of environmental pollutants, such as pesticides with organophosphate

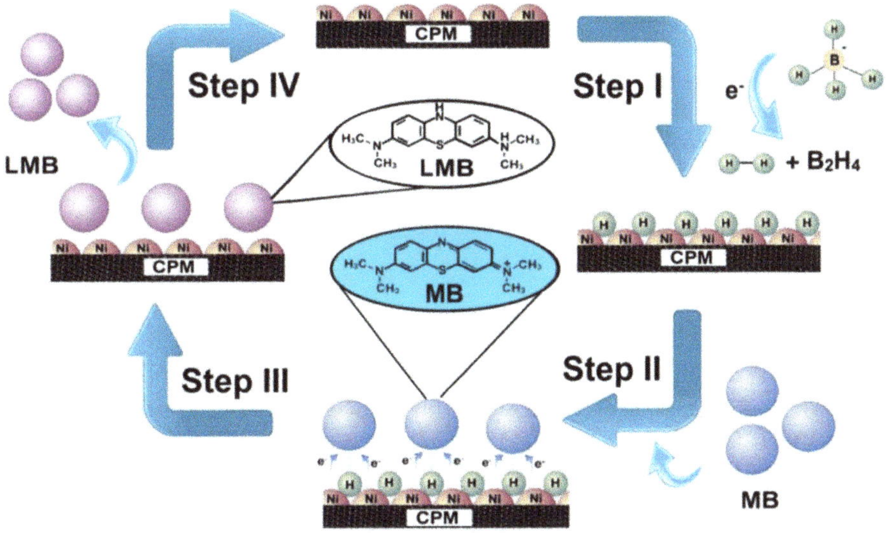

Figure 10.4. The suggested reaction process for using $NaBH_4$ to reduce MB over Ni/CPM catalysts. (Reproduced with permission from [108] . Copyright 2015 the American Chemical Society.)

containing compounds, phenolic chemicals, nitroaromatics, and organic dyes, has proven to be highly effective when fabricating with advanced functional materials for sensing platforms. Thus, the examined research hightlights a future pathway through the functionalization of advanced multifunctional nanocomposites toward various interdisciplinary applications in healthcare diagnostics and industrial and environmental monitoring.

Abbreviations

SWV	square wave voltammetry
DPV	differential pulse voltammetry
CV	cyclic voltammetry
CA	chronoamperometry
GCE	glassy carbon electrode
SPCE	screen printed carbon electrode
AChE	acetylcholinesterase
FNT	fenitrothion
NP	nanoparticle
MOF	metal–organic framework
LOD	limit of detection
CNT	carbon nanotube
DNT	dinitrotoluene
TNT	2,4,6-trinitrotoluene
NBz	nitrobenzene
4-NP	4-nitrophenol
2-NP	2-nitrophenol
SMZ	simazine

MP	methyl parathion
PQ	paraquat
IMI	imidacloprid
Au NP	gold nanoparticle
MWCNT	multi-walled carbon nanotube
LOQ	limit of quantitation
DNP	2,4-dinitrophenol
TNP	2,4,6-trinitrophenol

Acknowledgments

The author PA would like to express their gratitude to the FONDECYT/ANID (2022) Post-doctoral Project No. 3220176, funded by the Government of Chile, Santiago. The FONDECYT/ANID (2022) Post-doctoral Project No. 3220357, Government of Chile, Santiago, is gratefully acknowledged by the author PS. The author ME is grateful for the basic research support from the Chalmers University of Technology, Sweden. The author BC is grateful to the Department of Chemistry, Saveetha Engineering College (Autonomous), Chennai 602 105, Tamil Nadu, India, for research support.This work was supported by the Basic Science Research Program and Engineering and Academic Research (R&D) Program through the National Research Foundation of Korea (NRF) and funded by the Ministry of Education (2022R1I1A3072535 and RS202300249778).

References

[1] Zhang B, Li B and Wang Z 2020 Creation of carbazole-based fluorescent porous polymers for recognition and detection of various pesticides in water *ACS Sens.* **5** 162–70

[2] Zahran M, Khalifa Z, Zahran M A and Azzem M A 2021 Gum arabic-capped silver nanoparticles for electrochemical amplification sensing of methylene blue in river water *Electrochim. Acta* **394** 139152

[3] Arulraj A, Thachnatharen N, Rajkumar C and Mangalaraja R V 2022 Engineering 2D chalcogenides for energy and environmental remediation *Nanostructured Materials for Sustainable Energy and Environmental Remediation* (Bristol: IOP Publishing) pp 2-1–28

[4] Preethi S, Arulraj A, Mangalaraja R V, Ravichandran V and Subramanian N 2022 Biomass-derived carbon-based materials for microbicidal applications *Biomass-Derived Carbon Materials* (New York: Wiley) pp 63–92

[5] Tajik S, Beitollahi H, Nejad F G, Kirlikovali K O, Van Le Q, Jang H W, Varma R S, Farha O K and Shokouhimehr M 2020 Recent electrochemical applications of metal–organic framework-based materials *Cryst. Growth Des.* **20** 7034–64

[6] Matés J M, Segura J A, Alonso F J and Márquez J 2010 Roles of dioxins and heavy metals in cancer and neurological diseases using ROS-mediated mechanisms *Free Radical Biol. Med.* **49** 1328–41

[7] Nie D, Li P, Zhang D, Zhou T, Liang Y and Shi G 2010 Simultaneous determination of nitroaromatic compounds in water using capillary electrophoresis with amperometric detection on an electrode modified with a mesoporous nano-structured carbon material *Electrophoresis* **31** 2981–8

[8] Wang Q, Ma S, Huang H, Cao A, Li M and He L 2016 Highly sensitive and selective spectrofluorimetric determination of nitrite in food products with a novel fluorogenic probe *Food Control* **63** 117–21
[9] Baig S A, Akhtera N A, Ashfaq M and Asi M R 2009 Determination of the organophosphorus pesticide in vegetables by high-performance liquid chromatography *Am.-Eurasian J. Agri. Environ. Sci.* **6** 513–9
[10] Meng J, Shi C, Wei B, Yu W, Deng C and Zhang X 2011 Preparation of Fe_3O_4@ C@ PANI magnetic microspheres for the extraction and analysis of phenolic compounds in water samples by gas chromatography–mass spectrometry *J. Chromatogr.* A **1218** 2841–7
[11] Safavi A and Karimi M A 2002 Flow injection chemiluminescence determination of hydrazine by oxidation with chlorinated isocyanurates *Talanta* **58** 785–92
[12] Ghaedi M, Ahmadi F and Shokrollahi A 2007 Simultaneous preconcentration and determination of copper, nickel, cobalt and lead ions content by flame atomic absorption spectrometry *J. Hazard. Mater.* **142** 272–8
[13] Bakker E and Telting-Diaz M 2002 Electrochemical sensors *Anal. Chem.* **74** 2781–800
[14] Hanrahan G, Patil D G and Wang J 2004 Electrochemical sensors for environmental monitoring: design, development and applications *J. Environ. Monit.* **6** 657–64
[15] Chokkiah B, Eswaran M, Wabaidur S M, Alothman Z A, Lee S C and Dhanusuraman R 2022 An efficient amperometric sensor for chloride ion detection through electroactive e-spun PVA-PANi-g-C_3N_4 nanofiber *J. Mater. Sci., Mater. Electron.* **33** 9425–37
[16] Feng X, Han G, Cai J and Wang X 2022 Au@carbon quantum dots-MXene nanocomposite as an electrochemical sensor for sensitive detection of nitrite *J. Colloid Interface Sci.* **607** 1313–22
[17] Janssen L and Koene L 2002 The role of electrochemistry and electrochemical technology in environmental protection *Chem. Eng. J.* **85** 137–46
[18] Chokkiah B, Eswaran M, Alothman A A, Alsawat M, Ifseisi A A, Alqahtani K N and Dhanusuraman R 2022 Facile fabrication of hollow polyaniline/carbon nanofibers-coated platinum nanohybrid composite electrode as improved anode electrocatalyst for methanol oxidation *J. Mater. Sci., Mater. Electron.* **33** 8768–76
[19] Chokkiah B, Wabaidur S M, Siddiqui M R, Islam M A, Dhanusuraman R and Ponnusamy V K 2022 Easy fabrication of a novel electro-spun PVDF-g-C_3N_4-Pd nanocomposite material as improved anode electrocatalyst for direct alcohol fuel cell *Fuel* **324** 124496
[20] Walsh F 2001 Electrochemical technology for environmental treatment and clean energy conversion *Pure Appl. Chem.* **73** 1819–37
[21] Bavatharani C, Muthusankar E, Lee S C, Johan M R and Ragupathy D 2020 Excellent cyclic retention and supercapacitive performance of electrochemically active nanocomposite electrode *Sens. Lett.* **18** 395–400
[22] Eswaran M, Dhanusuraman R, Chokkiah B, Tsai P-C, Wabaidur S M, Alothman Z A and Ponnusamy V K 2022 Poly (diphenylamine) and its nanohybrids for chemicals and biomolecules analysis: a review *Curr. Anal. Chem.* **18** 546–62
[23] Howell K A, Achterberg E P, Braungardt C B, Tappin A D, Turner D R and Worsfold P J 2003 The determination of trace metals in estuarine and coastal waters using a voltammetric *in situ* profiling system *Analyst* **128** 734–41
[24] Janjani P, Bhardwaj U, Gupta R and Kushwaha H S 2022 Bimetallic Mn/Fe MOF modified screen-printed electrodes for non-enzymatic electrochemical sensing of organophosphate *Anal. Chim. Acta* **1202** 339676

[25] Hryniewicz B M, Orth E S and Vidotti M 2018 Enzymeless PEDOT-based electrochemical sensor for the detection of nitrophenols and organophosphates *Sensors Actuators* B **257** 570–8

[26] Wei Y, Xu R-X, Gao C, Liu J-H and Huang X-J 2012 Polishing-activated nano α-Al_2O_3: adsorption and electrochemical behavior toward organophosphate pesticides *Electrochem. Commun.* **18** 78–80

[27] Gopi P K, Ngo D B, Chen S-M, Ravikumar C H and Surareungchai W 2022 High-performance electrochemical sensing of hazardous pesticide Paraoxon using $BiVO_4$ nano dendrites equipped catalytic strips *Chemosphere* **288** 132511

[28] Wang J, Chen D, Gan Y, Sun X and Jin Y 2011 High sensitive self-assembled monolayer modified solid mounted resonator for organophosphate vapor detection *Appl. Surf. Sci.* **257** 4365–9

[29] Song D, Li Q, Lu X, Li Y, Li Y, Wang Y and Gao F 2018 Ultra-thin bimetallic alloy nanowires with porous architecture/monolayer MoS_2 nanosheet as a highly sensitive platform for the electrochemical assay of hazardous omethoate pollutant *J. Hazard. Mater.* **357** 466–74

[30] Wu S, Huang F, Lan X, Wang X, Wang J and Meng C 2013 Electrochemically reduced graphene oxide and Nafion nanocomposite for ultralow potential detection of organophosphate pesticide *Sensors Actuators* B **177** 724–9

[31] Song D, Wang Y, Lu X, Gao Y, Li Y and Gao F 2018 Ag nanoparticles-decorated nitrogen-fluorine co-doped monolayer MoS_2 nanosheet for highly sensitive electrochemical sensing of organophosphorus pesticides *Sensors Actuators* B **267** 5–13

[32] Mukherjee S, Pal S, Paria P, Bhattacharyya S, Ghosh K, Pal A, Ghosh D, Singh O K, Sarkar P and Behera B K 2021 On-spot biosensing device for organophosphate pesticide residue detection in fruits and vegetables *Curr. Res. Biotechnol.* **3** 308–16

[33] Ravi A K, Punnakkal N, Vasu S P, Nair B G and T G S B 2020 Manganese dioxide based electrochemical sensor for the detection of nitro-group containing organophosphates in vegetables and drinking water samples *J. Electroanal. Chem.* **859** 113841

[34] Zhao Y, Li X, Chen J, Lu X, Yang Y, Song D and Gao F 2022 Porous hierarchical peony-like cobalt-based bimetallic oxides structured by ultrathin nanosheets for highly sensitive electrochemical pesticides detection *Sensors Actuators* B **352** 131072

[35] Guan H, Zhang F, Yu J and Chi D 2012 The novel acetylcholinesterase biosensors based on liposome bioreactors–chitosan nanocomposite film for detection of organophosphates pesticides *Food Res. Int.* **49** 15–21

[36] Zhang P, Sun T, Rong S, Zeng D, Yu H, Zhang Z, Chang D and Pan H 2019 A sensitive amperometric AChE-biosensor for organophosphate pesticides detection based on conjugated polymer and Ag-rGO-NH_2 nanocomposite *Bioelectrochemistry* **127** 163–70

[37] Dzudzevic Cancar H, Soylemez S, Akpinar Y, Kesik M, Göker S, Gunbas G, Volkan M and Toppare L 2016 A novel acetylcholinesterase biosensor: core–shell magnetic nanoparticles incorporating a conjugated polymer for the detection of organophosphorus pesticides *ACS Appl. Mater. Interfaces* **8** 8058–67

[38] Liu G and Lin Y 2005 Electrochemical sensor for organophosphate pesticides and nerve agents using zirconia nanoparticles as selective sorbents *Anal. Chem.* **77** 5894–901

[39] Chen A, Du D and Lin Y 2012 Highly sensitive and selective immuno-capture/electrochemical assay of acetylcholinesterase activity in red blood cells: a biomarker of exposure to organophosphorus pesticides and nerve agents *Environ. Sci. Technol.* **46** 1828–33

[40] Shalini Devi K, Anusha N, Raja S and Senthil Kumar A 2018 A new strategy for direct electrochemical sensing of a organophosphorus pesticide, triazophos, using a coomassie brilliant-blue dye surface-confined carbon-black-nanoparticle-modified electrode *ACS Appl. Nano Mater.* **1** 4110–9
[41] Zhao Y, Zuo X, Lu X, Li Z and Gao F 2022 Hierarchical porous hollow N-doped Cu-based MOF derivatives as highly sensitive electrochemical sensing platform for pesticides detection *Sensors Actuators* B **362** 131749
[42] Ruiyi L, Hui J, Nana L, Dan X and Zaijun L 2022 Electrochemical detection of chlorpyrifos in fruits with gold-histidine functionalized graphene quantum dot-graphene hybrid and target-induced DNA cycle amplification *Sensors Actuators* B **355** 131314
[43] Mahmoudi E, Fakhri H, Hajian A, Afkhami A and Bagheri H 2019 High-performance electrochemical enzyme sensor for organophosphate pesticide detection using modified metal-organic framework sensing platforms *Bioelectrochemistry* **130** 107348
[44] Nesakumar N, Suresh I, Jegadeesan G B, Rayappan J B B and Kulandaiswamy A J 2022 An efficient electrochemical sensing platform for profenofos detection *Measurement* **202** 111807
[45] Singh M, Kashyap H, Singh P K, Mahata S, Rai V K and Rai A 2019 AuNPs/neutral red-biofunctionalized graphene nanocomposite for nonenzymatic electrochemical detection of organophosphate via NO_2 reduction *Sensors Actuators* B **290** 195–202
[46] Poudyal D C, Dhamu V N, Samson M, Muthukumar S and Prasad S 2023 Pesticide analytical screening system (PASS): a novel electrochemical system for multiplex screening of glyphosate and chlorpyrifos in high-fat and low-fat food matrices *Food Chem.* **400** 134075
[47] Zhang D, Lang X, Hui N and Wang J 2022 Dual-mode electrochemical biosensors based on chondroitin sulfate functionalized polypyrrole nanowires for ultrafast and ultratrace detection of acetamiprid pesticide *Microchem. J.* **179** 107530
[48] Song D, Jiang X, Li Y, Lu X, Luan S, Wang Y, Li Y and Gao F 2019 Metal–organic frameworks-derived MnO_2/Mn_3O_4 microcuboids with hierarchically ordered nanosheets and Ti_3C_2 MXene/Au NPs composites for electrochemical pesticide detection *J. Hazard. Mater.* **373** 367–76
[49] Duan S, Wu X, Shu Z, Xiao A, Chai B, Pi F, Wang J, Dai H and Liu X 2022 Curcumin-enhanced MOF electrochemical sensor for sensitive detection of methyl parathion in vegetables and fruits *Microchem. J.* **184** 108182
[50] Qiu L, Lv P, Zhao C, Feng X, Fang G, Liu J and Wang S 2019 Electrochemical detection of organophosphorus pesticides based on amino acids conjugated nanoenzyme modified electrodes *Sensors Actuators* B **286** 386–93
[51] Wan Y, Chen X, Li S, Yin M, Zhang W, Mao G, Wu C and Wang H 2022 A selective electroanalysis and photocatalytic removal strategy for pesticide residues using urchin-like $LaPO_4$@ Ag *Electrochim. Acta* **410** 140039
[52] Singh A P, Balayan S, Gupta S, Jain U, Sarin R and Chauhan N 2021 Detection of pesticide residues utilizing enzyme-electrode interface via nano-patterning of TiO_2 nano-particles and molybdenum disulfide (MoS_2) nanosheets *Process Biochem.* **108** 185–93
[53] Kunpatee K, Kaewdorn K, Duangtong J, Chaiyo S, Chailapakul O, Kalcher K, Kerr M and Samphao A 2022 A new disposable electrochemical sensor for the individual and simultaneous determination of carbamate pesticides using a nanocomposite modified screen-printed electrode *Microchem. J.* **177** 107318

[54] Ehzari H, Safari M, Samimi M, Shamsipur M and Gholivand M B 2022 A highly sensitive electrochemical biosensor for chlorpyrifos pesticide detection using the adsorbent nanomatrix contain the human serum albumin and the Pd:CdTe quantum dots *Microchem. J.* **179** 107424

[55] Zheng D, Chen M, Chen Y and Gao W 2022 *In-situ* preparation of hollow $CdCoS_2$ heterojunction with enhanced photocurrent response for highly photoelectrochemical sensing of organophosphorus pesticides *Anal. Chim. Acta* **1212** 339913

[56] Lei W, Wu Q, Si W, Gu Z, Zhang Y, Deng J and Hao Q 2013 Electrochemical determination of imidacloprid using poly (carbazole)/chemically reduced graphene oxide modified glassy carbon electrode *Sensors Actuators* B **183** 102–9

[57] Suresh I, Nesakumar N, Jegadeesan G B, Jeyaprakash B, Rayappan J B B and Kulandaiswamy A J 2022 Real-time detection of imidacloprid residues in water using f-MWCNT/EDTA as energetically suitable electrode interface *Anal. Chim. Acta* **1235** 340560

[58] Ajermoun N, Aghris S, Ettadili F, Alaoui O T, Laghrib F, Farahi A, Lahrich S, Bakasse M, Saqrane S and El Mhammedi M 2022 Phytotoxic effect of the insecticide imidacloprid in *Phaseolus vulgaris* L. plant and evaluation of its bioaccumulation and translocation by electrochemical methods *Environ. Res.* **214** 113794

[59] Li R, Zhe T, Li M, Li J, Li F, Bai F, Li L, Ma K, Shen S and Wang L 2022 Simple high-temperature annealing affords commercial carbon cloth with enhanced electrochemical performance for highly sensitive detection of imidacloprid *J. Pharm. Biomed. Anal.* **219** 114963

[60] Rashed M A, Faisal M, Alsareii S, Alsaiari M, Jalalah M and Harraz F A 2022 Highly sensitive and selective electrochemical sensor for detecting imidacloprid pesticide using novel silver nanoparticles/mesoporous carbon/hematite ore ternary nanocomposite *J. Environ. Chem. Eng.* **10** 108364

[61] Elbaz G A, Zaazaa H E, Monir H H, Abd El Halim L M and Atty S A 2022 Nano eco-friendly voltammetric determination of pesticide, imidacloprid and its residues in thyme and guava leaves *Sustain. Chem. Pharm.* **29** 100799

[62] Surucu O, Bolat G and Abaci S 2017 Electrochemical behavior and voltammetric detection of fenitrothion based on a pencil graphite electrode modified with reduced graphene oxide (RGO)/poly (E)-1-(4-((4-(phenylamino) phenyl) diazenyl) phenyl) ethanone (DPA) composite film *Talanta* **168** 113–20

[63] Kalinic B, Girardi L, Ragonese P, Faramawy A, Mattei G, Frasconi M, Baretta R, Bogialli S, Roverso M and Rizzi G 2022 Diffusion-driven formation of Co_3O_4 nanopetals layers for photoelectrochemical degradation of organophosphate pesticides *Appl. Surf. Sci.* **596** 153552

[64] Sakthivel K, Akilarasan M, Chen S-M, Ye Y-T, Chen T-W and Ali M A 2019 A novel synthetic approach to tungsten carbide polyhedrons: an effective electrocatalyst for the detection of organophosphate pesticide (fenitrothion) residues in environmental samples *Mater. Chem. Phys.* **233** 52–9

[65] Rajaji U, Chinnapaiyan S, Chen T-W, Chen S-M, Mani G, Mani V, Ali M A, Al-Hemaid F M and El-Shikh M S 2021 Rational construction of novel strontium hexaferrite decorated graphitic carbon nitrides for highly sensitive detection of neurotoxic organophosphate pesticide in fruits *Electrochim. Acta* **371** 137756

[66] Kilele J C, Chokkareddy R and Redhi G G 2021 Ultra-sensitive electrochemical sensor for fenitrothion pesticide residues in fruit samples using IL@ $CoFe_2O_4$NPs@ MWCNTs nanocomposite *Microchem. J.* **164** 106012

[67] Zeng Y, Yu D, Yu Y, Zhou T and Shi G 2012 Differential pulse voltammetric determination of methyl parathion based on multiwalled carbon nanotubes–poly (acrylamide) nanocomposite film modified electrode *J. Hazard. Mater.* **217** 315–22

[68] Xu M, Zhu J, Su H, Dong J, Ai S and Li R 2012 Electrochemical determination of methyl parathion using poly (malachite green)/graphene nanosheets–nafion composite film-modified glassy carbon electrode *J. Appl. Electrochem.* **42** 509–16

[69] Kokulnathan T, Wang T-J, Kumar E A and Ahmed F 2022 Construction of nickel cobalt-layered double hydroxide/functionalized–halloysite nanotubes composite for electrochemical detection of organophosphate insecticide *Chem. Eng. J.* **433** 133639

[70] Devi R K, Ganesan M, Chen T-W, Chen S-M, Lin K-Y, Akilarasan M, Al-Onazi W A, Rasheed R A and Elshikh M S 2022 Tailored architecture of molybdenum carbide/iron oxide micro flowers with graphitic carbon nitride: an electrochemical platform for nano-level detection of organophosphate pesticide in food samples *Food Chem.* **397** 133791

[71] Nagabooshanam S, John A T, Wadhwa S, Mathur A, Krishnamurthy S and Bharadwaj L M 2020 Electro-deposited nano-webbed structures based on polyaniline/multi walled carbon nanotubes for enzymatic detection of organophosphates *Food Chem.* **323** 126784

[72] Gong J, Wang L, Song D, Zhu X and Zhang L 2009 Stripping voltammetric analysis of organophosphate pesticides using Ni/Al layered double hydroxides as solid-phase extraction *Biosens. Bioelectron.* **25** 493–6

[73] Subbiah R and Tiwari R R 2021 The herbicide paraquat-induced molecular mechanisms in the development of acute lung injury and lung fibrosis *Crit. Rev. Toxicol.* **51** 36–64

[74] Nasir T, Herzog G, Hébrant M, Despas C, Liu L and Walcarius A 2018 Mesoporous silica thin films for improved electrochemical detection of paraquat *ACS Sens.* **3** 484–93

[75] Laghrib F, Bakasse M, Lahrich S and El Mhammedi M 2020 Electrochemical sensors for improved detection of paraquat in food samples: a review *Mater. Sci. Eng.* C **107** 110349

[76] Jeyapragasam T, Raju R, Chen S-M, Saraswathi R, Hatamleh A A, Chen T-W and Rwer S 2019 Poly (o-phenylenediamine)—multiwalled carbon nanotube nanocomposite based electrochemical sensing platform for paraquat detection *Int. J. Electrochem. Sci.* **14** 8326–39

[77] Niu Z, Liu Y, Li X, Yan K and Chen H 2022 Electrochemical sensor for ultrasensitive detection of paraquat based on metal-organic frameworks and para-sulfonatocalix [4] arene-AuNPs composite *Chemosphere* **307** 135570

[78] Kavazoi H S, Martin C S and Alessio P 2022 Comparative study of tetrasulfonated phthalocyanine modified screen-printed electrodes in paraquat *Synth. Met.* **284** 116988

[79] Li M, Wang X, Zhu Y, Jia X, Zhang S, Wang H, Li Y and Hu G 2023 Fe_2O_3-decorated boron/nitrogen-co-doped carbon nanosheets as an electrochemical sensing platform for ultrasensitive determination of paraquat in natural water *Chin. Chem. Lett.* **34** 107299

[80] Zhang J, Wang C, Niu Y, Li S and Luo R 2017 Electrochemical sensor based on molecularly imprinted composite membrane of poly (o-aminothiophenol) with gold nanoparticles for sensitive determination of herbicide simazine in environmental samples *Sensors Actuators* B **249** 747–55

[81] Akyol H, Riciputi Y, Capanoglu E, Caboni M F and Verardo V 2016 Phenolic compounds in the potato and its byproducts: an overview *Int. J. Mol. Sci.* **17** 835

[82] Ren J, Li L, Cui M, Zhai M, Yu C and Ji X 2016 Nitrobenzene electrochemical sensor based on silver nanoparticle supported on poly-melamine functional multi-walled carbon nanotube *Ionics* **22** 1937–45

[83] Boakye A, Yu K, Asinyo B K, Chai H, Raza T, Xu T, Zhang G and Qu L 2022 A portable electrochemical sensor based on manganese porphyrin-functionalized carbon cloth for highly sensitive detection of nitroaromatics and gaseous phenol *Langmuir* **38** 12058–69

[84] Xu G, Li B, Wang X and Luo X 2014 Electrochemical sensor for nitrobenzene based on carbon paste electrode modified with a poly (3, 4-ethylenedioxythiophene) and carbon nanotube nanocomposite *Microchim. Acta* **181** 463–9

[85] Yi Y, Kingsford O J, Ma Y, Wu Y and Zhu G 2020 Simultaneous electrochemical sensing of 1-chloro-4-nitrobenzene and N-(4-hydroxyphenyl) acetamide based on nitrogen-doped carbon black *Microchem. J.* **159** 105346

[86] Kokulnathan T, Jothi A I, Chen S-M, Almutairi G, Ahmed F, Arshi N and AlOtaibi B 2021 Integrating graphene oxide with magnesium oxide nanoparticles for electrochemical detection of nitrobenzene *J. Environ. Chem. Eng.* **9** 106310

[87] Vinoth S, Rajaitha P M and Pandikumar A 2020 *In-situ* pyrolytic processed zinc stannate incorporated graphitic carbon nitride nanocomposite for selective and sensitive electrochemical determination of nitrobenzene *Compos. Sci. Technol.* **195** 108192

[88] Cotchim S, Thavarungkul P, Kanatharana P and Limbut W 2015 A new strategy for 2,4,6-trinitrotoluene adsorption and electrochemical reduction on poly (melamine)/graphene oxide modified electrode *Electrochim. Acta* **184** 102–10

[89] Moon S, Charyulu D K, Lee J, Kim T, Kim J, Lee W and Lee K 2022 Chemical-bath-deposited rutile TiO_2 film for electrochemical detection of 2,4,6-trinitrotoluene *Thin Solid Films* **748** 139172

[90] Sağlam Ş, Üzer A, Tekdemir Y, Erçağ E and Apak R 2015 Electrochemical sensor for nitroaromatic type energetic materials using gold nanoparticles/poly (o-phenylenediamine–aniline) film modified glassy carbon electrode *Talanta* **139** 181–8

[91] Lima A P, Almeida P L, Sousa R M, Richter E M, Nossol E and Munoz R A 2019 Effect of alumina supported on glassy-carbon electrode on the electrochemical reduction of 2,4,6-trinitrotoluene: a simple strategy for its selective detection *J. Electroanal. Chem.* **851** 113385

[92] Zhang R, Zhang C, Zheng F, Li X, Sun C-L and Chen W 2018 Nitrogen and sulfur co-doped graphene nanoribbons: a novel metal-free catalyst for high performance electrochemical detection of 2, 4, 6-trinitrotoluene (TNT) *Carbon* **126** 328–37

[93] Salaria K, Mehta N, Krishna C and Mehta S K 2021 Electrochemical detection of TNT using CdS nanoparticles via cyclic voltammetry and amperometry *Curr. Res. Green Sustain. Chem.* **4** 100166

[94] Dubey M, Kumar R, Srivastava S, Verma A, Mathur A, Chawla V, Rubahn H-G, Joshi M and Mishra Y 2022 $ZnO/\alpha\text{-}MnO_2$ hybrid 1D nanostructure-based sensor for point-of-care monitoring of chlorinated phenol in drinking water *Mater. Today Chem.* **26** 101098

[95] Yang C-Y, Chen S-M, Tsai T-H and Unnikrishnan B 2012 Poly (diphenylamine) with multi-walled carbon nanotube composite film modified electrode for the determination of phenol *Int. J. Electrochem. Sci.* **7** 12796–807

[96] Padmanaban A, Murugadoss G, Venkatesh N, Hazra S, Kumar M R, Tamilselvi R and Sakthivel P 2021 Electrochemical determination of harmful catechol and rapid decolorization of textile dyes using ceria and tin doped ZnO nanoparticles *J. Environ. Chem. Eng.* **9** 105976

[97] Du Y, Wang B, Kang K, Ji X, Wang L, Zhao W and Ren J 2022 Signal synergistic amplification strategy based on functionalized CeMOFs for highly sensitive electrochemical detection of phenolic isomers *Microchem. J.* **177** 107285
[98] Tian F, Li H, Li M, Li C, Lei Y and Yang B 2017 Synthesis of one-dimensional poly (3,4-ethylenedioxythiophene)-graphene composites for the simultaneous detection of hydroquinone, catechol, resorcinol, and nitrite *Synth. Met.* **226** 148–56
[99] Ranjith K S, Vilian A E, Ghoreishian S M, Umapathi R, Hwang S-K, Oh C W, Huh Y S and Han Y-K 2022 Hybridized 1D–2D $MnMoO_4$–MXene nanocomposites as high-performing electrochemical sensing platform for the sensitive detection of dihydroxybenzene isomers in wastewater samples *J. Hazard. Mater.* **421** 126775
[100] Cheng D and Kan X 2020 Simultaneous determination of dihydroxybenzene isomers based on gold dendritic/pEDOT electrochemical sensor *J. Electroanal. Chem.* **857** 113741
[101] Wei W, Yang S, Hu H, Li H and Jiang Z 2021 Hierarchically grown $ZnFe_2O_4$-decorated polyaniline-coupled-graphene nanosheets as a novel electrocatalyst for selective detecting p-nitrophenol *Microchem. J.* **160** 105777
[102] Tang J, Zhang L, Han G, Liu Y and Tang W 2015 Graphene-chitosan composite modified electrode for simultaneous detection of nitrophenol isomers *J. Electrochem. Soc.* **162** B269
[103] Thangavelu D, Chen Y, Annamalai P, Ramadoss M and Narayanan V 2022 Rationally designed Ag@ polymer@ 2-D LDH nanoflakes for bifunctional efficient electrochemical sensing of 4-nitrophenol and water oxidation reaction *ACS Appl. Mater. Interfaces* **14** 6518–27
[104] Padmanaban A, Dhanasekaran T, Manigandan R, Kumar S P, Gnanamoorthy G, Stephen A and Narayanan V 2017 Facile solvothermal decomposition synthesis of single phase $ZnBi_{38}\,O_{60}$ nanobundles for sensitive detection of 4-nitrophenol *New J. Chem.* **41** 7020–7
[105] Fan Y, Liu J-H, Yang C-P, Yu M and Liu P 2011 Graphene–polyaniline composite film modified electrode for voltammetric determination of 4-aminophenol *Sensors Actuators* B **157** 669–74
[106] Kumar S P, Giribabu K, Manigandan R, Munusamy S, Muthamizh S, Padmanaban A, Dhanasekaran T, Suresh R and Narayanan V 2016 Simultaneous determination of paracetamol and 4-aminophenol based on poly (chromium Schiff base complex) modified electrode at nanomolar levels *Electrochim. Acta* **194** 116–26
[107] Zhang R, Liu J and Li Y 2019 MXene with great adsorption ability toward organic dye: an excellent material for constructing a ratiometric electrochemical sensing platform *ACS Sens.* **4** 2058–64
[108] Veerakumar P, Chen S-M, Madhu R, Veeramani V, Hung C-T and Liu S-B 2015 Nickel nanoparticle-decorated porous carbons for highly active catalytic reduction of organic dyes and sensitive detection of Hg (II) ions *ACS Appl. Mater. Interfaces* **7** 24810–21

Chapter 11

Recent trends and materials used for environmental monitoring and applications

C Rajkumar, Kazi Hasibur Rahman, P V Chandrasekar and Kuan-Chung Chen

The water–energy–environment nexus serves as a vital hub to resolve the major problems associated with human, economic, and environmental sustainability. Awareness of these complex correlations is important to balance the needs of distinct stakeholders and local and international policy decisions, and to conserve valuable resources in water management, the environment, and energy. Recently, extremely effective technologies for sustainable energy, water, and environmental interconnection have been developed, as well as future energy-efficient options to protect the environment and save energy, which is an important way to gain sustainable advancement. There are two temporal concerns: one is very crucial and immediate, which is the need to find a way to end energy generation from fossil fuel resources; and the second is to develop alternative energies and more sustainable access in the future. Improved designs of renewable materials have been established to replace petroleum-based materials, such as carbonized wood for energy storage and generation, white wood that enables radiation cooling, super wood with fine mechanical properties for advanced manufacturing, and transparent wood for solar cells and optoelectronics. In this chapter we discuss the new methodologies used in developing recent design technologies in the fields of energy, environment, and water, and the related potential in the fields of biodegradation and biodeterioration. Further, the increase in ultraviolet (UV) radiation due to global warming is also a major threat to the environment. Thus, this chapter also descibes the recent materials used in the detection of UV radiation and also those used as photocatalysts for effective environmental applications.

doi:10.1088/978-0-7503-5729-6ch11

11.1 Introduction

It is well known that water is one of the most important molecules on planet Earth, and is very important for the performance of Earth's ecosystem. A high volume of fresh water is now required due to the growth of the world's population, modern technological civilization, burgeoning industrialization, environmental and everyday global changes, and agricultural activities. Access to affordable and fresh water will remain a great challenge for people across the world and this issue is worsening every year due to uneven distribution, population growth, and climate change [1–3]. The United Nations (UN) reported that in 2050 approximately six billion people will suffer from a shortage of clean water [4]. Unified steps are essential to meet the challenge of increasing the global water reservoir and innovative materials play a vital role in removing the different toxic contaminants present in polluted water sources, improving the extraction of energy and chemicals from toxic water, and effectively detoxifying the water obtained from seawater, which is controlled by desalination. Membranes and polymers have been employed extensively for wastewater treatment and distillation of water, to remove salts, particles, metals, pathogens, and other constituents [5]. Rahman *et al* reported the detoxification of wastewater from toxic organic compounds and dyes of up to 90% using a conducting polymer which was composited with various amounts of TiO_2 nanoparticles [6]. Transition metal-doped TiO_2 nanoparticles are one of the key materials for the elimination of dye compounds in contaminated water [7]. Similarly, graphite and carbon materials are also candidates for wastewater treatment from a very early stage. Although these materials are stable, very cheap, and user-friendly, their action is restricted because of their intrinsic properties, and the manufacturing and disposal management of non-renewable materials, which are energy and environmental issues of concern [8, 9]. Due to this, in recent decades wood as well as other biomass materials have represented a suitable substitute for water engineering and energy applications. There are an estimate of over three trillion trees on Earth, which signifies over 380 trees per capita [10]. Wood panels are one of the important wood materials, and for the generation procedure, water is employed to immerse logs for the production of plywood, for cooling equipment and materials, and for cleaning. In the production systems, the industry sectors based on wood panels vary with respect to their usage of water. The direct wood based production process is well known to be a dry field, while panel production based on wood can create ca. 600 M m^3 of wastewater each year. This wastewater consists of a broad range of substances, including heavy metals, wood extracts, wood degradation products, and even surfactants, which are employed in cleaning procedures. Studies indicate that wood industry wastewaters are highly toxic to the environment even after substantial dilution [11]. During panel production, one of the most water-consuming techniques is wood hydrothermal treatment, in which logs are soaked in water basins, which increases wood plasticity prior to veneer cutting and debarking. Because of the hydrothermal technique, which is mostly carried out at 50 °C–70 °C, lignin, cellulose, and products of their degradation as well as wood extracts (e.g. tannins, resin acids, etc) are avoided under such conditions [12]. Wood extracts are usually formed in trees to protect them against attacks from pathogens such as bacteria or fungi, but their presence in wastewater creates a toxic environment [13].

11.2 Structural properties of wood

Wood is a structurally fibrous and porous tissue, which is found in the roots and trunks of trees. The components and structure of wood have been described widely in the literature [14–16]. The cell walls of wood mostly consist of ~10–30 wt% hemicelluloses, ~20–30 wt% lignin, and ~40–50 wt% cellulose. These components which are intercorrelated with everyone, confirming a better wood mechanical property [16]. Due to their geographical and type variation, different trees exhibit a broad range of wood microstructures. In particular, softwoods (e.g. spruce, cedar, pine, etc) usually have a homogeneous and simple structure with several fibers defined in a periodic manner; they show more porosity due to their rapid rate of growth. Due to the improvement of nanotechnology, nanomaterials based on wood have been employed in various emerging areas, including energy and the environment [15, 16]. In most circumstances, the choice of wood species is based on its chemical and physical properties for several applications. The procurement cost is also a main aspect in choosing the wood species. The hardwoods (e.g. balsam poplar and basswood) usually exhibit broader application prospects in devices for energy storage and conversion compared to softwoods, due to their denser porous structure, high mechanical strength, and low density [17]. However, no absolute model is used for choosing the wood species for many applications because it depends on the particular treatment methods and uses [18]. It is predictable that the use of nanotechnologies for treatment will create several unique properties in wood.

11.2.1 Transparent wood

By eliminating the light-absorbing components from wood and then introducing a refractive index material (such as epoxy resin), transparent wood was created. Additionally, useful transparent wood has been created. As an example, luminescent transparent wood was created by mixing Si and CdSe/ZnS core/shell quantum dots (QDs), and the light scattering of the transparent wood allowed for adjustment of the luminescence diffusivity. Rhodamine 6G, a fluorescent dye, and poly(methyl methacrylate), both of which were introduced into translucent wood, caused the wood to emit laser light more broadly thanks to the aligned cellulose fibers acting as optical resonators. By using Cs_xWO_3 nanoparticles, which absorb near-infrared radiation, Yu *et al* demonstrated a translucent wood with regulated thermal energy. The outcome was a high transparency of 86% at 550 nm for visible light transmittance and a high near-infrared absorption from 780 to 2500 nm for heat shielding.

11.2.2 Flexible wood

Flexible wood would be quite distinctive and have a promising future in flexible personal gadgets. However, as wood is often a bulky material with a rigid character, it is difficult to fabricate flexible wood. Creating superflexible wood has recently been achieved by a simple and scalable top-down method [19]. This superflexible wood is completely different to natural wood due to a change in chemical and physical characteristics. To assess biocompatibility and cell attachment, HEK293 cells were attached to the obtained flexible wood slice which was used as a bioscaffold.

11.2.3 Plasmonic wood

In addition, a wood slice modified with $CuFeSe_2$ NPs has also been developed for highly efficient solar steam generation and water purification (figure 11.1) [23]. Metallic nanoparticles (NPs) having plasmonic characteristics are utilized to change wood in order to broaden the applications of wood. By depositing metal nanoparticles (e.g. Ag NPs, Au, and Pd) inside of wood micro/mesostructures, plasmonic

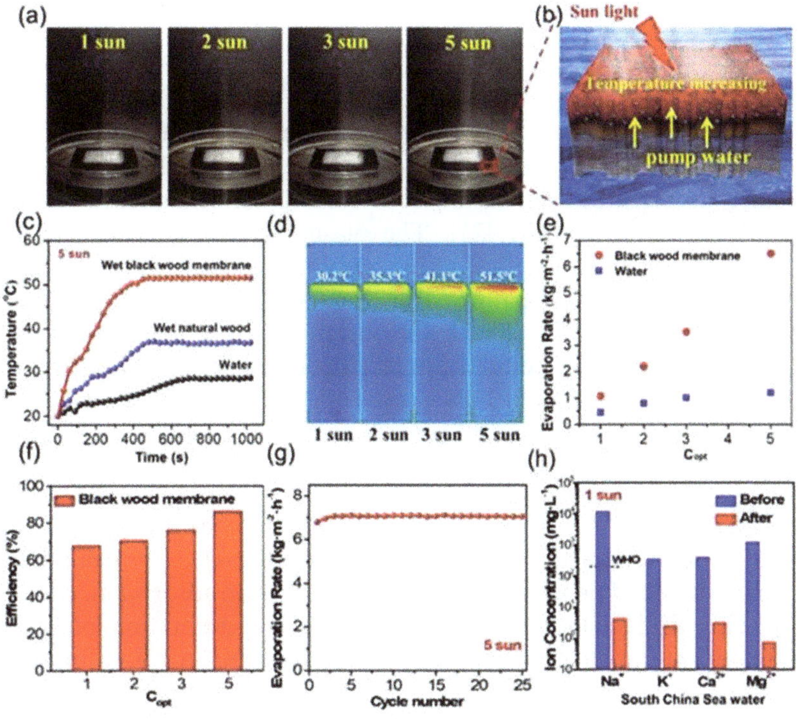

Figure 11.1. The performance of black wood membrane solar desalination. (a) Vapor columns at several illuminations. (b) Black wood membrane schematic diagram for effective generation of solar steam. (c) Under 5 kW ■ m^{-2} ■ h^{-1}, the wet natural and black wood of the temperature increasing process. (d) IR images of the black wood membrane temperature distribution at several illumination intensities. (e) The black wood membrane evaporation rate depends on different C_{opt}. (f) The black wood membrane solar steam efficiency for different values of C_{opt}. (g) The black wood membrane water evaporation stability in sunlight for 5 over 25 h (1 h for every cycle). (h) From the South China Sea, the measured Na^+, K^+, Ca^{2+}, and Mg^{2+} concentrations in actual seawater samples, each with ~1.05 wt% average salinity, after and before desalination using the black wood membrane at 1 kW ■ m^{-2}. (Reproduced with permission from [23]. Copyright 2018 the Royal Society of Chemistry.)

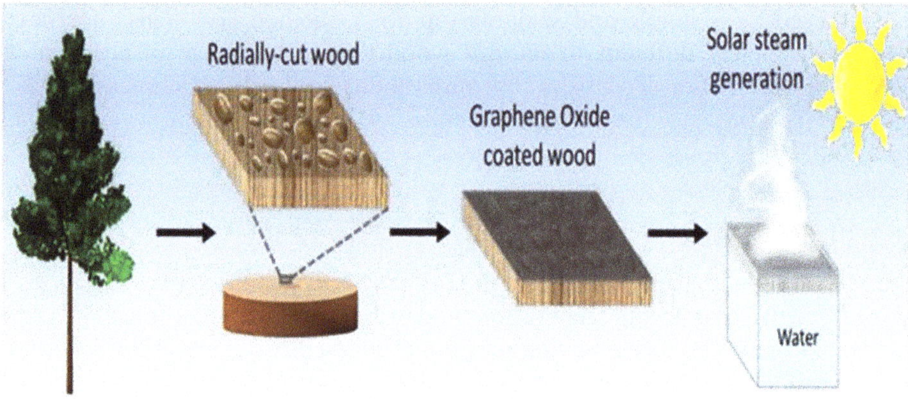

Figure 11.2. Scheme depicting the wood–graphene oxide composite preparation and solar steam production set-up with the bilayered composite structure. (Reproduced with permission from [22]. Copyright 2017 American Chemical Society.)

wood has been created successfully [20]. Due to the plasmonic effect, the Pd NP-coated plasmonic wood could transform the incoming light into heat energy. Additionally, the heat restricted on the wood's micro/nanochannel surface generates a lot of solar steam effectively [21]. The plasmonic wood, which has excellent hydrophobicity, could also be efficient in transporting water, maintaining a steady supply of water for vapor formation. The wood–GO evaporator separated bilayer structure (figure 11.2) contrasts with the integrated structure of such plasmonic wood [22]. For very effective solar steam production as well as water purification, a wood slice altered with $CuFeSe_2$ NPs has also been created (figure 11.1) [23].

11.3 Engineered nanowood for water and wastewater treatment and energy recovery

Engineered wood materials can perform a variety of tasks. For example, delignification can make wood pores more accessible for particle adsorption, carbonization can create electrically conductive nanowood for the generation of energy, and chemical treatment can alter the hydrophobicity of material to allow for the selective removal of contaminants in water or the distillation of water. The modification techniques, material characteristics, and the performance of applications are very crucial and are discussed in this chapter. A good overview of the use of cellulose nanoparticles, particularly cellulose derivatives for the filtration of water, is presented in the review in [24].

11.4 Cellulose-based membranes for water filtration

The fabrication methods for cellulose-based membranes include common vacuum filtration, blending, and phase inversion, as well as spin coating and interfacial polymerization. The base materials can be nanocellulose, such as cellulose nanofibrils, nanocrystals, or cellulose derivative-based nanofibers. The name 'nanocellulose' refers to a collection of materials that includes bacterial nanocellulose (BNC), nanocellulose crystals, and nanocellulose fibers (CNFs). CNFs are microfibrils detached from the plant cell wall. Mechanical and chemical treatments are used to break down the complex fiber matrix to prepare CNFs. The width of CNFs is dependent on the botanical source and corresponds to the microfibril width in the initial plant. CNFs made from wood possess a diameter of 10–30 nm. CNFs possess both amorphous and crystalline regions. Cellulose nanomaterials are being employed as coating materials or as additives in polymer matrices. Such alterations can result in extremely porous structures for smoother surfaces, high hydrophilicity, and better permeability to increase the mechanical properties to tolerate extreme pressures as well as to lessen membrane fouling. We primarily concentrate on current developments of cellulose-based membranes for water filtration, including forward osmosis (FO), nanofiltration (NF), and ultrafiltration (UF) membranes. Previous reviews have covered the early investigations of materials based on cellulose in membrane filtration. To achieve desired functions, several fabrication techniques can be used to create cellulose-based membranes. Common vacuum filtering, phase inversion, mixing, interfacial polymerization, and spin coating are a few examples.

11.5 UV photodetectors for environmental monitoring

Green plants growth processes are inhibited by excessive ultraviolet (UV)-B radition. Therefore, ozone depletion is a concern because it may lead to plant species loss and decrease the global food supply. Since all life on the planet is interconnected, changes in the balance of plant species can have major effects. Plants create the base of the food web, and prevent soil erosion as well as water loss. They are the prime creators of oxygen and the main sink (storage site) of carbon dioxide.

Terrestrial and aquatic biogeochemical cycles could be affected by excessive UV-B radiation, thus altering the chemically essential trace gases (e.g. carbon monoxide, carbon dioxide, ozone, carbonyl sulfide, etc) and both the sink and sources of greenhouse gases. These changes would create biosphere–atmosphere feedback that amplifies or mitigates the concentration of these gases in the atmosphere.

Currently, researchers and industry are showing more interest in photodetection in the UV region because of its application in environmental monitoring. Light in the UV region is divided into four regions, namely UV-A, UV-B, UV-C, and far UV. The wavelength of the UV-A region is between 400 and 320 nm. The wavelength of the UV-B region between 320 and 280 nm, the wavelength of the UV-C region is between 280 and 200 nm. The wavelength of the far UV region is between 200 and 10 nm, which touches the low energy frontier of the x-ray region. The atmospheric ozone layer mostly absorbs the UV radiation which is observed in solar radiation. In solar radition, wavelengths which are very long compared to 280 nm can puncture the atmosphere and reach the Earth. As UV photodetectors (PDs) have very high sensitivity in the region of UV compared to the visible region (400 to 800 nm), they are also called visible-blind UV photodetectors. The best UV photodetector should fulfill the 5S requirements, namely a high signal-to-noise ratio, speed, sensitivity, spectral selectivity, and stability. UV photodetecors have been prepared using various type of materials, such as zero-dimensional (0D), one-dimenstional (1D), two-dimensional (2D), as well as three-dimensional (3D) materials. The materials used to prepare UV photodetectors are also divided into metal oxides, metal chalcogenides, perovskite-structured materials, etc. Among these, metal oxides are more the focus to prepare UV photodetectors because of their compositions, physical and chemical properties, and special shapes. Moreover, they are among the more common minerals on Earth. Some of the metal oxides used to prepare UV photodetectors are SnO_2 (tetragonal, 3.6 eV), ZnO (hexagonal, 3.37 eV), In_2O_3 (cubic, 3.6 eV), and Ga_2O_3 (monoclinic, 4.2–4.9 eV). The UV photodetectors prepared using metal oxides offer many advantages compared to SiC, GaN, etc. For example, the UV photodetectors based on metal oxides do not oxidize easily and show a good sensitivity response. Therefore, they can be made small in size and easy to operate. Hence, researchers have shown more interest in preparing UV photodetectors using metal oxides.

The wide bandgap β-Ga_2O_3 semiconductor is gaining significant interest for preparing UV photodetectors because of its intrinsic solar-blind capability, large bandgap, and high thermal and chemical stability. Recently, researchers have also focused on Ga_2O_3 due to its unique advantages compared to other wide bandgap semiconductors, which include high breakdown voltage, a Baliga figure of merit, and intrinsic UV-C spectral selectivity. High-resistivity *n*-type κ-Ga_2O_3 epilayers were used to prepare UV PDs. The very high photo-gain was dependent on the applied bias linearly up to a specific critical bias voltage. Beyond this specific critical voltage, the photo-gain performance reached saturation at larger applied bias from the linear. The photo-gain is also high due to the collection of excess majority carriers (electrons) per unit time as a result of self-trapping/hole trapping in the conductive channel. The diffusion length of the minority carrier was estimated based on the saturation of the photo-gain voltage value. The calculation of the hole diffusion length in κ-Ga_2O_3 was first reported in this work. The UV PD has high stability and reproducibility over a long time under 250 nm illumination. It also has very high sensitivity and a high rejection ratio of $RS > 10^4$. Therefore, κ-Ga_2O_3 may be an appropriate material to fabricate solar-blind UV-C photodetectors [25]. A high performance UV PD was fabricated using Sn-doped Ga_2O_3 microwire. The chemical vapor deposition (CVD) technique was used to fabricate Sn-doped β-Ga_2O_3 microwires. An external quantum efficiency (EQE) could be reached of up to 58.87%, and responsivity could be reached up to 12 A W^{-1} under light illumination (254 nm) with a 100 μW cm^{-2} intensity at 10 V. Furthermore, it had a good I_{light}/I_{dark} ratio (2.5×10^3), low dark current (1.7 nA), and a fast decay time and rise time of 0.25 and 0.37 s, respectively. This is a feasible and economical method for preparing UV PDs (figure 11.3) [26]. A UV PD was fabricated on a p-type Si substrate using β-Ga_2O_3 nanowires (NWs) by a glancing angle deposition method inside a chamber with RF/DC sputtering. The β-Ga_2O_3 NWs were fabricated with an average length of ~270 nm, and had good photoabsorption and photoemission. The PD prepared employing β-Ga_2O_3 NW had a good responsivity of detectivity of

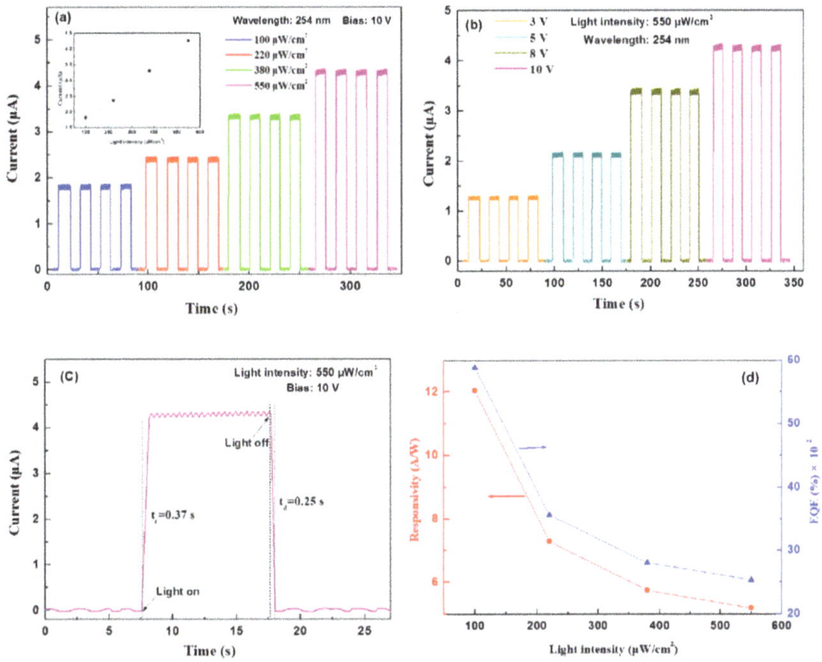

Figure 11.3. (a) The time-dependent photoresponse of a single Sn-doped β-Ga_2O_3 microwire photodetector under 254 nm illumination with various light intensities at 10 V. (b) The UV photodetector response time at various biases. (c) The expanded view of the UV PD rise/decay time. (d) Responsivity based on light intensities at 10 V. (Reproduced with permission from [26]. Copyright 2022 Elsevier.)

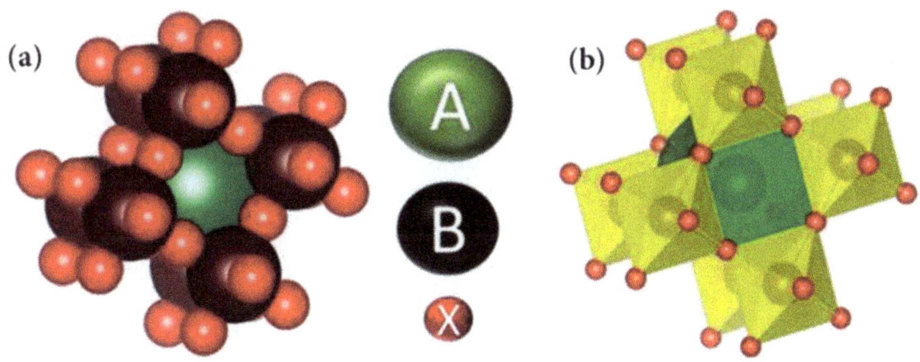

Figure 11.4. The crystal structure of perovskite with the generic chemical formula ABX_3. (a) Ball and stick configuration; (b) polyhedran configuration. Created with VESTA.3.5.8. (Reproduced with permission from [130]. Copyright 2011 IUCr Journals.)

1.37×10^{12} jones, 31.71 A W^{-1}, and an NEP of 2.44×10^{-12} W under 275 nm illumination at +5 V. Moreover, the prepared PD was worked at a very small power intensity of 6.37 μW cm^{-2}. It had fall and rise times of 0.25 and 0.18 s, respectively.

The photogenerated carriers were separated effectively by a strong p–n junction, which increased the performance of the PDs [27]. Jia *et al* constructed a deep UV photodetector using a NiO/β-Ga_2O_3 heterojunction. Magnetron sputtering was used to prepare the β-Ga_2O_3 layer. After annealing, the β-Ga_2O_3 layer had a selective orientation along the crystal plane (201) family. Under illumination (27 μW cm^{-2}) of 245 nm, the photodetector showed a maximum detectivity (*D**) (3.14 × 10^{12} cmHz^{1}/2W^{-1}) and high responsivity (*R*) (27.43 A W^{-1}) due to the p–NiO/n-β-Ga_2O_3 heterojunction [28].

Yadav *et al* improved the performace of a UV photodetector by incorporating lithium (Li) into nickel oxide (NiO). The presence of Ni vacancies was revealed by x-ray photoelectron spectroscopy (XPS). In lithium incorporated nickel oxide (LNO), the Li^+ is substitutional and incorporated in Ni^{2+} sites. As a result, a photodetector prepared using LNO showed high photocurrent (35.34 μA) in the UV region at a power density of 1.72 mW cm^{-2}. Moreover, it showed better detectivity (0.467 × 10^{12} jones) and responsivity (0.218 A W^{-1}) at a lower power density (0.31 mW cm^{-2}) as well [29].

Ren *et al* synthesized large-scale Bi_2O_3 nanotube (NT) arrays with the help of a ZnO nanorod template on a ITO substrate at room temperature. The as-prepared Bi_2O_3 NTs were used to fabricate a UV photodetector. Under UV irradiation, the Bi_2O_3 NT UV PD had a ~28.8 ms fast rise time, ~40.9 ms decay time, and 40.4 μA cm^{-2} average current density. Moreover, the Bi_2O_3 NTs UV PD was operated without any external bias and it has good repeatability and stability [30].

Among the several oxides, tungsten trioxide (WO_3) is also one of the favorable oxides for preparing UV photodetectors. UV photodetectors prepared using WO_3 have high sensitivity, high stability, and high responsivity. The WO_3 film thickness was reduced from 225 to 95 nm with the pressure of the sputter increasing from 10 to 20 mTorr owing to a low deposition rate due to the accumulation of argon ions. The deposited WO_3 film had high crystallinity and surface roughness at 10 mTorr due to the influence of Ostwald ripening. To enhance the photocurrent, the ohmic-contact between the WO_3 and Ti electrode and the surface recombination was inhibited due to passivation created by the SiO_2 layer on Si substrates. By minimizing the overall impedance, the tailored grain size, higher crystallinity, and WO_3 film surface roughness assisted in reaching a high EQE and responsivity [31].

Because of its high photoactive properties, titanium dioxide (TiO_2) is also among the important materials for application in photodetection. At different niobium (Nb) concentrations, Nb-doped TiO_2 (TNO) films were deposited on a substrate of Si/SiO_2 to fabricate highly sensitive UV PDs with a high EQE and high detectivity. The photocurrent was enhanced due to the contribution of an excess electron per Nb atom and Ti vacancy as a result of lower doping concentrations (1% of Nb) of Nb^{5+} [32]. TiO_2/GaN nanorods (NRs) decorated with platinum (Pt) NPs were used to prepare hybrid heterojunction UV PDs. Under UV illumination (λ = 382 nm, 1.32 mW cm^{-2}) at 0 V, the prepared PD had a high photoelectric performance, including high detectivity (1.57 × 10^{14} jones), high responsivity (44.6 A W^{-1}), short recovery/response times of 200 ms/180 ms, and large electronic quantum efficiency (8.65 × 10^{3}%). The performance of the fabricated photodetector was higher than for its pristine counterparts. The improvement of this hybrid UV PD photoresponse is due to the large photon absorption favored by the Pt NP localized surface plasmon resonance (LSPR) effect and the sufficient built-in potential at the interface of the GaN NRs and TiO_2 heterojunction, which provides excellent splitting and transportion of photogenerated charge carriers [33]. Moreover, TiO_2 and its composites were used to prepare UV PD [34–40]. Dip coated TiO_2 was used to fabricate UV PDs. The fabricated UV PD showed ohmic current–voltage (*I–V*) characteristics. The UV PD prepared using film deposited at eight cycles showed the best photocurrent (0.64 μA) at 365 nm illumination. Moreover, the UV PD prepared using a C_8 sample shows the highest photoresponsivity of 2.15 A W^{-1} at 5 V under UV illumination (365 nm). It also showed fast fall and rise times of 19 s and 17 s, respectively. Also, it shows good stability as well as good reproducibility [40]. Self-powered UV PDs were prepared using TiO_2 nanorod arrays/Au nanoparticles/polyterthiophene (TiO_2 NRs/Au/PTTh)), and TiO_2 nanorod arrays/polyterthiophene (TiO_2 NRs/PTTh) by combining the techniques of hydrothermal and electrochemical deposition. The UV PD prepared using TiO_2 NRs/PTTh had good stability, high sensitivity, and fast response. The UV PD performance was further enhanced using gold (Au) NPs as the middle layer owing to the pyro-phototronic effect between TiO_2 NRs and Au NPs. The detectability (D^*) of TiO_2 NRs/Au/PTTh and NRs/PTTh UV photodetectors were, respectively, 1.666×10^{10} jones and 2.155×10^9 jones and the responsivities (R) were 0.0774 mA W^{-1} and as 1.894 mA W^{-1}, respectively. The fall time (τ_f) and rise time (τ_r) of the UV PD prepared using TiO_2 NRs/Au/PTTh were shorter than those of the TiO_2 NRs/PTTh UV PD. These results confirmed that the TiO_2 NRs/Au/PTTh UV PD is appropriate for self-powered UV PD [39]. Dalal *et al* chose magnesium (Mg) doped chemically stable TiO_2 thin film for deep UV photodetection. The fabricated photodetectors used 110 nm size Mg doped TiO_2 thin film and undoped TiO_2 thin film on substrates of p-type silicon. A photodetector prepared using undoped TiO_2 thin film exhibited deep UV as well as good responsivity in the visible region. The Mg doped TiO_2 thin film had good hydrophobicity. Therefore, under 200 nm, the light sensitivity of the photodetector prepared using Mg doped TiO_2 thin film was four times higher than that of the photodetector prepared using undoped TiO_2 thin film. Moreover, the Mg doped TiO_2 thin film photodetector had a very small response in the visible region [41]. MgO capping layer radial heterostructures (RHs) on top of TiO_2 nanowires (RHNWs) were used to prepare PDs. Under UV illumination, the RHNW PD had a higher sensitivity than PD prepared using TiO_2-MgO axial heterostructure nanowires (AHNWs). At −1.5 V, the RHNW PD had a very low leakage current (~1.78×10^{-9} A) compared to AHNWs (~3.84×10^{-9} A) PDs. The RHNW PD had an extreme photoresponsivity of ~18 A W^{-1} with a fast rise time of ~0.12 s as well as a fast fall time of ~0.13 s under 200 nm illumination (–9 V) [37]. The TiO_2-reduced graphene oxide (rGO) was also used as a very efficient material for UV photodetection. A green synthesis technique was used to prepare TiO_2-rGO composites, including indifferent ratios. The TiO_2-rGO composite was used to fabricate PD on a SiO_2-Si wafer. The TiO_2-rGO composite XRD patterns showed the existence of both rutile and anatase phases of TiO_2. Under UV illumination of 370 nm, the UV PD had a high detectivity (7.92×10^{13} cm Hz.5W^{-1}), high responsivity (7.712 A W^{-1}), low response time (43 ms), and a power density of 10 μW cm^{-2} as a result of slow recombination in the device. It also has a very small dark current (7.3 nA), which indicates that it has reduced noise equivalent power [38]. Heterojunction arrays of TiO_2 nanorods (TNRs)/graphene (G)/$ZnIn_2S_4$ (ZIS)/ electrolyte were prepared through a three-step synthesis technique and they were used to construct self-powered UV PD. At 0 V the UV PD showed 13.79 mA W^{-1} responsivity in UV light as well as 0.39 mA W^{-1} responsivity in visible light. The UV PD showed a 5 ms short rise time and 10 ms decay time under UV illumination. For electron transport, an effective pathway was offered by the TNRs. ZIS stacking formed a type-II energy band structure which significantly improved the response of the TNRs under visible and UV light. Moreover, the separation and transfer of carriers were accelerated by the intermediate graphene layer because of good charge collection ability. Thus this UV PD may be good for UV–visible detection [34]. In a TiO_2 NRs/Au NPs heterojunction UV PD, a donor unit was created using 3,4-propanedioxythiophene and an acceptor unit was created using a novel D-A-D type conducting polymer (PProDOT-Py) with pyridine. In UV photodetection the effect of PProDOT-Py was explained in detail. The PProDOT-Py was unevenly dispersed on TiO_2 NRs in a cross-linked network form. Moreover, coordination bonds between the TiO_2 NRs and PProDOT-Py were present. The hybrid heterojunction was formed due to the energy levels matching between PProDOT-Py and TiO_2. The I_{photo}/I_{dark} and detectivity of the UV PD fabricated using TiO_2 NRs/Au NPs/ PProDOT-Py were 17 857.14 (2 V) and 1.067×10^{13} jones, respectively. At 0 V the UV PD had an high speed response time of 0.06 s [36].

UV PDs exhibit good performance and flexibility due to integration of $Ca_2Nb_3O_{10}$ nanosheets (FM@CNO) and filter membranes. In filter membrane pores, by using a simple vacuum filtration method, the $Ca_2Nb_3O_{10}$ nanosheets were integrated. At 5 V bias under 300 nm light illumination, the FM@CNO UV PD has good responsivity (0.08 A W^{-1}), a good UV/visible rejection ratio of 3.86×10^3, good detectivity of 1.1×10^{12} jones, and a high speed of 0.12/1.24 ms. Moreover, the FM@CNO UV PD showed good flexibility after several bending cycles [42].

One-dimensional (1D) nanostructures including nanorods and nanowires exhibit higher photoresponse compared to their bulk counterparts owing to the unidirectional transport that drastically decreases the scattering as well as photo-generated charge carrier recombination. The CuO/Cu_2O core–shell nanowire arrays were fabricated using chemical bath deposition, where the Cu_2O layer was grown thermally on the CuO nanowires. The PD prepared using CuO/Cu_2O core–shells via photolithography employing a Pd metal contact show better detectivity and responsivity, faster response and recovery times, and higher external quantum efficiency compared to that prepared using bare CuO nanowires. This is due to the formation of ohmic junctions at the interface of the Pd–Cu_2O which guarantees barrierless hole transport in the PD prepared using core–shells [43]. The UV PD prepared using the SnO_2/NiO heterojunction showed photo-responsivity of 3.3 mA W^{-1}, an ON/OFF ratio of 4.16×10^4, and detectivity of 7.4×10^{10} jones. The device, which has large area, showed a response time of 89 ms at low incident light intensity [44].

Similarly, a heterojunction of ZrO_2-RGO NFs/n-Si was used to fabricate a UV PD. The ZrO_2-RGO nanofibers (NFs) were fabricated using the electrospinning method. To investigate the photodetection, ZrO_2-RGO NFs were coated onto the n-Si wafers. A good responsivity (0.63 A W^{-1}) and detectivity (1.04×10^{10}) were attained in the UV region at 0 V [45]. $ZnGa_2O_4$ was also used to fabricate a high response UV PD. Au Schottky electrodes as asymmetric pairs were fabricated on the film of $ZnGa_2O_4$ to construct a novel self-powered solar-blind $ZnGa_2O_4$ UV PD. The *I–V* curve of the constructed device showed that the prepared device had a rectifying characteristic at ±8 V with a rectification ratio of 11.3. The $ZnGa_2O_4$ UV PD had good responsivity (22.2 mA W^{-1}) as well as a good UV–vis rejection ratio (1.3×10^4) at 0 V bias. Moreover, the UV PD decay and rise times are 30 ns and 10 ns, respectively, which are very high compared to other previous reports on the self-powered solar-blind UV PDs. The high speed of the UV PD is attributed to the high crystalline quality of the $ZnGa_2O_4$ film with low trap density and asymmetric Schottky junctions [46]. Similarly, several researchers used $ZnGa_2O_4$ for fabricating UV PDs [47, 48]. Young *et al* fabricated flexible UV PDs using 1D gallium (Ga)-doped zinc oxide nanorods using chemical bath deposition. The 1D ZnO nanorods were prepared on a flexible PEN (poly(ethylene 2,6-naphthalate)) substrate. The flexible substrate supports the improved performance of the UV photodetecors using Ga doping in ZnO. The UV photodetecors prepared using Ga doped ZnO showed good a light-to-dark current ratio of 570.2 at a 1 V bias compared to that (light-to-dark current ratio of 8.5) prepared using bare ZnO [49]. Ga incorporated ZnO (GZO) was used to improve the UV PD performance. The GZO thin films were fabricated using an easy two-step chemical bath deposition method. A structural study indicated that the film crystallinity was enhanced by an increased Ga doping level. An optical study confirmed that the UV absorption was higher for the 3% Ga doped ZnO film. The calculated EQE, responsivity (*R*), and photodetectivity (D^*), of the UV PD prepared using 3% Ga doped ZnO film were 125.10%, 0.38 A W^{-1}, and 1.24×10^{10} jones, respectively, which were higher than the prepared bare ZnO. The rise and decay times of the UV PD prepared using 3% Ga doped ZnO film were 1 and 3 s, respectively, which were faster than in previous reports [50]. Wang *et al* prepared self-powered solar-blind UV PDs (PPDs) using the heterojunction of Ga_2O_3/ZnO:V through a pulsed laser deposition (PLD) technique. They regulated the photoelectric process by introducing ferroelectric materials with spontaneous polarization which alter the depolarizing field direction when an external electric field was applied. In ZnO:V, the induced field of ferroelectric polarization could modulate the photoresponse behaviors of Ga_2O_3/ZnO:V PDs. Under 260 nm illumination, the Ga_2O_3/ZnO:V PD showed a photodetectivity of 1.18×10^{12} jones and a photoresponsivity of 95.53 mA W^{-1} in the downward poling state, which were 3.2 times higher compared to those of the unpoled state. Moreover, the speed of the photodetector, namely the rise/decay time, was increased from 10.3 μs/1.76 ms to 6.2 μs/0.43 ms. The Ga_2O_3/ZnO:V PD performance was improved due to widening of the depletion region and enlarging the built-in electric field under the influence of a depolarization field. This helps to split as well as transport the photogenerated carriers at the heterojunction interface [51]. Guo *et al* prepared a ZnO NW-based UV sensor employing a newly designed sensing layer which reduced the interference of humidity in the range of humidity (RH) of 10%–70%. This new technique was proposed by regulating the ZnO NWs properties to the hydrophobic state from the hydrophilic state using novel design of mixed materials. For example, hydrophobic silica NPs were covered on the ZnO NW surface. The prepared UV sensor showed good repeatability, less hysteresis (less than 1.65%), and good sensitivity (9.53 ppm (mW $cm^{-2})^{-1}$), with good linearity (the *R*2 value was 0.999 04). The UV photodetecor was used to monitor various environmental humidity conditions [52]. Ferhati *et al* fabricated low-noise solar-blind UV PDs using heterostructures of all-amorphous ZnO/SiC using a radio frequency (RF) magnetron sputtering method. After annealing, the properties of the device were enhanced. The device had a small dark current, good detectivity, and good photoresponse. The ZnO/SiC UV PD annealed at 500 °C showed a 5.9×10^3 UV–vis rejection ratio, a good signal-to-noise ratio (2.64×10^5), and good detectivity (2.4×10^{12} jones) [53]. The ZnO nanoclusters were deposited on an iron (Fe) coated glass substrate using laser assisted chemical bath deposition to fabricate the UV PD, which showed a good response [54]. ZnO nanostructures were deposited on silicon substrate using a thermal evaporation technique to fabricate a UV PD. It showed a fast response time and good photosensitivity [55]. The UV PD was prepared using a layer-by-layer polyaniline and ZnO arrangement. It had repeatable and high photocurrent, long-term stability, significant photoresponsivity, low rise/fall times, and considerable UV on/off switching at 0 V. The outcomes indicated that the UV PD with a PANI top layer had better responsivity than others [56]. UV PD based polymer–ZnO nanocomposites were prepared based on three polymers, poly(vinylidene difluoride-trifluoroethylene (P(VDF-TrFE)), polystyrene (PS), and poly(methyl methacrylate) (PMMA), which showed better performance compared to other devices because of its ferroelectric property [57]. Mandavkar *et al* fabricated a hybrid UV PD using MoS_2/ZnO/HNP architecture which improved the photocarrier injection in the PD. Because of its architecture, the photoresponsivity, photodetectivity, and photocurrent were improved under 385 nm illumination with an intensity of 54.9 mW mm^{-2} at 10 V. This UV PD performance was higher than for similar ZnO-based photodetectors. The photocurrent is improved owing to hot carrier injection via powerful LSPR as a result of PdAg HNPs. The photocurrent was also improved due to photocarriers from 2D MoS_2 nanoflakes which were placed on photo-induced carriers in the layer of ZnO QD [58]. Moreover, a lot of research was also done using ZnO and its composites and heterostructures to fabricate UV PDs because of its properties including wide bandgap and high exciton binding energy [35, 50, 51, 53, 54, 57–72].

11.6 Perovskite based photodetectors for UV detection

The innovation of functional materials with tunable optoelectronic properties will occupy important positions in the development of fundamental and applied research fields. In particular, metal halide perovskite materials with a typical crystal structure such as $CaTiO_3$ are emerging as a prominent semiconductor counterpart, outperforming all conventional materials in the optoelectronic field [73, 74]. The chemical formula of perovskites is ABX_3, where 'A' and 'B' are cations of different sizes and 'X' is an anion that bonds to both cations (figure 11.4). Progress has been achieved in the synthesis, structural characterization, and investigation of the physical properties of perovskite compounds in the form of 3D bulk crystals, 2D nanosheets, 1D nanorods or nanowires, and 0D quantum dots or nanocrystals [75]. The perovskite materials exhibit intriguing and unusual physical properties that have been studied extensively for both practical applications and theoretical modeling [76]. In this sense, the potential applications of perovskite materials are diverse and include applications in sensors, fuel cells, solar cells, photodetectors, lasers, storage devices, and spintronic applications [77, 78]. Among the above, photodetectors are essential optoelectronic components in industrial and scientific applications, including image capture, optical communications, and environmental monitoring [79, 80].

Monitoring UV radiation with high precision is one of the important applications of photodetectors. With an exciting performance in terms of sensitivity, specific detection ability, and response speed, the perovskite-based photodetectors have emerged as the best choice for UV photodetection and some recent reports strongly confirm the same. To date, many successful UV photodetectors have been fabricated with wide bandgap materials such as GaN and ZnO, which are very expensive for developing and fabricating the photodetector [60, 81] (figure 11.5).

Recently, many exciting results on UV detection using perovskite PDs have been reported [82]. Perovskite photodetectors with various device configurations have been developed and demonstrated unprecedented advances in photodetection performance. Various strategies (figure 11.6) have been applied to synthesize perovskites with different structural forms (polycrystalline film, single crystals, and low-dimensional nanostructures), including solution processing, evaporation-induced self-assembly growth, chemical vapor transport process, template method, etc. There are already several comprehensive review articles that mainly focus on the growth techniques and mechanisms [83–85]. Although perovskites have found widespread applications in many fields, this chapter mainly focuses on the unique

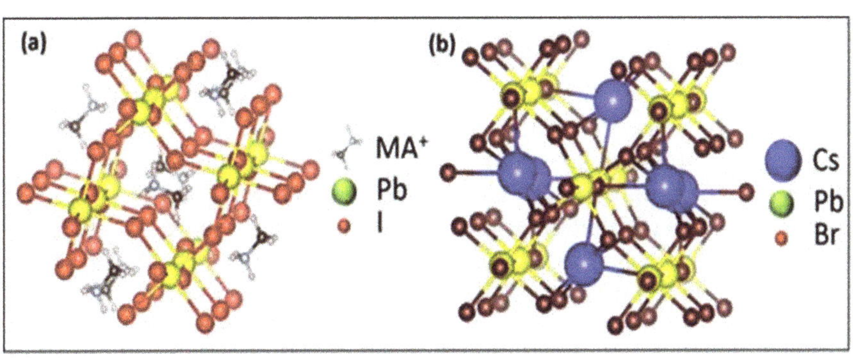

Figure 11.5. Schematic crystal structure of (a) $CH_3NH_3PbI_3$ and (b) $CsPbBr_3$. Created with VESTA.3.5.8. (Reproduced with permission from [130]. Copyright 2011 IUCr Journals.)

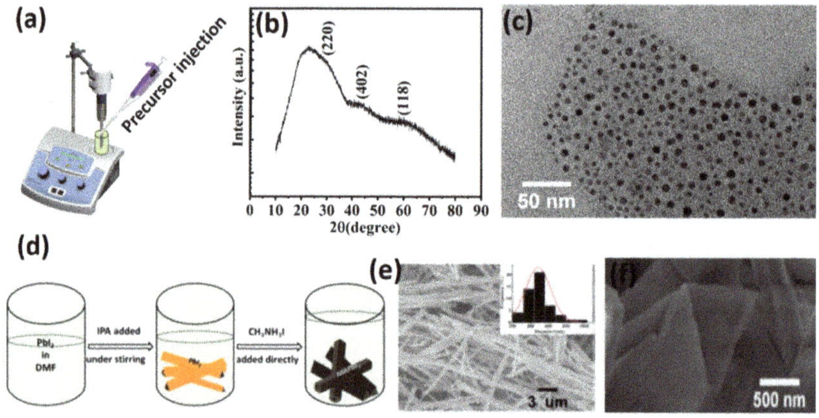

Figure 11.6. Various strategies to synthesize perovskite nanostructures. (a) $CH_3NH_3PbI_3$ quantum dots synthesized via an ultrasonic precipitation technique. (b) XRD pattern of $CH_3NH_3PbI_3$ QD thin films. (c) HRTEM image of $CH_3NH_3PbI_3$ QDs. (Reproduced with permission from [96]. Copyright 2020 Wiley.) (d) Schematic illustration of the synthesis strategy of $CH_3NH_3PbI_3$ nano/micro tubes. The nano/micro tubes are synthesized by an antisolvent crystallization technique. (b) SEM image of nano/micro tubes (FESEM image of single nanotubes). (Reproduced with permission from [103]. Copyright 2019 Royal Society of Chemistry.)

properties and performances of perovskite materials as UV photodetectors and also summarizes the most interesting work in this important field.

11.7 Synthesis of perovskite structures

The schematic crystal structure of $CH_3NH_3PbI_3$ and $CsPbBr_3$ is depicted in figure 11.5. The large optical absorption coefficient of perovskites ranges from 300 to 800 nm, making it a suitable material for broadband photodetector applications. Moreover, low charge recombination was observed in perovskites due to low trap density within the bandgap. In addition to the above advantages, a thin layer of perovskite can absorb full light radiation due to its high absorptivity [86]. As a result, the movement distance of photogenerated carriers is small, resulting in a rapid response to light. These properties strongly suggest that the perovskite materials are ideal building blocks for building powerful photodetectors

There are currently three main structures of perovskites utilized to fabricate photodetectors: polycrystalline film, single crystals, and low-dimensional nanostructures [87]. A precisely controlled crystallization process is used to fabricate the polycrystalline thin film structure, allowing the surface morphology, crystal size, and thickness to be tailored to the application requirements [88]. Despite their high defect density, polycrystalline perovskite thin films have significant advantages for optoelectronic device applications [89]. On the other hand, single crystal perovskites have been used to fabricate high performance photodetectors due to their absence of grain boundaries, high carrier mobility, and low trap density [90]. In addition to single crystal and polycrystalline perovskite materials, nanostructured perovskites (nanowires, nanorods, and nanosheets) have also been synthesized successfully and have attracted increasing research attention, mainly due to their intriguing optical and electronic properties arising from the quantum size effect [91–99].

Perovskite-inspired materials (PIMs) have been synthesized and fabricated using various strategies and have been reported previously [100, 101]. The various strategies include solution-phase recrystallization growth [102, 103], vapor-phase conversion [104], direct vapor-phase growth [105], colloidal nanowire synthesis [106], space-confined nanowire growth [107], nanowire growth via intermediate adducts [108], and ion exchange of existing perovskite nanowires [109]. In the following section, representative recent works on the development of perovskite-based UV photodetectors are summarized.

11.8 Perovskite based UV photodetectors

The comparative optical absorption spectra of perovskite materials are shown in figure 11.7. It should be noted that the absorption coefficient of perovskite materials is 50% higher compared to other solar cell materials, which makes them highly recommended for applications in optoelectronic devices.

Recently, Li *et al* reported a Cs_4PbI_6 single crystal based photodetector with planar $Au/Cs_4PbI_6/Au$ structure, showing a high photoresponsivity of 5.07 mA W^{-1} with excellent response and recovery times of 11.65 and 12.80 μs, respectively [126]. The performance of the device is directly related to the unique physical properties of the perovskite material, for example, the single crystal Cs_4PbI_6 has a large mobility value of 9.5×10^{-4} cm^2 V^{-1} and a low trap density of 1.7×10^{10} cm^{-3} with an optical bandgap energy of 3.46 eV, which is almost similar to the traditional ZnO material. A highly sensitive UV-A photodetector has been fabricated using $CH_3NH_3PbCl_3/Ga_2O_3$ hybrid polycrystalline structure [127]. Interestingly, the device can

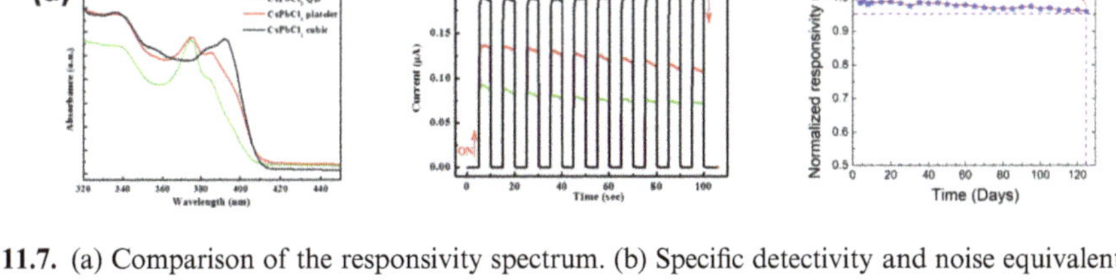

Figure 11.7. (a) Comparison of the responsivity spectrum. (b) Specific detectivity and noise equivalent power of the hybrid $CsPbBr_3$–Si photodetection scheme in the UV wavelength region. (Reproduced from [110]. CC BY 4.0.) (c) Schematic device structure of planar $CsPbCl_3$ QDs. (d) Absorption spectra of $CsPbCl_3$ perovskite material in the form of QDs, platelets, and cubic crystals. (e) Photostability of $CsPbCl_3$ UV PDs with different structural form. (Reproduced from [128]. CC BY 3.0.) (f) Time-dependent responsivity stability test. (Reproduced with permission from [124]. Copyright 2020 American Chemical Society.)

be operated with a high specific detectivity of 5.4 × 10^{10} jones without bias (self-powered). In addition, the response speed of the device was found to be 6/67 ms, which is perfect for the atmospheric UV monitoring process. In the another approach, in order to overcome the low responsivity of Si-based PDs (less than 0.1 A W^{-1} below 400 nm), highly luminescent $CsPbBr_3$ nanocrystals (NCs) were integrated with Si-PDs, from which responsivity and detectivity was increased significantly (figures 11.7(a) and (b)) [110]. A comparative photodetector performance of $CsPbCl_3$ in the UV range was demonstrated by Zhang *et al* [128]. The solution processing technique produced a simple and transparent planar PD (figure 11.7(c)) that integrates well with conventional systems. As shown in figure 11.7(d), well defined absorption characteristics were observed for cubic crystals of $CsPbCl_3$. The as-fabricated device exhibited a high photoresponsivity of 1.89 A W^{-1} and small response time of 41 ms. The visible-blind UV PDs are fabricated using a Cs_2SnCl_6 all-inorganic perovskite photoactive layer, which showed a high specific detectivity of 6.3 × 10^{15} jones comparable with photomultiplier tube (PMT) detectivity. In addition, the photosensitivity of the device was stable for a long period of time, allow its effective use in practical applications, as shown in figure 11.7(f). A high photosensitivity up to 10^5 A W^{-1} was observed in $CH_3NH_3PbCl_3$-based sensors as reported by Wu *et al* [125]. The hybrid device consisted of a transistor-configured $CH_3NH_3PbCl_3$/SCNT composite structure, in which maximum photosensitivity and better detection ability were observed under a light intensity as low as 1 W cm^{-2}, as shown in table 11.1. Fabricated on a flexible polyethylene naphthalate (PEN) substrate, the device exhibited a photoresponsivity of up to 1.8 × 10^5 A W^{-1} (365 nm, 1 μW cm^{-2}), which is consistent with the above responsivity value for the silicon-substrate device. Thus, it was proved that the negative effect of the PEN substrate on the performance of the photosensor is limited.

Zhou *et al* reported a self-powered perovskite-TiO_2 photodetector with $R = 0.26$ A W^{-1} under 364 nm ultraviolet light [129]. The device consists of a hybrid structure of a $CH_3NH_3PbI_3$ layer on vertically aligned TiO_2 nanorods. It was concluded that the UV response is mainly due to the wide bandgap of TiO_2, however, the perovskite plays the role of carrier separation and transport.

Recently, energy self-sufficient optoelectronic systems with an integrated power supply unit have become increasingly popular, since their power module converts environmental energy into electrical energy in order to operate the functional device without additional energy costs. Su *et al* designed and fabricated PDs that exploit the combined properties of the photoelectric effect and triboelectric effect of $MAPbI_3$ perovskite. The photodetector device consist of two friction layers, where the top layer is a Cu electrode with an elastic buffer layer and bottom layer composed of $MAPbI_3/TiO_2$/FTO. The device is characterized by high sensitivity (7.5 V W^{-1}), a fast response time (< 80 ms), and wide detection ranges from UV to visible. The mechanism of this photodetector is mainly based on a change in light-induced surface charge density.

Table 11.1. Summary of the figures of merit of perovskite-based UV photodetectors.

Device structure	Response wavelength (nm)	R (mA W^{-1})	D^* (jones)	τ_r/τ_f	References
$CsPbBr_3$ NC	278 nm				[110]
Au/$CH_3NH_3PbI_3$/Al	1 Sun illumination	240	Not		
mentioned	71/112 μs	[111]			
ITO/ZnO:$CsPbBr_3$/Ag	405 nm	11.5	Not		
mentioned	0.409/0.017 s	[66]			
ZnO NRs/$CsPbBr_3$	Visible light	300	1.15×10^{13}		[112]
PVP-$CsPbI_3$	400–700 nm	100	1×10^{12}	5.7/6.2 μs	[113]
P_3HT-PDPP$_3$T/$CsPbBr_3$/ SnO_2	300–950 nm	250	1.2×10^{13}	111/306 μs	[114]
$CH_3NH_3PbI_3$	400 nm	250	1.53×10^{11}	110/72 ms	[115]
ITO/ZnO/CdS/$CsPbBr_3$/Au	405 nm	86	6.2×10^{11}	0.3/0.25 s	[116]
Au/$CH_3NH_3PbI_3$/Au	400 nm	160	5.89×10^{11}	150/50 ms	[117]
FTO/TiO_2/$CsPbBr_3$/carbon	405 nm	350	3.83×10^{13}	–/1.46 μs	[118]
(4-AMP)$(MA)_2Pb_3Br_{10}$/ $MAPbBr_3$	405 nm	1.19	1.26×10^{12}	600/600 μs	[119]
Au/$CH_3NH_3PbI_3$/Au	White light	196 V mW^{-1}-cm^{-2}	—	—	[120]
$CH_3NH_3PbI_3$	UV–visible	7.5 V W^{-1}		<80 ms/–	[121]
Au/$CH_3NH_3PbI_3$ NRs/Au	300 nm	2.2	1.76×10^{11}	27.2/26.2 ms	[122]
ITO/TiO_2/$CsPbBr_3$/ SpiroOMeTAD/Au	405 nm	10.1×10^3	9.35×10^{13}	8.0/2.3 s	[123]
ITO/NiO/Cs_2SnCl_6:Nd^{3+} /BCP/Al	372 nm		6.3×10^{15}	2.5 μs/1.8 μs	[124]
$CH_3NH_3PbCl_3$/SCNT	365 nm	2×10^5	6×10^{14}	0.25 s/0.65 s	[125]

There are still a number of problems and unanswered questions regarding the commercialization of perovskite-based photodetectors. The toxicity of lead, which endangers human health and pollutes the environment, has been criticized repeatedly. Recently, some research on lead-free perovskites has surfaced. Although toxicity has been reduced by lead-free perovskites, device performance still needs to be improved. Among the perovskite components, lead (Pb) acts as a core skeleton, which plays an important role in generating and transporting charges. The charge carriers are transferred along the Pb–halogen bond, resulting in poor performance and lower stability after replacing Pb with Sn or Bi. Many researchers have shown that the role of lead is irreplaceable, so the toxicity treatment of perovskite needs to be examined further. Overall, the focus should be on the crystal growing process, defect engineering, and device fabrication engineering to provide an in-depth understanding of perovskite properties and device physics.

11.9 Conclusion

In summary, the different kinds of wood used to treat water and wastewater, in energy recovery, filtering water, and contaminant removal are discussed. The metal oxide and perovskite based UV photodetectors have also been discussed. Mostly wide bandgap materials are used to fabricate UV photodetectors. In recent years, perovskite-structured materials have been used to fabricate UV photodetectors and as composites to metal oxides. Perovskite-structured materials were shown to have good photodetecting performance, however, the stability of the device and the materials remain poor. Thus, novel materials for the effective detection of UV still need to be explored.

References

[1] Mekonnen M M and Hoekstra A Y 2016 Four billion people facing severe water scarcity *Sci. Adv.* **2** e1500323

[2] Preethi S, Arulraj A, Mangalaraja R V, Ravichandran V and Subramanian N 2022 Biomass-derived carbon-based materials for microbicidal applications *Biomass-Derived Carbon Materials* (New York: Wiley) pp 63–92

[3] Arulraj A, Thachnatharen N, Rajkumar C and Mangalaraja R V 2022 Engineering 2D chalcogenides for energy and environmental remediation *Nanomaterials for Sustainable Energy and Environmental Remediation* (Bristol: IOP Publishing) pp 2-1–2-28

[4] UNESCO 2018 World water assessment programme *The United Nations World Water Development Report: Nature-Based Solutions for Water WWAP (United Nations World Water Assessment Programme)/UN-Water. Paris, UNESCO.*

[5] Hou D, Jassby D, Nerenberg R and Ren Z J 2019 Hydrophobic gas transfer membranes for wastewater treatment and resource recovery *Environ. Sci. Technol.* **53** 11618–35

[6] Rahman K H and Kar A K 2020 Effect of band gap variation and sensitization process of polyaniline (PANI)-TiO_2 p–n heterojunction photocatalysts on the enhancement of photocatalytic degradation of toxic methylene blue with UV irradiation *J. Environ. Chem. Eng.* **8** 104181

[7] Rahman K H, Kar A K and Chen K -C 2022 Oxidation-induced catalytic performance of heterostructured Ni-TiO_2 nanoparticles and formation of leuco-methylene blue *Res. Chem. Intermed.* **48** 4475–501

[8] Lu L, Guest J S, Peters C A, Zhu X, Rau G H and Ren Z J 2018 Wastewater treatment for carbon capture and utilization *Nat. Sustain.* **1** 750–8

[9] Arulraj A, Mehana Usmaniya U, Senguttuvan G, Sivakumar V and Khalid M 2021 Chalcogenides nanocrystals and its applications In: Mubarak, N.M., Khalid, M., Walvekar, R., Numan, A. (eds) Contemporary Nanomaterials in Material Engineering Applications. Engineering Materials. (Cham: Springer) pp 201–27

[10] Crowther T W *et al* 2015 Mapping tree density at a global scale *Nature* **525** 201–5

[11] Laohaprapanon S, Kaczala F, Salomon P S, Marques M and Hogland W 2012 Wastewater generated during cleaning/washing procedures in a wood-floor industry: toxicity on the microalgae *Desmodesmus subspicatus Environ. Technol.* **33** 2439–46

[12] Hedenström E, Fagerlund Edfeldt A, Edman M and Jonsson B-G 2016 Resveratrol, piceatannol, and isorhapontigenin from Norway spruce (*Picea abies*) debarking wastewater

as inhibitors on the growth of nine species of wood-decaying fungi *Wood Sci. Technol.* **50** 617–29

[13] Vek V, Oven P and Poljanšek I 2016 Review on lipophilic and hydrophilic extractives in tissues of common beech *Drv. Ind.* **67** 85–96

[14] Jiang F, Li T, Li Y, Zhang Y, Gong A, Dai J, Hitz E, Luo W and Hu L 2018 Wood-based nanotechnologies toward sustainability *Adv. Mater.* **30** 1703453

[15] Klemm D, Heublein B, Fink H-P and Bohn A 2005 Cellulose: fascinating biopolymer and sustainable raw material *Angew. Chem. Int. Ed.* **44** 3358–93

[16] Li Y, Vasileva E, Sychugov I, Popov S and Berglund L 2018 Optically transparent wood: recent progress, opportunities, and challenges *Adv. Opt. Mater.* **6** 1800059

[17] Sjöström E 1993 *Wood Chemistry* (Amsterdam: Elsevier)

[18] Song J *et al* 2018 Processing bulk natural wood into a high-performance structural material *Nature* **554** 224–8

[19] Song J *et al* 2017 Superflexible wood *ACS Appl. Mater. Interfaces* **9** 23520–7

[20] Zhu M *et al* 2018 Plasmonic wood for high-efficiency solar steam generation *Adv. Energy Mater.* **8** 1701028

[21] Li T *et al* 2018 Scalable and highly efficient mesoporous wood-based solar steam generation device: localized heat, rapid water transport *Adv. Funct. Mater.* **28** 1707134

[22] Liu K -K, Jiang Q, Tadepalli S, Raliya R, Biswas P, Naik R R and Singamaneni S 2017 Wood–graphene oxide composite for highly efficient solar steam generation and desalination *ACS Appl. Mater. Interfaces* **9** 7675–81

[23] Liu H, Chen C, Wen H, Guo R, Williams N A, Wang B, Chen F and Hu L 2018 Narrow bandgap semiconductor decorated wood membrane for high-efficiency solar-assisted water purification *J. Mater. Chem.* A **6** 18839–46

[24] Chen C, Zhang Y, Li Y, Dai J, Song J, Yao Y, Gong Y, Kierzewski I, Xie J and Hu L 2017 All-wood, low tortuosity, aqueous, biodegradable supercapacitors with ultra-high capacitance *Energy Environ. Sci.* **10** 538–45

[25] Borelli C, Bosio A, Parisini A, Pavesi M, Vantaggio S and Fornari R 2022 Electronic properties and photo-gain of UV-C photodetectors based on high-resistivity orthorhombic κ-Ga_2O_3 epilayers *Mater. Sci. Eng.* B **286** 116056

[26] Feng Q, Dong Z, Liu W, Liang S, Yi Z, Yu C, Xie J and Song Z 2022 High responsivity solar-blind UV photodetector based on single centimeter-sized Sn-doped β-Ga_2O_3 microwire *Micro Nanostruct.* **167** 207255

[27] Pedapudi M C and Dhar J C 2022 Ultrasensitive p–n junction UV-C photodetector based on p-Si/β-Ga_2O_3 nanowire arrays *Sensors Actuators* A **344** 113673

[28] Jia M, Wang F, Tang L, Xiang J, Teng K S and Lau S P 2020 High-performance deep ultraviolet photodetector based on NiO/β-Ga_2O_3 heterojunction *Nanoscale Res. Lett.* **15** 47

[29] Yadav P V K, Ajitha B, Annapureddy V, Reddy Y A K and Sreedhar A 2021 Improved UV photodetector performance of NiO films by substitutional incorporation of Li *Mater. Lett.* **301** 130296

[30] Ren S, Gao S, Lu H, Rong P, Zhang X, Li L, Jiao S, Wang D, Zhang Y and Wang J 2022 A facile template-assisted synthesis of large-scale Bi_2O_3 nanotube arrays for self-powered UV photodetector *Appl. Surf. Sci.* **590** 3–5

[31] Karthik Yadav P V, Ajitha B, Reddy Y A K, Minnam Reddy V R, Reddeppa M and Kim M D 2021 Effect of sputter pressure on UV photodetector performance of WO_3 thin films *Appl. Surf. Sci.* **536** 147947

[32] Yadav P V K, Ajitha B, Ahmed C M A, Reddy Y A K and Minnam Reddy V R 2022 Superior UV photodetector performance of TiO_2 films using Nb doping *J. Phys. Chem. Solids* **160** 110350
[33] Pasupuleti K S, Chougule S S, Jung N, Yu Y J, Oh J E and Kim M D 2022 Plasmonic Pt nanoparticles triggered efficient charge separation in TiO_2/GaN NRs hybrid heterojunction for the high performance self-powered UV photodetectors *Appl. Surf. Sci.* **594** 153474
[34] Wang X, Bai Z, Zhang Y, Zhang Z, Liu J, Gao Y and Zhang Q 2022 High performance self-powered UV–visible photodetectors based on TiO_2/graphene/$ZnIn_2S_4$/electrolyte heterojunctions *Ceram. Int.* **48** 13090–9
[35] Yin Z, Shan Y, Yu M, Yang L, Song J, Hu P and Teng F 2022 Enhanced performance of UV photodetector based on ZnO nanorod arrays via TiO_2 as electrons trap layer *Mater. Sci. Semicond. Process.* **148** 106813
[36] Haile L T *et al* 2022 TiO_2 nanoarrays/Au nanoparticles/PProDOT-Py hybrid heterojunction UV photodetector *Org. Electron.* **110** 106644
[37] Dalal A, Samajdar D P and Mondal A 2022 Hetero-radial MgO capped TiO_2 nanowire arrays as a deep UV and self-powered photodetector *J. Alloys Compd.* **928** 167120
[38] Phukan P and Sahu P P 2020 High performance UV photodetector based on metal–semiconductor–metal structure using TiO_2-rGO composite *Opt. Mater.* **109** 110330
[39] Zhang H, Abdiryim T, Jamal R, Li J, Liu H, Kadir A, Zou D, Che Y and Serkjan N 2022 Self-powered TiO_2 NRs UV photodetectors: heterojunction with PTTh and enhanced responsivity by Au nanoparticles *J. Alloys Compd.* **899** 163279
[40] Kumbhar S M, Shevate S S, Patil A R, Shaikh S K and Rajpure K Y 2020 Dip coated TiO_2 based metal-semiconductor-metal ultraviolet photodetector for UV A monitoring *Superlattices Microstruct.* **141** 106490
[41] Dalal A, Mishra M, Chakrabarti S, Gupta R K and Mondal A 2022 Characteristics of Mg doped TiO_2 thin film based deep UV photodetector *Vacuum* **201** 111115
[42] Zhang Y, Cao F, Li S, Liu X, Kang L, Wu L and Fang X 2022 Integration of filter membrane and Ca_2Nb_3O10 nanosheets for high performance flexible UV photodetectors *J. Mater. Sci. Technol.* **129** 108–14
[43] Kajli S K, Ray D and Roy S C 2022 Efficient UV–visible photodetector based on single CuO/Cu_2O core-shell nanowire *J. Alloys Compd.* **895** 162546
[44] Athira M, Bharath S P and Angappane S 2022 SnO_2–NiO heterojunction based self-powered UV photodetectors *Sensors Actuators* A **340** 113540
[45] Yıldırım F, Khalili S, Orhan Z, Chenari H M and Aydoğan S 2023 Highly sensitive self-powered UV–visible photodetector based on ZrO_2-RGO nanofibers/n-Si heterojunction *J. Alloys Compd.* **935** 168054
[46] Han D, Liu K, Hou Q, Chen X, Yang J, Li B, Zhang Z, Liu L and Shen D 2020 Self-powered solar-blind $ZnGa_2O_4$ UV photodetector with ultra-fast response speed *Sensors Actuators* A **315** 112354
[47] Ling K, Li K, Bai R, Zhao B and Liu X 2022 Fabrication and performance characterization of deep UV photodetectors based on magnetron sputtered $ZnGa_2O_4$ thin films *Mater. Sci. Semicond. Process.* **152** 107096
[48] Horng R H, Li Y S, Lin K L, Tarntair F G, Nittayakasetwat S and Hsiao C L 2022 Effects of UV-ozone treatment on the performance of deep-ultraviolet photodetectors based on $ZnGa_2O_4$ epilayers *Mater. Chem. Phys.* **292** 126847

[49] Young S J *et al* 2020 Flexible ultraviolet photodetectors based on one-dimensional gallium-doped zinc oxide nanostructures *ACS Appl. Electron. Mater.* **2** 3522–9
[50] Kumar K D A, Mele P, Murahari P, Abdeltawab A A, Mohammady S Z, Ubaidullah M, Samdani M S and AlFaify S 2022 Enhancement of performance of Ga incorporated ZnO UV photodetectors prepared by simplified two step chemical solution process *Sensors Actuators* A **333** 113217
[51] Wang H, Ma J, Chen H, Wang L, Li P and Liu Y 2022 Ferroelectricity enhanced self-powered solar-blind UV photodetector based on Ga_2O_3/ZnO:V heterojunction *Mater. Today Phys.* **30** 100929
[52] Guo Y, Zhou J, Ji Z, Liu Y, Cao R, Zhuo F, Tan K, Duan H and Fu Y 2022 A new strategy to minimize humidity influences on acoustic wave ultraviolet sensors using ZnO nanowires wrapped with hydrophobic silica nanoparticles *Microsyst. Nanoeng* **8** 121
[53] Ferhati H, Djeffal F, Bendjerad A, Saidi A and Benhaya A 2021 Post-annealing effects on RF sputtered all-amorphous ZnO/SiC heterostructure for solar-blind highly-detective and ultralow dark-noise UV photodetector *J. Non. Cryst. Solids.* **574** 121168
[54] Lahewil A S Z, Zyoud S H, Ahmed N M, Omar A F and Azman N Z N 2022 Synthesis ZnO nanoclusters micro active area using continues wave blue laser-assisted chemical bath deposition based on UV photodetector *Optik* **260** 169099
[55] Rajkumar C 2022 Effect of annealing temperature on response time of ZnO photoconductor fabricated using thermal evaporation technique *Phys. Scr.* **97** 115806
[56] Nasirian S and Hadizadeh F 2022 A cheap self-powered UV-photodetector based on layer-by-layer arrangement of polyaniline and ZnO *Polymer* **245** 124699
[57] Hanna B *et al* 2022 UV photodetectors using a polymer/ZnO nanocomposite thin film *Sensors Actuators* A **338** 113495
[58] Mandavkar R *et al* 2022 Significantly improved photo carrier injection by the MoS_2/ZnO/HNP hybrid UV photodetector architecture *Appl. Surf. Sci.* **574** 151739
[59] Mohammad S M *et al* 2021 Self-powered UV photodetector performance optimization based on Ag nanoparticles-encapsulated-ZnO nanorods by photo-deposition method *Sensors Actuators* A **331** 113032
[60] Inamdar S I and Rajpure K Y 2014 High-performance metal–semiconductor–metal UV photodetector based on spray deposited ZnO thin films *J. Alloys Compd.* **595** 55–9
[61] Pallavolu M R, Maddaka R, Viswanath S K, Banerjee A N, Kim M D and Joo S W 2023 High-responsivity self-powered UV photodetector performance of pristine and V-doped ZnO nano-flowers *Opt. Laser Technol.* **157** 108776
[62] Bhardwaj R, Singh R and Mukherjee S 2021 Advances in ZnO-material based UV photodetectors *Encycl. Smart Mater.* **3** 200–23
[63] Zou D, Jamal R, Abdiryim T, Liu L, Liu H and Fan N 2022 Construction and properties of ZnO/poly(EDOT-pyridine-EDOT) core/shell heterostructure nanoarrays for UV photodetector *Synth. Met.* **291** 117192
[64] Lin S, Mandavkar R, Habib M A, Burse S, Khalid T, Joni M H, Li M -Y, Kunwar S and Lee J 2023 Investigation on the AgPt and AgPd hybrid alloy nanoparticles (HANPs) for the hybrid MoS_2/ZnO/HANP UV photodetector application *Appl. Surf. Sci.* **611** 155559
[65] Guan H, Mao G, Zhong T, Zhao T, Liang S, Xing L and Xue X 2021 A self-powered UV photodetector based on the hydrovoltaic and photoelectric coupling properties of ZnO nanowire arrays *J. Alloys Compd.* **867** 159073

[66] Li C, Han C, Zhang Y, Zang Z, Wang M, Tang X and Du J 2017 Enhanced photoresponse of self-powered perovskite photodetector based on ZnO nanoparticles decorated $CsPbBr_3$ films *Sol. Energy Mater. Sol. Cells* **172** 341–6
[67] Zheng H, Jiang Y, Yang S, Zhang Y, Yan X, Hu J, Shi Y and Zou B 2020 ZnO nanorods array as light absorption antenna for high-gain UV photodetectors *J. Alloys Compd.* **812** 152158
[68] Podder S, Basumatary B, Gogoi D, Bora J and Pal A R 2021 Pyro-phototronic application in the Au/ZnO interface for the fabrication of a highly responsive ultrafast UV photodetector *Appl. Surf. Sci.* **537** 147893
[69] Deka Boruah B 2019 Zinc oxide ultraviolet photodetectors: rapid progress from conventional to self-powered photodetectors *Nanoscale Adv.* **1** 2059–85
[70] Abbasi F, Zahedi F and Yousefi M H 2021 Fabricating and investigating high photoresponse UV photodetector based on Ni-doped ZnO nanostructures *Opt. Commun.* **482** 126565
[71] Zyoud S H, Ahmed N M, Lahewil A S Z and Omar A F 2022 Micro spot ZnO nanotubes using laser assisted chemical bath deposition: a low-cost approach to UV photodetector fabrication *Sensors Actuators* A **338** 113485
[72] Wang H, Ma J, Cong L, Zhou H, Li P, Fei L, Li B, Xu H and Liu Y 2021 Piezoelectric effect enhanced flexible UV photodetector based on Ga_2O_3/ZnO heterojunction *Mater. Today Phys.* **20** 100464
[73] Yu J, Liu G, Chen C, Li Y, Xu M, Wang T, Zhao G and Zhang L 2020 Perovskite $CsPbBr_3$ crystals: growth and applications *J. Mater. Chem.* C **8** 6326–41
[74] Murali B, Kolli H K, Yin J, Ketavath R, Bakr O M and Mohammed O F 2020 Single crystals: the next big wave of perovskite optoelectronics *ACS Mater. Lett.* **2** 184–214
[75] Mei F, Sun D, Mei S, Feng J, Zhou Y, Xu J and Xiao X 2019 Recent progress in perovskite-based photodetectors: the design of materials and structures *Adv. Phys.* X **4** 1592709
[76] Wang H and Kim D H 2017 Perovskite-based photodetectors: materials and devices *Chem. Soc. Rev.* **46** 5204–36
[77] Saparov B and Mitzi D B 2016 Organic–inorganic perovskites: structural versatility for functional materials design *Chem. Rev.* **116** 4558–96
[78] Green M A, Ho-Baillie A and Snaith H J 2014 The emergence of perovskite solar cells *Nat. Photonics* **8** 506–14
[79] Konstantatos G and Sargent E H 2010 Nanostructured materials for photon detection *Nat. Nanotechnol.* **5** 391–400
[80] Koppens F H L, Mueller T, Avouris P, Ferrari A C, Vitiello M S and Polini M 2014 Photodetectors based on graphene, other two-dimensional materials and hybrid systems *Nat. Nanotechnol.* **9** 780–93
[81] Li D, Sun X, Song H, Li Z, Chen Y, Jiang H and Miao G 2012 Realization of a high-performance GaN UV detector by nanoplasmonic enhancement *Adv. Mater.* **24** 845–9
[82] Zhou Y *et al* 2022 Halide-exchanged perovskite photodetectors for wearable visible-blind ultraviolet monitoring *Nano Energy* **100** 107516
[83] Huang H, Polavarapu L, Sichert J A, Susha A S, Urban A S and Rogach A L 2016 Colloidal lead halide perovskite nanocrystals: synthesis, optical properties and applications *NPG Asia Mater.* **8** e328–8

[84] Cao L, Liu X, Li Y, Li X, Du L, Chen S, Zhao S and Wang C 2020 Recent progress in all-inorganic metal halide nanostructured perovskites: materials design, optical properties, and application *Front. Phys.* **16** 33201
[85] Shi Z, Guo J, Chen Y, Li Q, Pan Y, Zhang H, Xia Y and Huang W 2017 Lead-free organic–inorganic hybrid perovskites for photovoltaic applications: recent advances and perspectives *Adv. Mater.* **29** 1605005
[86] Yin W -J, Shi T and Yan Y 2014 Unique properties of halide perovskites as possible origins of the superior solar cell performance *Adv. Mater.* **26** 4653–8
[87] Chen Y, He M, Peng J, Sun Y and Liang Z 2016 Structure and growth control of organic–inorganic halide perovskites for optoelectronics: from polycrystalline films to single crystals *Adv. Sci.* **3** 1500392
[88] Zhao Y and Zhu K 2014 Solution chemistry engineering toward high-efficiency perovskite solar cells *J. Phys. Chem. Lett.* **5** 4175–86
[89] Brandt R E, Stevanović V, Ginley D S and Buonassisi T 2015 Identifying defect-tolerant semiconductors with high minority-carrier lifetimes: beyond hybrid lead halide perovskites *MRS Commun.* **5** 265–75
[90] Chen G, Li W, Yu Y and Yang Q 2015 Fast and low-temperature synthesis of one-dimensional (1D) single-crystalline SbSI microrod for high performance photodetector *RSC Adv.* **5** 21859–64
[91] Bekenstein Y, Koscher B A, Eaton S W, Yang P and Alivisatos A P 2015 Highly luminescent colloidal nanoplates of perovskite cesium lead halide and their oriented assemblies *J. Am. Chem. Soc.* **137** 16008–11
[92] Hassan Y, Song Y, Pensack R D, Abdelrahman A I, Kobayashi Y, Winnik M A and Scholes G D 2016 Structure-tuned lead halide perovskite nanocrystals *Adv. Mater.* **28** 566–73
[93] Spina M, Bonvin E, Sienkiewicz A, Náfrádi B, Forró L and Horváth E 2016 Controlled growth of $CH_3NH_3PbI_3$ nanowires in arrays ofopen nanofluidic channels *Sci. Rep.* **6** 19834
[94] Li X *et al* 2017 Constructing fast carrier tracks into flexible perovskite photodetectors to greatly improve responsivity *ACS Nano* **11** 2015–23
[95] Fu Y, Meng F, Rowley M B, Thompson B J, Shearer M J, Ma D, Hamers R J, Wright J C and Jin S 2015 Solution growth of single crystal methylammonium lead halide perovskite nanostructures for optoelectronic and photovoltaic applications *J. Am. Chem. Soc.* **137** 5810–8
[96] Perumal Veeramalai C, Yang S, Zhi R, Sulaman M, Saleem M I, Cui Y, Tang Y, Jiang Y, Tang L and Zou B 2020 Solution-processed, self-powered broadband $CH_3NH_3PbI_3$ photodetectors driven by asymmetric electrodes *Adv.Opt. Mater.* **8** 2000215
[97] Yang D *et al* 2019 Interfacial synthesis of monodisperse $CsPbBr_3$ nanorods with tunable aspect ratio and clean surface for efficient light-emitting diode applications *Chem. Mater.* **31** 1575–83
[98] Lin C -C, Yeh S -Y, Huang W -L, Xu Y -X, Huang Y -S, Yeh T -H, Tien C -H, Chen L -C and Tseng Z -L 2020 Using thermally crosslinkable hole transporting layer to improve interface characteristics for perovskite $CsPbBr_3$ quantum-dot light-emitting diodes *Polymers* **12** 2243
[99] Zhang P, Zhang G, Liu L, Ju D, Zhang L, Cheng K and Tao X 2018 Anisotropic optoelectronic properties of melt-grown bulk $CsPbBr_3$ single crystal *J. Phys. Chem. Lett.* **9** 5040–6

[100] Fu Y, Zhu H, Chen J, Hautzinger M P, Zhu X Y and Jin S 2019 Metal halide perovskite nanostructures for optoelectronic applications and the study of physical properties *Nat. Rev. Mater.* **4** 169–88

[101] Perumal Veeramalai C, Feng S, Zhang X, Pammi S V N, Pecunia V and Li C 2021 Lead–halide perovskites for next-generation self-powered photodetectors: a comprehensive review *Photonics Res* **9** 968–91

[102] Liang K, Mitzi D B and Prikas M T 1998 Synthesis and characterization of organic –inorganic perovskite thin films prepared using a versatile two-step dipping technique *Chem. Mater.* **10** 403–11

[103] Chandrasekar P V, Yang S, Hu J, Sulaman M, Shi Y, Saleem M I, Tang Y, Jiang Y and Zou B 2019 Solution-phase, template-free synthesis of PbI_2 and $MAPbI_3$ nano/microtubes for high-sensitivity photodetectors *Nanoscale.* **11** 5188–96

[104] Ha S T, Liu X, Zhang Q, Giovanni D, Sum T C and Xiong Q 2014 Synthesis of organic–inorganic lead halide perovskite nanoplatelets: towards high-performance perovskite solar cells and optoelectronic devices *Adv. Opt. Mater.* **2** 838–44

[105] Chen J, Fu Y, Samad L, Dang L, Zhao Y, Shen S, Guo L and Jin S 2017 Vapor-phase epitaxial growth of aligned nanowire networks of cesium lead halide perovskites ($CsPbX_3$, X = Cl, Br, I) *Nano Lett.* **17** 460–6

[106] Zhang D, Eaton S W, Yu Y, Dou L and Yang P 2015 Solution-phase synthesis of cesium lead halide perovskite nanowires *J. Am. Chem. Soc.* **137** 9230–3

[107] Ashley M J, O'Brien M N, Hedderick K R, Mason J A, Ross M B and Mirkin C A 2016 Templated synthesis of uniform perovskite nanowire arrays *J. Am. Chem. Soc.* **138** 10096–9

[108] Horváth E, Spina M, Szekrényes Z, Kamarás K, Gaal R, Gachet D and Forró L 2014 Nanowires of methylammonium lead iodide ($CH_3NH_3PbI_3$) prepared by low temperature solution-mediated crystallization *Nano Lett.* **14** 6761–6

[109] Dou L, Lai M, Kley C S, Yang Y, Bischak C G, Zhang D, Eaton S W, Ginsberg N S and Yang P 2017 Spatially resolved multicolor $CsPbX_3$ nanowire heterojunctions via anion exchange**114** 7216–21

[110] Kang C H *et al* 2019 High-speed colour-converting photodetector with all-inorganic $CsPbBr_3$ perovskite nanocrystals for ultraviolet light communication *Light: Sci. Appl.* **8** 94

[111] Ding J, Fang H, Lian Z, Li J, Lv Q, Wang L, Sun J -L and Yan Q 2016 A self-powered photodetector based on a $CH_3NH_3PbI_3$ single crystal with asymmetric electrodes *Cryst. Eng. Commun.* **18** 4405–11

[112] Xue M, Zhou H, Ma G, Yang L, Song Z, Zhang J and Wang H 2018 Investigation of the stability for self-powered $CsPbBr_3$ perovskite photodetector with an all-inorganic structure *Sol. Energy Mater. Sol. Cells* **187** 69–75

[113] Bansode U, Rahman A and Ogale S 2019 Low-temperature processing of optimally polymer-wrapped α-$CsPbI_3$ for self-powered flexible photo-detector application *J. Mater. Chem.* C **7** 6986–96

[114] Cao F, Tian W, Deng K, Wang M and Li L 2019 Self-powered UV–Vis–NIR photodetector based on conjugated-polymer/$CsPbBr_3$ nanowire array *Adv. Func. Mater.* **29** 1906756

[115] Ghosh J, Natu G and Giri P K 2019 Plasmonic hole-transport-layer enabled self-powered hybrid perovskite photodetector using a modified perovskite deposition method in ambient air *Org. Electron.* **71** 175–84

[116] Saleem M I, Yang S, Zhi R, Li H, Sulaman M, Chandrasekar P V, Zhang Z, Batool A and Zou B 2020 Self-powered, all-solution processed, trilayer heterojunction perovskite-based photodetectors *Nanotechnology* **31** 254001
[117] Zhang X, Dong X, Wang S, Liu H, Hu W and Li X 2020 A self-powered photodetector based on polarization-driven in $CH_3NH_3PbI_3$ single crystal (100) plane *Chem. Eng. J.* **404** 125957
[118] Zhu W, Deng M, Chen D, Chen D, Xi H, Chang J, Zhang J, Zhang C and Hao Y 2020 Sacrificial additive-assisted film growth endows self-powered $CsPbBr_3$ photodetectors with ultra-low dark current and high sensitivity *J. Mater. Chem.* C **8** 209–18
[119] Zhang X, Ji C, Liu X, Wang S, Li L, Peng Y, Yao Y, Hong M and Luo J 2020 Solution-grown large-sized single-crystalline 2D/3D perovskite heterostructure for self-powered photodetection *Adv. Opt. Mater.* **8** 2000311
[120] Fang H, Li Q, Ding J, Li N, Tian H, Zhang L, Ren T, Dai J, Wang L and Yan Q 2016 A self-powered organolead halide perovskite single crystal photodetector driven by a DVD-based triboelectric nanogenerator *J. Mater. Chem.* C **4** 630–6
[121] Su L, Zhao Z X, Li H Y, Yuan J, Wang Z L, Cao G Z and Zhu G 2015 High-performance organolead halide perovskite-based self-powered triboelectric photodetector *ACS Nano* **9** 11310–6
[122] Lim S, Ha M, Lee Y and Ko H 2018 Large-area, solution-processed, hierarchical $MAPbI_3$ nanoribbon arrays for self-powered flexible photodetectors *Adv. Opt. Mater.* **6** 1800615
[123] Shen K *et al* 2020 Flexible and self-powered photodetector arrays based on all-inorganic $CsPbBr_3$ quantum dots *Adv. Mater.* **32** 2000004
[124] Shao D, Zhu W, Liu X, Li M, Chen J, Sun Y -Y, Xin G, Lian J and Sawyer S 2020 Ultrasensitive UV photodetector based on interfacial charge-controlled inorganic perovskite–polymer hybrid structure *ACS Appl. Mater. Interfaces* **12** 43106–14
[125] Wu X, Zhou B, Zhou J, Chen Y, Chu Y and Huang J 2018 Distinguishable detection of ultraviolet, visible, and infrared spectrum with high-responsivity perovskite-based flexible photosensors *Small* **14** 1800527
[126] Li C, Nie J, Cai J, Zhou S, Huang J, Ouyang X, Wu Z and Xu Q 2022 0D structured Cs_4PbI_6 single crystals for highly performance UV photodetection *J. Alloys Compd.* **896** 163047
[127] Liu S, Jiao S, Lu H, Nie Y, Gao S, Wang D, Wang J and Zhao L 2022 Polycrystalline perovskite $CH_3NH_3PbCl_3$/amorphous Ga_2O_3 hybrid structure for high-speed, low-dark current and self-powered UVA photodetector *J. Alloys Compd.* **890** 161827
[128] Zhang J, Wang Q, Zhang X, Jiang J, Gao Z, Jin Z and Liu S 2017 High-performance transparent ultraviolet photodetectors based on inorganic perovskite $CsPbCl_3$ nanocrystals *RSC Adv.* **7** 36722–7
[129] Zhou H, Song Z, Tao P, Lei H, Gui P, Mei J, Wang H and Fang G 2016 Self-powered, ultraviolet-visible perovskite photodetector based on TiO_2 nanorods *RSC Adv.* **6** 6205–8
[130] Momma K and Izumi F 2011 VESTA 3 for three-dimensional visualization of crystal, volumetric and morphology data *J. Appl. Crystallogr.* **44** 1272–6

Chapter 12

An integrated approach of green nanotechnology in wastewater remediation using algae

Kanagam Nachiappan, Kaaviya Velmurugan, Harini Gurumurthy Vedavalli, Abirami Krishnan, Gopinath Sekar, Yamunadevi Balakrishnan, Sathishkumar Panneerselvam, Balaji Sundaramoorthy, Kalaivani Thiagarajan[1] and Rajasekaran Chandrasekaran[1]

The global economy is expanding due to industrialization and modernization in order to achieve socio-economic development quickly. However, rapid economic growth necessitates high energy use, which increases emissions and poses a significant risk to both the environment and human health. Algae biomass is used for adsorption or cell integration in wastewater bioremediation polluted with heavy metal. The removal or degradation of toxicants from contaminated locations using various types of plants, algae, and cyanobacteria is known as phytoremediation or phycoremediation. Apart from remediation, nano-biotechnology provides indirect solutions in all phases of growth, process, production, storage, packaging, and transportation of products from agriculture. By using nanoscale devices and improving sensing, treating, and remediating environmental pollutants, nanotechnology is developing crucial technological qualities for safeguarding the environment. The use of nanoremediation techniques can reduce the cost of cleaning up large contaminated areas as well as the time it takes. They can also do away with the need to treat and remove polluted soil and reduce some pollutant concentrations to almost zero under *in situ* conditions.

[1] Equally contributed.

doi:10.1088/978-0-7503-5729-6ch12

12.1 Introduction

The word 'globalization' derives from the verb 'globalize', which means 'to make worldwide in scope' [1, 2]. Because of globalization, industrialization and modernization have sped up the heavy metal accumulation in the environment causing toxicity in living organisms. Examples of this include the release of metal ions by functioning cellular units and reactive oxygen species, such as superoxide anions, hydrogen peroxide, hydroxyl radicals, and singlet oxygen. Due to the deactivation of enzymes, biomolecule oxidation, such as that of nucleic acids, proteins, and lipids, and disruption of the integrity of the cellular membrane occur, which damage the cellular system. Eutrophication, according to the Oslo and Paris Commission (OSPAR), 'may cause a variety of unfavorable changes in the marine ecosystem, including a change in the flora and fauna composition that impacts habitats and biodiversity and the oxygen depletion resulting in the fatality of fish and other species' [3].

Opportunistic plant species that have evolved to these new environmental conditions proliferated, changing the composition and operation of all life forms of the community (phytoplankton, zooplankton, benthic fauna, fish, etc). These blooms or proliferations produce a significant amount of biomass, hazardous emissions, hypoxia or anoxia in the aquatic environment, and oxygen loss resulting from their bacterial decomposition (CO_2, H_2S, and CH_4). Several of these proliferations could be dangerous. The first physiological/biochemical reactions to such a disturbance are exhibited at the individual level, which are then followed by behavioral or morphological reactions, and finally effects are seen at the population and community levels. The most obvious effects of eutrophication are anoxia, which can cause a huge number of deaths, and toxic vegetal blooms, which reduce biodiversity [4].

In freshwater habitats such as rivers and lakes, microalgae are very common. Microalgae are categorized according to their basic structures as photoautotrophs, and photoheterotrophs, lacking differentiation into leaves, stems, and roots, without any protective coatings on their reproductive organs. Blue-green algae, among other types of microalgae, are crucial to bioremediation. Microalgae play a substantial role in biogeochemical cycling and heavy metal transformation in the environment [5, 6]. Microalgae belonging to the phyla Cyanophyta and Chlorophyta serve as biological sensors for identifying heavy metal toxicity, as they are sensitive to changes in trace-level concentrations of organic and inorganic contaminants [7, 8]. Therefore, most of the microalgal species can be used for the *in situ* process of bioremediation in contaminated water bodies [9, 10]. Microalgal species can remove heavy metals by various mechanisms, such as metabolic and non-metabolic, which can be differentiated as absorption (metabolic) and adsorption (non-metabolic) [11]. The uptake of heavy metals inside microalgal species is prevented either by the adsorption capacity of its cell wall components or by the secretion of metal-binding pollutants into the environment around it [12]. Some of the heavy metals that enter into microalgae through micronutrient transport can be detoxified by metallothionein, a metal-binding protein [13], or chelated by organic molecules such as citrate, polyphosphate, and malate.

Additionally, the green synthesis of nanoparticles (NPs) has become a viable option for providing nanomaterials with improved physicochemical properties while also being environmentally benign. The nanomaterials synthesized using different methodologies reported in the literature have drawbacks such as poisonous by-products, high-priced instrumentation, and energy-intensive experimental procedures which significantly reduce their wide applicability [14]. The biogenic synthesis of nanoparticles has emerged as a possible prospect, providing a sustainable, economical, and energy-efficient method for the synthesis of a wide range of NPs. Numerous biological organisms, including bacteria, fungi, and algae, are used in the production of NPs [15]. However, because of their superior growth rates, noticeably higher rates of retention of carbon dioxide (CO_2), and excessive build-up of heavy metals, the distinctive characteristics of microalgae and cyanobacteria allow their potential application for NPs synthesis. The increased surface areas, compact dimensions, high adsorption capacity, catalytic activity, and sometimes microbiological effectiveness of nanoparticles undoubtedly contribute to their tremendous potential. [16].

The algal extract acts as a living cell-producing unit for the effective green production of nanoparticles since it is a natural capping and reducing agent. By controlling their metabolic processes, both physiological and organic approaches enable algal cells to absorb heavy metals and use them as a source of nutrients to produce biomass [17]. Despite having considerable potential, the literature lacks studies on the synthesis of nanoparticles from microalgae for the removal of toxic pollutants from wastewater. This chapter summarizes the current improvements and potential of algae nanoparticles in treating wastewater.

12.2 A growing industrial sector in the twenty-first century

Recently, researchers and policymakers have called attention to the significance of globalization and its impact on a variety of social, political, economic, and, in particular, environmental factors [18]. The economic upliftment of any nation is connected by trade openness, and foreign direct investment (FDI) has been expedited by globalization. Because the main sources of energy are fossil fuels (such as coal, gas, oil, and petroleum), the degradation of the environment has grown in importance in recent years. Emissions of the main greenhouse gas component, CO_2, have increased dramatically during the last few decades. By the middle of this century, human activity will have produced more CO_2 than in the previous 50 million years [18]. The Paris Agreement on Climate Change, which aims to improve environmental quality by tackling issues related to global warming and its detrimental effects, has been ratified by 144 out of 197 countries. World Health Organisation [19] research highlights the importance of building up investments in the healthcare industry that focus on low-emissions development and institution resilience.

12.3 Major industrial pollutants of the environment

12.3.1 Tannery industry

The chemical and organic substances used in tanning can have a negative impact on the environment. The tanning stage of the process makes use of substances including chromium, vegetable tannins, as well as aldehydes [20]. Some of the effluents released from the tannery industry include acids, alkalis, chromium salts, tannins, solvents, auxiliaries, surfactants, acids, natural and synthetic tanning agents, sulfonated oils, and salts.

12.3.2 Textile industry

Textile is any filament, fiber, or yarn that can be made into fabric or cloth. The pretreatment, dyeing, printing, and finishing processes are just a few of the different steps in the production of industrial textiles. In addition to using a lot of power and water, these manufacturing processes produce a significant amount of waste that affects the environment very badly [21], including starch, waxes, carboxymethyl cellulose (CMC), polyvinyl alcohol (PVA), wetting agents fats, pectins [22].

12.3.3 Petrochemical industry

Methanol, ethylene, propylene, butadiene, benzene, toluene, and xylenes are some of the basic petrochemicals. The removal of such chemicals from industrial effluents is crucial since it produces chemicals such as ethylene and its derivatives, benzene, and toluene, and additional effluents such as hydrogen, carbon monoxide, synthetic gases, and other pollutants [23].

12.3.4 Dairy industry

Due to its high water use, the dairy industry is one of the food industries that contributes the most pollution [24]. High levels of oil and grease, fats, sugar, colors, preservatives, dissolved oxygen, total solids, biological oxygen demand (BOD), chemical oxygen demand (COD), suspended particles, and generally nutrients such as nitrogen, ammonia, and phosphate [25] cause problems.

12.4 The adverse effect of contaminants on the ecosystem

12.4.1 Health risks—biomagnification

Through the transport of heavy metals from the abiotic environment into living creatures, they build up in the biota at different trophic levels and contaminate food chains and webs. The health of animals and people is significantly affected by the this trophic transmission, bioaccumulation, and biomagnification of dangerous heavy metals in the food chain. The bioavailability of numerous heavy metals can be determined based on their solubility, mobility, and bioavailability. The direct consumption of heavy metals found in many food crops and wild seafood is harmful to people's health. These heavy metals, especially those with a higher affinity for sulfur atoms, which thus interfere with the operation of numerous enzymes, have a deleterious effect on biochemical processes in humans. Pb causes nephrotoxicity, and when paired with Hg, heavy metals such as Cd can cause kidney damage. Inorganic arsenic can cause cancer, and the mixture of all these metals, even in small doses, may be more dangerous than any one metal by itself [26]. About 90% of the danger to human health was brought on by eating fish polluted with heavy metals, primarily mercury and selenium. Hussain *et al* [27] have reported that Hg accumulation in fish tissues in the Himalayan region has exceeded the permissible levels of the JECFA and EPA.

12.4.2 Eutrophication in aquatic systems

According to the IPCC's fifth assessment report, the average surface temperature of the entire world has increased by 0.4 °C. The 0.85 °C (0.65 °C to 1.06°C) increase from 1880 to 2012 is proof that there has been global warming. The hydrological cycle will be affected by warmer weather, which will also change atmospheric and meteorological features such as precipitation patterns, atmospheric water vapor, and evaporation. This will have an effect on the water quality by increasing various types of water pollution.

Water contamination comes in many forms, one of which is eutrophication [28]. In the scientific literature, the word 'eutrophication' describes a natural process of rising organic matter production that happens when an aquatic habitat develops over geological time, unless it eventually fills up completely. It can also be used to refer to a brief process (hours, days, months, or years) that is started by human effects. Anthropogenic eutrophication is the excessive manufacture of organic matter brought on by anthropogenic phosphorus and nitrogen inputs. The significant disruption of aquatic ecosystems has an impact on the associated goods and services, people's health, and economic activity in the areas where it happens. Green macro-algae blooms in coastal locations and poisonous cyanobacteria blooms in lakes and streams are two of its most well-known effects. The term 'eutrophication' has many definitions because it is used by both scientists and decision-makers in government [4]. Environmental factors can affect the form and strength of reactions: eutrophication is triggered by long water residence durations, high temperatures, and adequate light [4]. Due to the decomposition and settling of more organic waste, eutrophic freshwaters have low concentrations of dissolved oxygen (DO) in the bottom water and sediment pore water. Low DO circumstances, however, are probably advantageous for overall CH_4 emissions [29].

12.4.3 Soil pollution

The natural system's most important component, soil, provides the foundation for human growth and survival. The Earth naturally contains a lot of heavy metal elements, but human activities such as farming, tourism, and industrial manufacturing raise the pollution level. The majority of the heavy metals produced by these processes are released into the environment through sewage, waste gas, and solid debris after ending up in the soil [31]. To maintain regular life processes, life forms need trace levels of heavy metals. Although there are many contributing factors, which include the use of pesticides and fertilizers, improper storage of industrial waste, chemical manufacturing, metal plating, the tanning process, sewage irrigation, atmospheric deposition, etc, it was determined that human activity was the primary offender [30]. Heavy metal contamination in polluted soils deteriorates ecosystem functioning and in time it harms life forms via the food chain [31]. Metals are added in small amounts into agricultural soils along with both organic and mineral fertilizers (mainly calcium and phosphates). Fei *et al* [32] found that while the combined effect of two heavy metals present in the soil exhibited a nonlinear amplification, no single heavy metal showed a statistically significant impact on the spatial distribution of stomach cancer (SC) incidence. This was done by combining the analysis of SC incidence data with the processing of soil heavy metal data.

12.5 Algal bioremediation

12.5.1 Micro- and macroalgal remediation

Compared to the physicochemical methods, a variety of benefits are provided by microalgal species when employing them for the remediation of heavy metals. Potential microalgae are extremely prevalent in the natural freshwater environment and have amazing capacities for absorbing and accumulating heavy metals in a variety of environments, including domestic water, freshwater, and industrial effluents [33]. Additionally, several microalgal species have been found to be effective at phycoremediation, sequestering metal ions such as chromium, cadmium, copper, and nickel [34]. Both living and dead microalgal biomass are used to filter out heavy metals in contaminated water. Algae are autotrophic, which means they create more biomass and require less food than other microbial biosorbents. These sorts of biosorbents have additionally been employed to remove contaminants such as heavy metals [35]. Algae feature a range of chemical moieties on their surface, including hydroxyl, carboxyl, phosphate, and amide, that act as metal-binding sites. The authors of [35] used dead *Chlorella vulgaris* cells to extract Cd^{2+} and Cu^{2+}. Algae are photosynthesis-capable organisms that can survive in both freshwater and saltwater. Certain peptides produced by algae can both gather heavy metal ions and shield the algae from their toxicity [36]. According to previous studies, *Fucus vesiculosus* is very efficient at absorbing Pb(II), which can be explained via the Langmuir and Freundlich equations [37]. Applying the Langmuir and Freundlich isotherm models, the adsorption capabilities of *Cladophora fascicularis* were examined for the removal of Pb(II) from wastewater. The maximal Pb(II) adsorption capacity of *C. fascicularis* at pH 5.0 is 198.5 mg g^{-1}, and its biosorption kinetics are reliable. Because they need a lot of nitrogen and phosphorus to make proteins (which make up 45%–60% of microalgal dry weight), nucleic acids, and phospholipids, microalgae consume a lot of nutrients. Due to the change in pH brought on by photosynthesis, nutrient removal can also be boosted by NH_3 stripping or P precipitation [71]. In a continuous treatment process, the use of microalgae would provide a great advantage, because most industries are in dire need of implementing cost-effective continuous treatment processes. Algal species appear to be the ideal organisms for use in remediation research since they are reasonably simple to grow, adapt, and manipulate in a lab setting [39]. Due to its superior nutrient removal effectiveness and low implementation and servicing costs, phycoremediation is also preferable to other traditional physicochemical methods such as ion exchange, reverse osmosis, dialysis and electro-dialysis, membrane separation, activated carbon adsorption, and chemical reduction or oxidation [40].

12.5.2 Role of Blue Green Algae in Bioremediation

Bioaccumulation, bioremediation, and biosorption are heavy metal removal processes. The capacity of cyanobacteria to retain cell surface cationic metals through binding with phosphate and other anions contained in extracellular polysaccharides (EPS), makes biosorption the most prevalent method for heavy metal elimination. The biosorption mechanism is made simple by the presence of abiotic factors such as pH, temperature, and contact time, as well as anionic groups present on the extracellular surface. It has also been discovered that dead cyanobacterial biomass is more effective at biosorption than live cells, which causes the release of saxitoxin and other frequently occurring exotoxins from living cyanobacteria. By using a passive absorption process termed biosorption, cyanobacteria sequester heavy metals at physiological pH levels. Bacterial cell walls with negative charges cause them to interact ionically with positively charged heavy metals in the surroundings [41, 42]. All live and dead organisms possess negatively charged functional groups on their cell surface. Additionally, there are many ways for biosorption to happen [43, 44], via the functional classes of cell surface peptidoglycan receptors. Peptidoglycans' carboxyl groups provide a superb surface for heavy metal adsorption [45]. Metal adsorption can be triggered by metal binding to membrane proteins, lipids, lipopolysaccharides, and exopolysaccharides [5]. When there is a gradient in the concentration of metal ions between the external and intracellular environments, metal ions migrate through the cell membrane. Another method of active transport is binding to chelating proteins [47]. Cyanobacteria produce metallothioneins, chelating proteins with cysteine residues rich in thiols, in their cytoplasm. As metal ions reach the cytoplasm, they aid in their sequestration [48]. Organometallic complexes are moved to cellular spaces (such as vacuoles) for storage after they have formed. This procedure supports detoxification and the maintenance of cytoplasmic ion concentrations [49]. Enzymatic sequestration of metal ions in cells is a component of bioaccumulation. Bioaccumulation begins when intracellular metal ion transport begins and the extracellular metal ion gradient is greater than the intracellular metal ion gradient [50]. Porins carry heavy metal ions into the periplasm from the outside membrane. Metal ions are carried by three proteins from the periplasm to the cytoplasm: active primary transporters, secondary carriers, and channels. With the help of metal ion gradients, channels mediate energy-demanding passive transport. Secondary carriers rely on the protein's propulsion and are energy-dependent. Protein complexes, which transport metal ions by ATP hydrolysis, are the main active transporters. Metal ions combine with metal-binding moieties to form complexes when they are in the cytoplasm. These metallothioneins, photokeratitis, and polyphosphates are metal-binding moieties.

Once the metal has entered the cell, the metabolic activity transforms it into a less dangerous or harmless form. Cells are poisoned when this system malfunctions because of high metal concentrations [51].

12.6 Algae-coated nanoparticles in wastewater remediation

12.6.1 Advantages of nanotechnology in remediation

In order to create, use, and understand material structures, systems, and devices with basically novel properties and functions due to their microscopic structure, nanotechnology explores materials with dimensions of a few nanometers or less. Nanotechnology is a fantastic field to concentrate on because of the effective physiochemical and crystalline properties of nanoparticles. Due to their unique thermal, electrical, biological, optical, chemical, and physical features compared to their bulk-sized counterparts, they have drawn the interest or researchers [52]. Nanoparticles can be utilized for a variety of purposes, such as medical treatment, solar and fuel cell batteries for the generation of green energy, water and air filters to reduce pollution, and they can even be used as catalysts in contemporary manufacturing processes to do away with the use of toxic ingredients. Nanoparticles with their small size (1–100 nm) and enormous surface areas opens up opportunities for valuable and effective energy management and pollutant control strategies. The primary concept for controlling pollutants at the nano-scale is segregating specific factors and molecules from an aggregate of pollutants via nanofabrication, thermal partitioning, or nanostructuring. It is claimed that gram-positive and gram-negative bacteria are resistant to metal nanoparticles (gold, silver, and copper oxide) made from *Bifurcaria bifurcate, Galaxaura elongate*, *Sargassum plagiophyllum*, *Caulerpa racemose*, *Microcoleus* sp., and *Chlorococcum humicola*. Diatoms increase the amount of dissolved oxygen (DO) in aquatic life by absorbing carbon dioxide from the atmosphere and releasing oxygen [53]. Greater biodiversity, healthier marine life, and improved quality of water may result from this. Algal nanoparticles' capacity to pierce extrapolysaccharides (EPS) and cell membranes makes them potential anti-biofilm agents for the treatment of multidrug-resistant bacteria in the future. They can also be investigated extensively in the nanocomposite and biosensor applications [17].

12.6.2 Characterization and synthesis of nanoparticles to enhance remediation

Global interest is high in the biosynthetic production of nanoparticles (by microbes or plant extracts), which is a technology that is significantly simpler than physical and chemical approaches [54, 55]. Yeast, bacteria, and fungi are only a few examples of microorganisms that are crucial in the removal of metal ions. Microalgae are receiving the most attention among all the biological systems due to their capacity to biorevert hazardous metals and further change them into more sensitive variants. Microalgae perform photosynthesis, turning sunlight and carbon dioxide into usable biomass and storing nutrients such as nitrogen and phosphorus that cause eutrophication in aquatic environments.

The synthesis of nanoparticles utilizing algae may be completed in three crucial phases:

1. Heating or boiling water or an organic solvent for a predetermined amount of time.
2. Ionic metallic compound molar solution preparation.
3. Algal solutions as well as molar solutions of ionic metallic compounds are incubated continuously with stirring or without stirring for a predetermined amount of time under particular circumstances [56].

Nanoparticle production is dose-dependent and influenced by the type of algae employed. The reduction of metals is caused by a multitude of biomolecules, including pigments, peptides, and polysaccharides. By using amino groups, cysteine residues, and sulphated polysaccharides, proteins stabilize and cap the metal nanoparticles in aqueous solutions. Algae are used to synthesize nanoparticles, which takes a lot less time than other bio-synthesizing techniques [57]. So far, numerous seaweeds, particularly *Chaetomorphalinum*, *Enteromorpha flexuosa*, *Fucus vesiculosus*, *Turbinariaconoides*, *Sargassum wightii*, *Stoechospermum marginatum*, *Ulva fasciata*, and *Ulva reticulata* have been used for synthesizing AgNPs of various sizes and shapes. We merely investigate marine algae for NP synthesis. Ninety per cent of the algal-bound gold produced by *C. vulgaris* forms a strong binding to tetrachloro aurate ions, and the gold crystals accumulated in the inside and outside elements of cell surfaces with tetrahedral, decahedral, and icosahedral structures [58].

12.7 Adsorption mechanism of algal-based nanomaterials

The pseudo-first-order model and the pseudo-second-order model are two kinetic models that are frequently used to explore the dynamics of algae adsorption to nanoparticles (NPs). To fit the adsorption isotherm with the collected experimental data, the Langmuir and Freundlich isotherm models are utilized frequently [59]. In general, investigations of the adsorption kinetics and isotherm models reveal that a pseudo-second-order kinetic model and Langmuir isotherm model are the best fits for the adsorption of algae to NPs [60, 62]. The ability of algae to bind to NPs increased with an increase in initial concentration [61]. The initial stages of the adsorption process were very rapid, but after that the rate of adsorption gradually dropped until it achieved saturation [60]. With a shift in solution pH, the amount of NPs that algae absorbed from the water changed dramatically. Thus, the pH ranges with relatively strong acids can facilitate the adsorption of algae to NPs [62]. In addition, the shape, biomass dosage, and physicochemical characteristics of the algae all had an impact on how much of them were absorbed by NPs [59].

NPs need to penetrate two barriers, namely cell walls and plasma membranes, in order to be taken up by algae [65]. The rather strong and durable cell walls of algae are typically regarded as the initial barrier to stopping NPs from being internalized. The cell walls of algae are often porous and semipermeable. The pores have a diameter that falls between 5 and 20 nm. Only molecules with a size less than the maximal pore size may pass easily through the cell wall, thus the pores prevent the passage of larger molecules. In addition to glycoproteins and polysaccharides, cellulose is the primary component of the cell wall in algal cells. These elements can serve as binding sites to encourage the adsorption of NPs by algae [63]. The plasma membrane serves as the second barrier that must be overcome by NPs after they have passed through the cell wall. Passive diffusion and endocytosis are thought to be the two main mechanisms by which NPs travel through the bilayer lipid membrane. Additionally, during cell division, the permeability of the cell wall may change, making the newly formed cell walls more NP-permeable and enhancing the uptake of NPs by algal cells [64].

Without the use of any cross-linking agents, chitosan nanoparticles (chitosan NPs) were successfully integrated into cotton fiber surfaces during green adsorption. Various analytical approaches were used to study the interactions between the cotton fibers and chitosan NPs throughout the green adsorption process. The findings showed that intermolecular hydrogen bond connections between cotton fibers' hydroxyl groups and the chitosan NPs' amino groups existed and gradually improved as the chitosan NP mass concentrations increased. Because of the increased contacts upon chitosan NP adsorption, the acidity of the fibers increased while the crystallinity index decreased. Additionally, chitosan NP and crystalline cotton fibers interacted hydrophobically, causing preferential adsorption onto the hydrophobic crystallographic plane [66].

Wet procedures (physical and chemical) are the traditional and most often used methods to create nanoparticles (figure 12.1). These techniques fall into one of two categories: top-down methods or bottom-up methods. Chemically, stabilizing and reducing substances such as potassium bitartrate, sodium dodecyl benzyl sulfate, methoxy polyethylene glycol (PEG), polyvinyl pyrrolidone, or sodium borohydride are used to produce nanoparticles in a liquid medium. The physical techniques also include pyrolysis and attrition. Unfortunately, the physical and chemical techniques used in both systems have some harmful effects on the environment, require rigorous production, and are prohibitively expensive. According to research, nanoparticles' high surface energy causes them to attract atoms and molecules, further altering their surface characteristics. As a result, they cannot exist in the ecosystem in their natural state. Nanoparticles are unsuitable for therapeutic usage due to these environmental interactions. This method of action has caused concern to grow around the creation of non-toxic and environmentally friendly processes for the synthesis and assembly of nanoparticles [67].

Recent research has shown that polycyclic aromatic hydrocarbons can be accumulated and degraded by freshwater algae such *Scenedesmus quadricauda*, *C. vulgaris*, *Selenastrum capricornutum*, and *Scenedesmus platydiscus*. As a tertiary wastewater treatment step, algae systems have historically been employed. Additionally, they have been proposed as a viable secondary therapy method. Algae can be employed in purification processes for a variety of objectives, including the removal of heavy metals, the reduction of chemical and biological oxygen demand, the removal of *Escherichia coli*, and the removal of N and/or P [40].

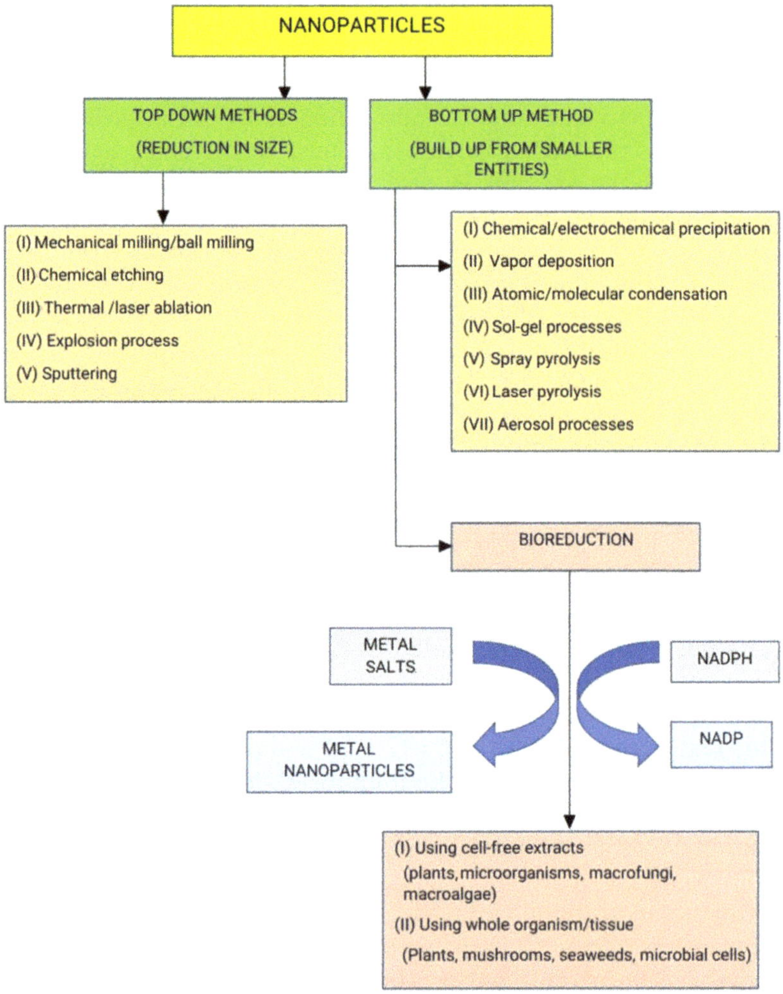

Figure 12.1. The traditional and most popular techniques for producing nanoparticles.

12.8 Applications of algal nanotechnology in remediation

Eliminating environmental toxins is currently the most difficult undertaking due to the rapid development of urbanization. Although heavy metals are naturally existing components in the environment, their careless and indiscriminate release into the ecosystem as a result of anthropogenic and industrial activity has significantly affected the stability of the environment. As a result, the issue of heavy metallic pollutants appears to pose a serious risk because of their persistent and resistant nature. The standard treatment method has some strengths, but the remediation system is complicated because of a number of linked but unrelated limitations [68]. The degradation and antimicrobial property of algal nanoparticle has a wide application in various industries as tabulated in Table 12.1.

Table 12.1. Nanotechnology applications of Algae across various industries

S. No.	Algal species	Nanoparticle	Industry	Action	Effect	References
1	*Oscillatoria* sp.	OG-AgNPs	Medicine	Degradation of H_2O_2 at trace ppm levels	Exhibit potent phycotoxic effect	[69]
2	*Sargassum myriocystum*	AgNPs	Textiles and dyeing	Degradation organic dyes	98% potential photocatalytic activity	[70]
3	*Oscillatoria* sp.	Ag_2O/AgO-NPs	Water treatment	Antibacterial and antifungal properties	More efficient than chemically synthesized NPs after 24 h treatment	[38]
4	*Bifurcaria bifurcate*	CuONPs	Water treatment	Significant antibacterial activity	Their cell surface contains amines and carboxyl groups, and copper ions have a higher affinity for them	[57]
5	*Sargassum wighitii*	MgONPs	Biological wastewater	Antifungal and antibacterial action against human diseases	Higher antimicrobial activity than the usual synthetic antibiotics	[72]

12.9 Future prospects and challenges

Micro- and macro-algae are stunning biotechnological tools for the elimination of hazardous waste in the environment. Environmentally hazardous waste products are dangerous and present serious problems for both the environment and living beings. Through many mechanisms, including biodegradation, accumulation, and immobilization, algal biomass has the capacity to remove these contaminants from contaminated locations [20]. Algal growth is less expensive since it requires light and carbon dioxide; as a result, the global microalgal environmental remediation market shows that this industry had grown significantly over the previous few years and further rapid expansion is anticipated. Additionally, research is needed to improve the remediation process so that it is more beneficial, affordable, practicable, widely accepted by the public, and also more effective at reducing contaminants at several stages [73].

Modern techniques for removing heavy metals from wastewater have recently attracted interest. Biological methods have been investigated and used because they had been proven to be easy, eco-friendly, and sustainable. Microalgal species use particular extracellular and intracellular processes to eliminate heavy metals. As a result, attempts to enhance the remediation mechanisms that are already in place are increasing in order to improve and render this system more efficient and long-lasting. This chapter also elaborates on recent developments in the use of microalgae for heavy metal cleanup [75]. Along with this, difficulties with the phycoremediation of heavy metals and ways to overcome them, that might make such systems a more effective, economical, sustainable, and clean technology were examined. With consideration of the bioeconomic potential, phytoremediation can play a significant role in the remediation of heavy metals in an environmentally friendly manner [74, 76].

Environmental biotechnological applications offer a huge range of potential due to the improper disposal of toxic heavy metals in commercial effluents and their effects on the environment and human health. With the help of such organisms, pollutants can be digested, broken down into less toxic molecules, or concentrated with the help of adsorption onto the surface of dead biomass. Microalgae, which have earned the title of 'wonder organisms', are capable of carrying out bioremediation successfully with the help of two mechanisms: biosorption and bio-assimilation. They have the capacity to grow as 'algal blooms' in polluted water and absorb a variety of contaminants. After harvesting and lipid/protein extraction, the algal biomass can be utilized as a powerful biosorbent.

12.10 Conclusion

The global warming and climate change exceed national boundaries. To assist nations in transitioning to a low-emission economy, these issues require coordinated solutions at all scales and international cooperation. The first physiological/biochemical responses to such disturbances are seen at the individual level, which is then followed by behavioral or morphological responses, and eventually there are changes at the population and community levels. Anoxia, which can result in a large number of deaths, and toxic vegetal blooms, which decrease biodiversity, are two of eutrophication's most notable impacts. The most successful front-runner in creating nanoparticles that can successfully supply a number of flexible applications is algae (macro/micro). A low-cost microalgae-based biomass with high bioremediation potential to neutralize and adsorb heavy metallic ions can be developed with careful attention to parameters such as increasing conditions, cell shape, pre-treatment of biomass, etc The use of algae in the production of NPs has encouraged the development of straightforward, environmentally friendly, affordable, and quick processes, therefore reducing the use of solvents and chemical compounds.

The method of biosorption by algal NPs, when compared to other techniques, offers a number of benefits, including a high removal rate, simplicity in desorption, reduction in sludge formation, selective removal of HM species, and cheap operating costs. Once the metal enters the cell, metabolic processes within the cell convert it into a less dangerous or harmless form. For suitable evaluation of nano-remediation strategies, significant research on unique environments is required to avoid any feasible unwanted environmental effects. The new nanosubstances and their production methods are required to pose negligible harm to human health or the environment. The microalgal production technique has certain significant drawbacks, including being time-consuming, involving arduous tasks, microalgae harvesting, and considerable usage of arable land. Microalgal nanoparticles are often a novel research topic for improved removal efficiency to avoid negative effects. As a result, the advancement of nanomaterials has demonstrated that they have fantastic qualities that favor their use in environmentally friendly remediation and, in particular, in the removal of heavy metals from the environment, which overcomes the drawbacks of other NPs.

Acknowledgments

The authors would like to extend their profound sense of gratitude to our management, Vellore Institute of Technology, and the Dean SBST for their constant support and encouragement in all our endeavors.

References

[1] Agba A M O and Odu E 2013 Globalization and the challenge of industrialization in developing nations: the Nigeria experience *IOSR J. Human. Soc. Sci.* **12** 41–7

[2] Zafar M W, Saud S and Hou F 2019 The impact of globalization and financial development on environmental quality: evidence from selected countries in the Organization for Economic Co-operation and Development (OECD) *Environ. Sci. Pollut. Res.* **26** 13246–62

[3] Ménesguen A and Lacroix G 2018 Modelling the marine eutrophication: a review *Sci. Total Environ.* **636** 339–54

[4] Le Moal M *et al* 2019 Eutrophication: a new wine in an old bottle? *Sci. Total Environ.* **651** 1–11

[5] Fomina M and Gadd G M 2014 Biosorption: current perspectives on concept, definition and application *Bioresour. Technol.* **160** 3–14

[6] Manikandan A S, Babu P, Shyamalagowri S, Kamaraj M, Muthukumaran P and Aravind J 2022 Emerging role of microalgae in heavy metal bioremediation *J. Basic Microbiol.* **62** 330–47

[7] Ramakrishnan B, Megharaj M, Venkateswarlu K, Naidu R and Sethunathan N 2010 The impacts of environmental pollutants on microalgae and cyanobacteria *Crit. Rev. Environ. Sci. Technol.* **40** 699–821
[8] Sicard C, Perullini M, Spedalieri C, Coradin T, Brayner R, Livage J, Jobbágy M and Bilmes S A 2011 CeO_2 nanoparticles for the protection of photosynthetic organisms immobilized in silica gels *Chem. Mater.* **23** 1374–8
[9] Gadd G M 2009 Biosorption: critical review of scientific rationale, environmental importance and significance for pollution treatment *J. Chem. Technol. Biotechnol.: Int. Res. Proc., Environ. Clean Technol.* **84** 13–28
[10] Ateia M, Yoshimura C and Nasr M 2016 *In-situ* biological water treatment technologies for environmental remediation: a review *J. Bioremed. Biodeg.* **7** 2
[11] Kaplan D 2013 Absorption and adsorption of heavy metals by microalgae *Handbook of Microalgal Culture: Applied Phycology and Biotechnology* (Hoboken, NJ: Wiley) pp 602–11
[12] Monteiro C M, Castro P M and Malcata F X 2012 Metal uptake by microalgae: underlying mechanisms and practical applications *Biotechnol. Progr.* **28** 299–311
[13] Arunakumara K K I U and Zhang X 2007 Heavy metal bioaccumulation and toxicity with special reference to microalgae *J. Ocean Univ. China* **7** 60–4
[14] Baker S, Rakshith D, Kavitha K S, Santosh P, Kavitha H U, Rao Y and Satish S 2013 Plants: emerging as nanofactories towards facile route in synthesis of nanoparticles *BioImpacts* **3** 111
[15] Khan F, Shahid A, Zhu H, Wang N, Javed M R, Ahmad N, Xu J, Alam M A and Mehmood M A 2022 Prospects of algae-based green synthesis of nanoparticles for environmental applications *Chemosphere* **293** 133571
[16] De Luca A and Ferrer B B 2017 Nanomaterials for water remediation: synthesis, application and environmental fate *Nanotechnologies for Environmental Remediation* (Cham: Springer) pp 25–60
[17] Agarwal P, Gupta R and Agarwal N 2019 Advances in synthesis and applications of microalgal nanoparticles for wastewater treatment *J. Nanotechnol.* **2019** 7392713
[18] Le H P and Ozturk I 2020 The impacts of globalization, financial development, government expenditures, and institutional quality on CO_2 emissions in the presence of the environmental Kuznets curve *Environ. Sci. Pollut. Res.* **27** 22680–97
[19] WHO 2015 *Joint Water Supply and Sanitation Monitoring Programme, 2015. Progress on Sanitation and Drinking Water: 2015 Update and MDG Assessment* (Geneva: World Health Organization)
[20] Abdel-Raouf N, Al-Homaidan A A and Ibraheem I 2012 Microalgae and wastewater treatment *Saudi J. Biol. Sci.* **19** 257–75
[21] Madhav S, Ahamad A, Singh P and Mishra P K 2018 A review of textile industry: wet processing, environmental impacts, and effluent treatment methods *Environ. Qual. Manage.* **27** 31–41
[22] Roy C A K 2014 Environmental impacts of the textile industry and its assessment through life cycle assessment *Roadmap to Sustainable Textiles And Clothing: Environmental and Social Aspects of Textiles and Clothing Supply Chain* (Cham: Springer) pp 1–39
[23] Balasubramani K and Sivarajasekar N 2018 A short account on petrochemical industry effluent treatment *Int. J. Petrochem. Sci. Eng.* **3** 12–3

[24] Raghunath B V, Punnagaiarasi A, Rajarajan G, Irshad A, Elango A and Mahesh Kumar G 2016 Impact of dairy effluent on environment—a review *Integrated Waste Management in India: Status and Future Prospects for Environmental Sustainability* (Cham: Springer) pp 239–49

[25] Chokshi K, Pancha I, Ghosh A and Mishra S 2016 Microalgal biomass generation by phycoremediation of dairy industry wastewater: an integrated approach towards sustainable biofuel production *Bioresour. Technol.* **221** 455–60

[26] Ali H and Khan E 2019 Trophic transfer, bioaccumulation, and biomagnification of non-essential hazardous heavy metals and metalloids in food chains/webs—concepts and implications for wildlife and human health *Human Ecol. Risk Assess.: Inter. J.* **25** 1353–76

[27] Hussain M M *et al* 2021 Arsenic speciation and biotransformation pathways in the aquatic ecosystem: the significance of algae *J. Hazard. Mater.* **403** 124027

[28] Nazari-Sharabian M, Ahmad S and Karakouzian M 2018 Climate change and eutrophication: a short review. *Eng., Technol. Appl. Sci. Res.* **8** 3668

[29] Li Y, Shang J, Zhang C, Zhang W, Niu L, Wang L and Zhang H 2021 The role of freshwater eutrophication in greenhouse gas emissions: a review *Sci. Total Environ.* **768** 144582

[30] Shah A I 2017 Heavy metal impact on aquatic life and human health–an over view *IAIA17 Conf. Proc. IA's Contribution in Addressing Climate Change, 37th Annual Conf. of the Int. Association for Impact Assessment* pp 4–7

[31] Jia X, Fu T, Hu B, Shi Z, Zhou L and Zhu Y 2020 Identification of the potential risk areas for soil heavy metal pollution based on the source-sink theory *J. Hazard. Mater.* **393** 122424

[32] Fei H, Dong J, Chen D, Hu T, Duan X, Shakir I, Huang Y and Duan X 2019 Single atom electrocatalysts supported on graphene or graphene-like carbons *Chem. Soc. Rev.* **48** 5207–41

[33] Preethi S, Arulraj A, Mangalaraja R V, Ravichandran V and Subramanian N 2022 Biomass-derived carbon-based materials for microbicidal applications *Biomass-Derived Carbon Materials* (New York: Wiley) pp 63–92

[34] Ajayan K V, Selvaraju M, Unnikannan P and Sruthi P 2015 Phycoremediation of tannery wastewater using microalgae *Scenedesmus* species *Int. J. Phytorem.* **17** 907–16

[35] Abbas S H, Ismail I M, Mostafa T M and Sulaymon A H 2014 Biosorption of heavy metals: a review *J. Chem. Sci. Technol.* **3** 74–102

[36] Maqsood Q, Hussain N, Mumtaz M, Bilal M and Iqbal H M 2022 Novel strategies and advancement in reducing heavy metals from the contaminated environment *Arch. Microbiol.* **204** 478

[37] Demey H, Vincent T and Guibal E 2018 A novel algal-based sorbent for heavy metal removal *Chem. Eng. J.* **332** 582–95

[38] El-Sheekh M M, Hassan L H and Morsi H H 2021 Evaluation of antimicrobial activities of blue-green algae-mediated silver and gold nanoparticles *Rend. Lincei. Sci. Fis. Nat.* **32** 747–59

[39] Dresback K, Ghoshal D and Goyal A 2001 Phycoremediation of trichloroethylene (TCE). *Physiol. Mol. Biol. Plants* **7** 117–23

[40] Rao P, Kumar R R, Raghavan B G, Subramanian V V and Sivasubramanian V 2011 Application of phycoremediation technology in the treatment of wastewater from a leather-processing chemical manufacturing facility *Water SA* **37** 7–14

[41] Dvořák P *et al* 2015 Species concepts and speciation factors in cyanobacteria, with connection to the problems of diversity and classification *Biodivers. Conserv.* **24** 739–57

[42] Echlin P and Morris I 1965 The relationship between blue-green algae and bacteria *Biol. Rev.* **40** 143–84

[43] Fay P 1992 Oxygen relations of nitrogen fixation in cyanobacteria. *Microbiol. Rev.* **56** 340

[44] Fan T, Liu Y, Feng B, Zeng G, Yang C, Zhou M, Zhou H, Tan Z and Wang X 2008 Biosorption of cadmium (II), zinc (II) and lead (II) by *Penicillium simplicissimum*: isotherms, kinetics and thermodynamics *J. Hazard. Mater.* **160** 655–61

[45] Ferris M J and Hirsch C F 1991 Method for isolation and purification of cyanobacteria *Appl. Environ. Microbiol.* **57** 1448

[47] Gauthier P *et al* 2014 *Metal–PAH Mixtures in the Aquatic Environment: A Review of Co-toxic Mechanisms Leading to More-Than-Additive Outcomes* vol 154 (Amsterdam: Elsevier) pp 253–69

[48] Goodwin T W 1965 Chemistry and biochemistry of plant pigments *Chemistry and Biochemistry of Plant Pigments Q. Rev. Biol.* **52** 96 [preprint]

[49] Gupta P and Diwan B 2017 Bacterial exopolysaccharide mediated heavy metal removal: a review on biosynthesis, mechanism and remediation strategies *Biotechnol. Rep.* **13** 58–71

[50] Heindl J E *et al* 2014 Mechanisms and regulation of surface interactions and biofilm formation in agrobacterium *Front. Plant Sci.* **5** 176

[51] Iram S, Shabbir R, Zafar H and Javaid M 2015 Biosorption and bioaccumulation of copper and lead by heavy metal-resistant fungal isolates *Arab. J. Sci. Eng.* **40** 1867–73

[52] Ramirez-Merida L G, Zepka L Q, de Menezes C R and Jacob-Lopes E 2015 Microalgae as nanofactory for production of antimicrobial molecules *J. Nanomed. Nanotechnol.* **S6** 1

[53] Hafez E E and Kabeil S S 2013 Antimicrobial activity of nano-silver particles produced by micro algae *J. Pure Appl. Microbiol.* **7** 35–42

[54] Sharma A, Sharma S, Sharma K, Chetri S P, Vashishtha A, Singh P, Kumar R, Rathi B and Agrawal V 2016 Algae as crucial organisms in advancing nanotechnology: a systematic review *J. Appl. Phycol.* **28** 1759–74

[55] Madkour L H 2017 Ecofriendly green biosynthesized metallic nanoparticles: bio-reduction mechanism, characterization and pharmaceutical applications in biotechnology industry *Drugs Ther.* **3** 1–11

[56] Thakkar K N, Mhatre S S and Parikh R Y 2010 Biological synthesis of metallic nanoparticles *Nanomed. Nanotechnol. Biol. Med.* **6** 257–62

[57] Abboud Y, Saffaj T, Chagraoui A, El Bouari A, Brouzi K, Tanane O and Ihssane B 2014 Biosynthesis, characterization and antimicrobial activity of copper oxide nanoparticles (CONPs) produced using brown alga extract (*Bifurcariabifurcata*) *Appl. Nanosci.* **4** 571–6

[58] Vincy W, Mahathalana T J, Sukumaran S and Jeeva S 2017 Algae as a source for synthesis of nanoparticles-a review *Int. J. Latest Trends Eng. Technol* **5** 5–9

[59] Esmaeili A and Beni A A 2015 Novel membrane reactor design for heavy-metal removal by alginate nanoparticles *J. Ind. Eng. Chem.* **26** 122–8

[60] Harja M, Buema G, Bulgariu L, Bulgariu D, Sutiman D M and Ciobanu G 2015 Removal of cadmium (II) from aqueous solution by adsorption onto modified algae and ash *Korean J. Chem. Eng.* **32** 1804–11

[61] Chakraborty T W and Weber R K 1974 Pore and solid diffusion model for fixed bed adsorbents *J. Am. Inst. Chem. Eng* **20** 228

[62] Vijayaraghavan K, Mahadevan A, Sathishkumar M, Pavagadhi S and Balasubramanian R 2011 Biosynthesis of Au (0) from Au (III) via biosorption and bioreduction using brown marine alga *Turbinaria conoides Chem. Eng. J.* **167** 223–7

[63] Kim M S *et al* 2014 Using citrate-functionalized TiO_2 nanoparticles to study the effect of particle size on zebrafish embryo toxicity *Analyst* **139** 964–72
[64] Wang F, Guan W, Xu L, Ding Z, Ma H, Ma A and Terry N 2019 Effects of nanoparticles on algae: adsorption, distribution, ecotoxicity and fate *Appl. Sci.* **9** 1534
[65] Wang J and Wang W X 2014 Significance of physicochemical and uptake kinetics in controlling the toxicity of metallic nanomaterials to aquatic organisms *J. Zhejiang Univ. Sci.* A **15** 573–92
[66] Wang M, She Y, Xiao Z, Hu J, Zhou R and Zhang J 2014 The green adsorption of chitosan tripolyphosphate nanoparticles on cotton fiber surfaces *Carbohydr. Polym.* **101** 812–8
[67] Dahoumane S A, Wujcik E K and Jeffryes C 2016 Noble metal, oxide and chalcogenide-based nanomaterials from scalable phototrophic culture systems *Enzyme Microb. Technol.* **95** 13–27
[68] Mehrotra T, Sinha S and Singh R 2021 Application of nanotechnology in the remediation of heavy metal toxicity: a promising approach *New Trends in Removal of Heavy Metals from Industrial Wastewater* (Amsterdam: Elsevier) pp 359–73
[69] Abo-Elmagd R A, Hamouda R A and Hussein M H 2022 Phycotoxicity and catalytic reduction activity of green synthesized *Oscillatoria* gelatin-capped silver nanoparticles *Sci. Rep.* **12** 1–19
[70] Balaraman P, Balasubramanian B, Kaliannan D, Durai M, Kamyab H, Park S, Chelliapan S, Lee C T, Maluventhen V and Maruthupandian A 2020 Phyco-synthesis of silver nanoparticles mediated from marine algae *Sargassum myriocystum* and its potential biological and environmental applications *Waste Biomass Valoriz.* **11** 5255–71
[71] El-Sheekh M M, Farghl A A, Galal H R and Bayoumi H S 2016 Bioremediation of different types of polluted water using microalgae *Rend. Lincei.* **27** 401–10
[72] Pugazhendhi A, Prabhu R, Muruganantham K, Shanmuganathan R and Natarajan S 2019 Anticancer, antimicrobial and photocatalytic activities of green synthesized magnesium oxide nanoparticles (MgONPs) using aqueous extract of *Sargassum wightii J. Photochem. Photobiol.* B **190** 86–97
[73] Dwivedi S and Ahmad I Z 2021 Microalgal nanotechnology for the remediation of environmental pollutants *Bioprospecting Algae for Nanosized Materials* (Cham: Springer) pp 403–28
[74] Chugh M, Kumar L, Shah M P and Bharadvaja N 2022 Algal bioremediation of heavy metals: an insight into removal mechanisms, recovery of by-products, challenges, and future opportunities *Energy Nexus* **7** 100129
[75] Baghour M 2019 Algal degradation of organic pollutants *Handbook of Ecomaterials* (Cham: Springer) vol 1 pp 565–86
[76] Elahi A, Arooj I, Bukhari D A and Rehman A 2020 Successive use of microorganisms to remove chromium from wastewater *Appl. Microbiol. Biotechnol.* **104** 3729–43

IOP Publishing

Materials Technology for the Energy and Environmental Nexus, Volume 2

R V Mangalaraja, Arunachalam Arulraj, Radhamanohar Aepuru and Chan Siew Hwa

Chapter 13

Trace level detection of heavy metal ions using electrochemical sensors: mercury, lead, cadmium, and chromium

Bavatharani Chokkiah, Muthusankar Eswaran, Padmanaban Annamalai, Manikandan Ramalingam, Ivan Mijakovic, Bharathkumar Sivakumar, Vinoth Kumar Ponnusamy and Kiruthiga Ramakrishnan

Heavy metal ions (HMIs) are carcinogens, non-biocompatible, and pollute most natural resources, including environmental water samples. Some HMIs, such as cadmium (Cd), chromium (Cr), mercury (Hg), and lead (Pb) are extremely toxic and dangerous to human health, even in trace amounts. As a result, fast, precise, and dependable methods for detecting HMIs are essential. Compared to other optical and spectroscopic approaches, this chapter spotlights various electrochemical detection techniques for HMIs that are eco-friendly, affordable, and offer onsite and real-time monitoring. Various electrochemical methods are classified based on the different types of sensing signals produced by the presence of HMIs in the solution, such as current, electrical conductivity, impedance, and potential. Furthermore, current developments in the electrochemical detection of HMIs using different sensing technologies with the help of advanced composite materials, such as metals, metal oxide nanomaterials, metal films, graphene, carbon nanotubes, and polymers, are discussed and highlighted.

13.1 Introduction

Heavy metal ions (HMIs) are found in elements/ions or complexes in potable water, the Earth's crust, airborne particulate, and living organisms. HMIs are considered low-density and highly toxic chemical components. HMIs are commonly distributed, non-biodegradable, and lead to a high threat to human health and the

doi:10.1088/978-0-7503-5729-6ch13

environment. When an HMI enters a cell, the formation of the bond between the metal and thiol protein groups changes the biochemical lifecycle owing to its toxicity. Some of these HMIs, such as copper, cobalt, iron, manganese, zinc, etc, are essential for living organisms in trace amounts and their presence at high concentrations creates toxic effects. HMIs such as arsenic, cadmium, chromium, lead, and mercury are harmful elements even at lesser concentrations [1–4].

The scientific community has been forced to fabricate and develop specialized and extremely sensitive analytical instruments for metal pollutant detection in aqueous systems due to the harmful effects of HMIs on humans and other creatures. Significant work has gone into achieving these goals. As a result, a variety of analytical methods are now used for the trace-level detection of toxic metal ions, including electron spin resonance spectroscopy (ESRS) [5], extended x-ray absorption fine structure spectroscopy [6], Fourier transform-infrared spectroscopy (FT-IR) and ion chromatography [7], capillary electrophoresis [8], spectropolarimetry, and chemiluminescence [9]. Other methods for the identification of environmental HMIs are atomic absorption spectroscopy (AAS), experimental neutron activation analysis (ENAA), inductively coupled plasma mass spectroscopy (ICPMS), and x-ray fluorescence spectroscopy (XFS) [10]. Although these analytical methods have high sensitivity and accuracy, they are expensive and have labour-intensive sample preparation and lengthy measurement times, which has forced researchers to look for alternative analytical approaches that are straightforward, rapid, cost-effective, selective, and portable. Due to their practical field-based size, good selectivity, quick response, and relatively reduced cost, electrochemical sensors are particularly well-suited for the determination of toxic metal ions [11] and organic molecules [12–16].

13.2 Heavy metal ion detection using electrochemical sensors

In contrast to the aforementioned techniques, electrochemical sensors facilitate qualitative and quantitative measurement with ultra-high sensitivity and selectivity. Also, owing to their facile fabrication and processability, electrochemical sensors are simple to include in miniaturized systems [17]. Additionally, the selectivity for particular unmodified electrodes can identify HMIs without using any identification probe/molecular receptors since heavy metals have a fixed redox potential. Electrochemical sensing makes use of several techniques, including cyclic voltammetry (CV), chronoamperometry (CA), differential pulse voltammetry (DPV), potentiometry, impedimetry, and conductometry (figure 13.1).

The anodic stripping voltammetry (ASV) approach is very flexible in determining HMIs. On the electrode surface to preconcentrate the analyte, ASV analysis typically entails electrochemical accumulation or deposition of HMIs at a constant potential and stripping/dissolving the deposited analyte from the electrode surface. Due to its high sensitivity, excellent repeatability, and broad cathodic potential range for detecting HMIs, an electrode based on mercury was the first to consider electrochemical HMI detection [18]. The mercury-based electrode, which displays similar electrochemical behaviour in many ways, has been substituted by the more ecologically suitable bismuth electrode due to the high toxicity of mercury.

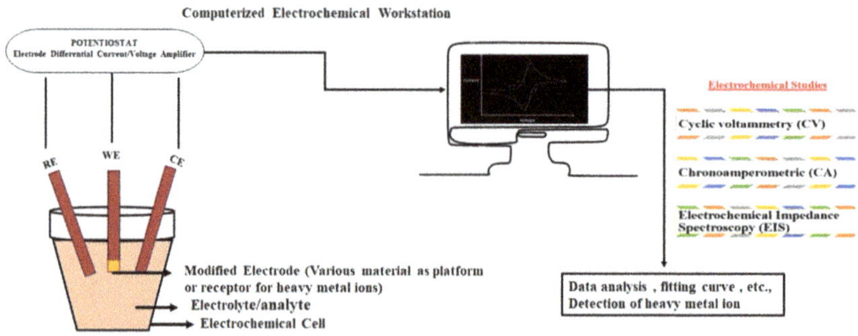

Figure 13.1. General electrochemical set-up for HMI detection.

Because of its inherent oxidation, the bismuth electrode is unstable in air and has a comparatively small cathodic potential range. Electrodes made of gold, glassy carbon (GCEs), and boron-doped diamond have also been employed in the ASV investigation of HMIs and electrodes based on mercury and bismuth. The DPV method was used with a boron-doped diamond electrode to achieve a level of detection ~68 nM in 0.1 M KNO_3 [17, 18].

While bulk or macro electrodes have been employed in the ASV investigation of HMIs, their limit of detection (LOD) and sensitivity are insufficient for the identification of tiny amounts of HMIs. Bulk electrodes have a number of drawbacks, including the high stripping potential necessary for the detection of an analyte and the other coexisting interference metals, as well as the supporting electrolyte background ions. Modifying bulk electrode surfaces with advanced nanoparticles is a practical method for solving these issues [19] and using micro/nanoelectrodes is an alternative method to address these issues. Thus, through understanding various research improvements, various advanced nanomaterials and nanocomposites the have been engaged toward detecting HMIs using versatile electrochemical sensors.

13.3 The detection of heavy metal ions

13.3.1 Cadmium

A non-essential metal, cadmium serves no recognized purpose in humans, but it is equivalently poisonous to lead. Cadmium was categorized as a 'human carcinogenic agent' in group 1 by the International Agency for Research on Cancer in 2018. It builds up in the soil as a result of numerous human activities. This cadmium is subsequently passed to the human body though eating tainted plants and foods. Smoking and occupational exposure may be some additional sources. A significant issue in terms of global health, cadmium exposure can cause cancer or have an impact on other bodily functions. Excess cadmium consumption or exposure causes fever, lung and prostate cancer, cardiovascular disease, and reproductive, urinary, skeletal, muscular, respiratory, and nervous problems, etc, and could be fatal. Not only is it toxic in humans, but it is toxic even at low dosages for many aquatic plants, animals, and terrestrial classes [20]. Therefore, it is desired and most important to produce HMI sensors to monitor the environment using easy, affordable, and effective methods in order to protect the environment.

Organic complexes, surfactants, and other species [21] can interact easily with the electrode; electrochemically, it is challenging to detect cadmium in seawater. A team has been working on electrochemically modulated preconcentration and matrix exclusion via adsorption to improve the performance of the ion-selective electrode to get around this limitation [22]. Despite the fact that mercury-based electrodes are frequently employed for cadmium and other HMIs detection, their employment was constrained by their toxicity [23]. Other electrodes, such as Au, Pt, Ir, GCE, and Ag, have drawn a lot of attention. Bismuth films were employed for the detection of cadmium [24], with methyl methacrylate and decyl methacrylate serving as the matrix (MMA-DMA). With good reproducibility even in seawater [22], this approach produced an LOD of 2×10^{-10} M.

Despite their exceptional strength, superior electrical conductivity, and quick charge transfer rates, carbon nanotubes (CNTs) suffer from excessive background current and a lack of reactive functional groups that would allow them to interact with HMIs. This can be avoided by applying epoxy to the sidewalls of the CNTs, which reduces current leakage [25]. Another study coated the array with bismuth and employed a CNT nanoarray to detect Cd. This strategy produced an LOD of $0.04\ \mu g\ l^{-1}$. In order to achieve high sensitivity and low LOD, graphene with another promising nanomaterial was investigated in three dissimilar forms, including expanded graphite, graphite (natural), and nanographene sheets [26]. The other most promising substance for detecting Cd^{2+} is carbon nanoparticles. In one study, carbon nanoparticles were employed to alter the screen-printed electrode (SPE) surface, which was then used to detect metal ions. The difference in binding energies results from variations in the enthalpy and free energy brought about by the interfacial interactions with regard to temperature. Therefore, with respect to temperature, the observable changes in the communications of HMIs will takes place with advanced nanoparticles [27].

Khan *et al* used the sol–gel polymerization technique to create a poly(2-methylaniline) (P2MA)-$Ce_2(WO_4)_3$@CNT nanocomposite (NC) with a mulberry form. Several characterization methods, such as scanning electron microscopy (SEM), x-ray diffraction (XRD), Fourier transform-infrared spectroscopy (FT-IR), and thermogravimetric (TGA) analysis, were used to describe this composite. A Nafion (NF) attached modified GCE was engaged in detecting Cd ions utilizing a fairly straightforward P2MA-$Ce_2(WO_4)_3$@CNT NC. A calibration curve was used to calculate the sensitivity, which was ~5.138 $\mu A\ \mu M^{-1}\ cm^{-2}$. Furthermore, the LOD was estimated to be around 0.11 nM, and the linear dynamic concentration of the P_2MA-$Ce_2(WO_4)_3$@CNT/NF/GCE electrochemical sensor was used from 1.0 to 0.1 M with a regression coefficient of $R^2 = 0.99$. Additionally, in comparison to the detection of other sensors of Cd^{2+} ions, the proposed NC sensor was highly selective [28, 29].

Nanoconjugate combinations of reduced graphene oxide (rGO)/gold nanoparticles (AuNPs) and tetraphenylporphyrin (TPP) are used to create a sensor for the electrochemical detection of Cd^{2+} ions. The incorporation of Cd ions and tetraphenylporphyrin is accelerated by rGO. Additionally, AuNPs improve the electrochemical detection sensitivity. This sensor had superior sensitivity, with a low LOD and excellent selectivity, from 50 nM to 300 μM. Up to five repeated cycles, the sensor's stability and reproducibility were preserved [30]. Using a one-step wet chemical process, rGO was mixed with a semiconductor and a gas-detecting material made of tin oxide (SnO_2). Hummer's approach was used to modify rGO, which resulted in a high current distinct peak revealing a substantial 2D electrical conductivity. Compared to other SnO_2 and GCE, the modified SnO_2/rGO NC electrode exhibits a high current. The Cd ions are detected using an electrode made of SnO_2 quantum dots (SnO_2/Nafion/Au). According to impedance studies, the customized SnO_2/Nafion/Au electrode has a lower impedance than the gold electrode. The customized electrode increases sensitivity to cadmium ions due to SnO_2 capability as an effective electron mediator in electrochemical detection. The customized SnO_2/Nafion/Au modified electrode used in the fabrication of the sensor has a superior sensitivity of ~77.5×10^2 nA ppm^{-1} cm^{-2}, a low LOD ~ 0.5 ppm, and a reponse time of < 2 s [31].

The progress of a consistent technique for detecting Cd (II) ions would diminish the counterpart of current existing methods. Sreekanth *et al* used ethyl green (EG) and MWCNTs to detect Cd (II) ions using an electrochemical technique based on DNA. The working electrode for DPV studies of Cd (II) ion detection was made up of GCE/MWCNTs. The functioning electrode was immobilized with the dsDNA. The reduced peak current of the indicator dye EG, which preferentially attaches to ssDNA, is substantially lower in the presence of dsDNA. After engaging with the dsDNA, the Cd (II) ions unwind it into ssDNA. The EG molecules then fix to the ssDNAs, increasing the reduction peak current. The linear sensing range reached by this approach is 2–10.0 nM with a sensitivity of ~5 nA nM^{-1}. The WHO's permitted limit for human exposure is 2 nM, less than the LOD. This study significantly expands the range of heavy metals detected in environmental samples using MCNT-modified electrodes as electrochemical biosensors [32]. Fakude *et al* reported an investigation on an electrochemical aptasensor towards Cd (II) on a modified SPCE with carbon black and Au nanoparticles (AuNPs) (figure 13.2) [33]. Electrodeposited AuNPs created the nanostructure while the cycling voltage was between −400 and 1100 mV at a rate of 50 mV s^{-1}. Upon interacting with Cd (II), the immobilized 36-base thiolated ssDNA aptamer on the SPCE-CB-AuNPs underwent a conformational shift that altered the electrochemical current response. Using square wave voltammetry, the aptasensor sensed Cd (II) in the linear dynamic range of 1–50 ppb with an LOD of 0.14 ppb after the incubation period, and aptamer concentration was optimized at 30 min and 2 μM, correspondingly. This aptasensor showed high selectivity for Cd (II), adequate stability, and outstanding repeatability, supporting its potential use in the environmental detection of cadmium.

A new kind of heterocyclic ligands/nanocellulose/Co_3O_4/CNT NC was prepared, physiochemically investigated, and employed as a novel electrochemical detection for the ultra-trace sensing of Cd ions using an impedimetric method directly in samples from tap water by Magar *et al.* Cellulosic nanoarchitectures combined with

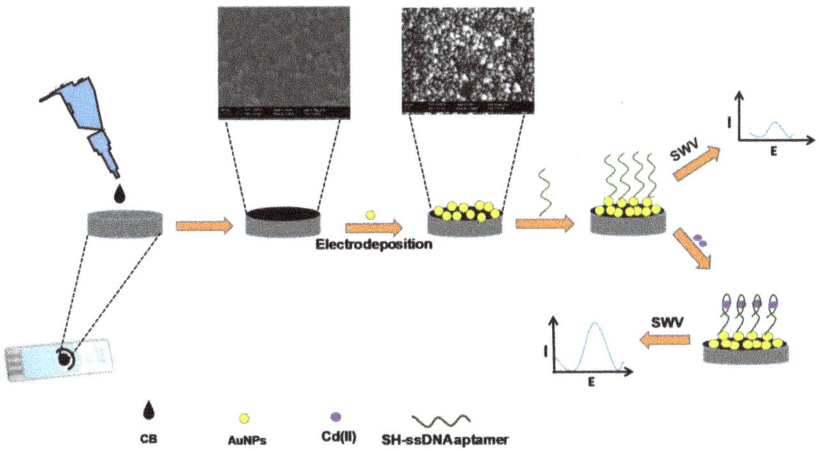

Figure 13.2. Schematic representation of the fabrication method and determination of the Cd (II) aptasensor. (Reproduced with permission from [33]. Copyright 2020 Elsevier.)

a number of heterocyclic ligands (L/nanocellulose) and employed as the sensor were created using a straightforward synthesis method. Metal oxides/CNTs were mixed with a chosen L/nanocellulose to create a CNT/Co_3O_4/L/cellulose NC to achieve strong electrocatalytic and quick electrochemical reactions. To achieve sub-picomolar rapid sensitivity with no fractious reactivity with other metal ions, the impedimetric technique was refined. An LOD of 1.5×10^{-13} M, $R^2 = 0.99$, good sensitivity, and a broad linear range were obtained over the dynamic concentrations (10^{-13}–10^{-2} M) from the calibration curve. A significant retrieval percentage (96%–120%) was achieved when the electrochemical sensor was used for the examination of an actual sample (tap water) [34].

13.3.2 Mercury

Compared with non-hybrid materials, using hybrid nanomaterials has the added benefit of improving analytical properties and increasing sensitivity. Martin *et al* reported research on AuNPs and hybrid materials made of CNTs and AuNPs and demonstrated a quicker response time and a higher detection limit [35]. Graphene oxide (GO) was utilized to detect Hg^{2+} more precisely, and 2.5×10^{-8} M was found to be the detection limit in this work. The communication between mercury ions with 1-octadecanethiol caused it to achieve a reaction time of < 10 s. Further, the surface alteration was performed via GO and thioglycolic acid. AuNPs then functionalized this combination to achieve high selectivity and a low detection limit. The chelation interface between mercury and thioglycolic acid caused this hybrid material to exhibit better qualities than graphene oxide alone [36].

For mercury detection in groundwater, SPEs coated with AuNPs have also been utilized extensively. When combined with micro-extraction using ionic liquids, they may identify trace amounts of mercury (even at levels below ppb) in several types of water [37] or urine [38]. Additionally, SPEs tailored with AuNPs have been employed with other nanomaterials, such as carbon black, to produce an effective electroactive surface area higher than that of normal graphite and employed efficiently to analyse samples of soil and river water [39]. SPEs are low-cost, disposable, and portable and have significant potential to be employed in point-of-care devices, remote or environmental monitoring, etc. These characteristics make them particularly intriguing for a potential commercial mercury detection sensor. Although they have not been employed widely for mercury detection, paper-based electrodes are also a reasonably priced substitute for the determination of several species. Only one publication over the past few years described the use of Au and selenium NP altered carbon paper electrodes for the determination of mercury and nitrate in samples from a lake [40]. The lowest limit of the linear range for this electroactive material, however, was only able to sense 14 ppb, which is over the specified permissible limits in potable water.

Likewise, carbon nanomaterials have been used to create electrode surfaces employed in mercury detection. To produce a more successful preconcentration, other materials with complex capabilities are required as these constituents do not exhibit a high attraction towards detecting mercury. As a fashionable material, graphene has recently been used in various structural arrangements (GO, rGO, graphitic carbon nitride, etc): with proteins behaving as the sorbent for solid-phase extraction, with cyclodextrin or pectin, grafted with carboimidazole, PEDOT nanorods, and ion-imprinted polymers [41, 42]. These materials were all used to detect mercury in samples of river, waste, and tap water by enabling mercury ion complexation on the facade of the electrode prior to electrochemical detection. It has also been found that similar environmental samples contain other carbon nanomaterials, such as single/multi-walled nanotubes or ultrathin g-C_3N_4 (containing –NH_2 and –NH functional groups) [43]. The fact that the latter is the only case of a 2D layered material (besides graphene) engaged in the electrochemical detection of mercury is unexpected. Current interest in 2D materials indicates that more examples of these materials for the detection of mercury may be reported in the years to come [44].

Recently, Zuo *et al* developed a poly(3,4-ethylene dioxythiophene) nanorod/GO NC-customized GCE (PEDOT/GO/GCE) electrochemical sensor for Hg^{2+} detection at trace levels [45]. Additionally, reduced Hg^{2+} concentrations on the PEDOT/GO/GCE have been found via differential pulse stripping voltammetry (DPSV). The experimental parameters have also been tuned, including the deposition potential (0.2 V), accumulation period (360 s), and pH values (pH 5.0, 0.1 M ABS). Based on the optimized settings, an adequate linear correlation between the peak current density and concentration range of Hg^{2+} amid 10.0 nM and 3.0 mM. The LOD was calculated as 2.78 nM. Ultimately, the efficacy of Hg^{2+} detection in tap water was accepted.

Eksin *et al* investigated mercury (II) ion (Hg^{2+}) detection using a one-use, environmentally friendly sensor technology based on herbal Ag nanoparticles. To do this, AgNPs and folic acid (FA) were employed to customize a pencil graphite electrode (PGE) surface. To effectively modify the surface of the PGE, the amounts of AgNPs and FA were initially tuned by the DPV technique. All surface alteration steps were examined using EIS and CV. Using the CV approach, the LOD for Hg^{2+} was determined and discovered to be 8.43 μM. The electrochemical sensor displayed remarkable Hg^{2+}selectivity compared to other heavy metal ions [46]. Singh *et al* carried out the hydrothermal synthesis of a Cu metal–organic framework (MOF), examining the electrochemical performance for Hg^{2+} electrochemical detection. The created electrochemical sensor was easy to use and did not require any elaborate set-up procedures before performing the detection of mercury. The affinity of Hg^{2+} for the N–H and O–H groups was exploited in the detection of Hg^{2+}. The O-H groups in the MOF developed more electroactive in an immediate environment, which increased the Hg^{2+} anodic peak currents and improved sensitivity.

Cu_2O acts as a conducting substance that enhances the Cu-charge MOF's transfer kinetics. In ideal circumstances, the LOD and linear dynamic range were discovered to be, respectively, 0.0633 and 0.1–50 nM. Hence, the electrochemical sensor has a better linear dynamic range, a lower detection threshold, greater stability, reproducibility, and selectivity for Hg^{2+}. The electrochemical results were discovered to be highly accurate and precise since the RSD value was observed to be relatively low. This study thoroughly assessed Cu-capability MOFs to find traces of Hg^{2+} in samples of tap water and canned tuna fish. Overall, the electrochemical sensor offers a practical methodology that is straightforward, excellently sensitive and selective, and provides a helpful device for assessing trace amounts of mercury in tap water and canned tuna fish [47].

Priya *et al* established a flexible electrode using bismuth molybdate/graphene (BiM/GR) NC for Hg^{2+} electrochemical sensing (figure 13.3). The high electrical conductivity, collaborative effect, and copious electroactive spots of the BiM/GR NC are responsible for its remarkable electrocatalytic activity. An important finding was that the BiM/GR NC demonstrated respectable electrochemical sensing behaviour for Hg^{2+} in the broad detection limit of 0.02–149 μM with an ultralow LOD of 5.0 nM and a high sensitivity of 9.5 μA μM^{-1} cm^2 which is significantly

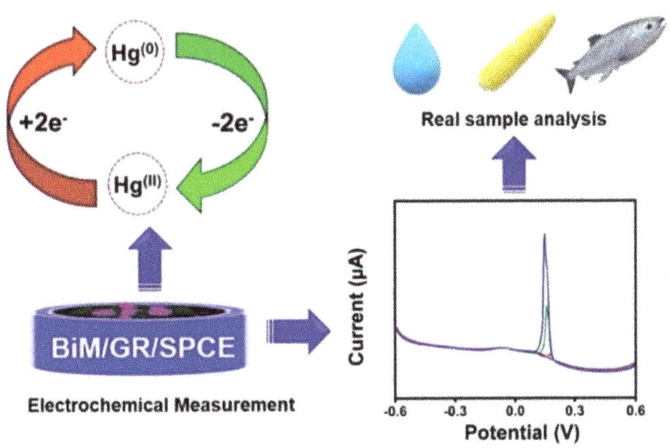

Figure 13.3. Illustration of the fabrication of bismuth molybdate/graphene nanocomposite for electrochemical determination of mercury. (Reproduced with permission from [48]. Copyright 2022 American Chemical Society.)

lower than the threshold value (30 nM) set by the United States Environmental Protection Agency and World Health Organization. Specifically, Hg^{2+} in samples of river water, corn, and fish was successfully detected using the BiM/GR nanocomposite-customized electrode with acceptable recovery values [48].

13.3.3 Lead

Lead is one of the most hazardous metals. Lead poisoning is the term used to describe its long-term toxicity which tends to cause serious issues to the cardiovascular, hematopoietic, and neurological systems. Pb (II) causes neurological, haematological, and renal problems in case of a large dose. It can result in psychological abnormalities, learning difficulties, and disorders of brain development in young age groups. In addition to neurological, cardiovascular, and gastrointestinal issues, lead poisoning reduces fertility and increases the foetal mortality rate, followed by miscarriages. Lead may also have carcinogenic and mutagenic effects, cause DNA damage and oxidative stress in animals, and affect calcium homeostasis in mice when exposed to lead metal ions [20].

Another electrochemical sensor group that falls under the potentiometric category is ion-selective electrodes. Ion-selective electrodes for the detection of Pb^{2+} were created employing oleic acid as an ion exchanger and polysulfoaminoathraquinone (PSA) nanoparticles as solid ionophores. Dioctyl phthalate (DOP) was used as a plasticizer because of its dielectric constant, good dissolution capacity to other membrane additives, and low inclination to exude from the matrix [49]. The potential was shown to rise as Pb^{2+} concentration did as well. The fabricated sensor exhibited good stability for up to 21 days and had a limit of detection for Pb^{2+} ions of 1.6×10^{-7}M.

The gold electrode surface was altered to create a DNAzyme sensor that could detect Pb^{2+} ions. DNAzyme typically combines complementary DNA strands, but in this case, it combines complementary DNA strands with gold nanoparticles to create a DNA-Au bio-bar code. Owing to the bulk solution reactions, the use of DNAzyme in Pb^{2+} ion detection is restricted. However, the attachment of AuNPs accelerated the planar processes in this analysis, and Ru $(NH_3)_6{}^{3+}$ redox mediator was engaged to improve the charge transfer process. DPV was employed to assess the modified electrode's electrochemical behaviour in the presence of Pb^{2+} ions. The hydrolytic cleavage of the catalytic DNA strands in the presence of Pb^{2+} broke the substrate. The electrochemical signals were then lowered as an outcome of the electrostatic communication between DNA and hexaamineruthenium (III) chloride (RuHex). Since the solution's ionic strength significantly affected the electrostatic interactions, its impact coupled with various RuHex concentrations was investigated. Comparing this approach to the other divalent metal ions, Pb^{2+} ions were shown to have a high degree of selectivity [3, 50].

Tarley *et al* used an NC as one of the recently developed electrochemically active and highly selective platforms for determining Pb^{2+} in water samples. It is based on an ion-imprinted polymer (IIP) (methacrylic acid)-1-(2-pyridylazo)-2-naphthol (PAN)/MWCNT [51]. Based on the results, the polymer applied the imprinting effect, ligand PAN, and MWCNTs synergistically to increase the sensitivity and selectivity of Pb^{2+} determination. Additionally, the LOQ and LOD were 0.50 and 0.16 $\mu g\ l^{-1}$, respectively. Moreover, for 10.0 and 5.0 $\mu g\ l^{-1}$ of Pb^{2+}, the technique's intraday precision was 2.53% and 1.68% and the intraday accuracy was 1.72% and 1.86% (RSD). The operative use of this fabricated electrochemical sensor for water and synthetic urine samples was finally observed, with good retrieval values ranging between 95% and 103%. Zuo *et al* prepared highly sensitive electrochemical sensors for Pb (II) using a modified GCE made of a Co_3O_4/reduced GO (rGO)/CHIT NC [52]. The Co_3O_4/rGO/CHIT/GCE procurement process and the Pb (II) electrochemical detection are shown in figure 13.4. Pb (II) shows a linear calibration curve in the 1.0–200.0 nM concentration range. Additionally, a sensitive Pb sensor with 0.35 nM as the LOD was obtained by combining the sensible conductivity and more significant active rGO surface areas, the potent Pb (II) absorption ability of the Co_3O_4 NPs, and the excellent Pb (II) compound-making ability of chitosan (CHIT). The Co_3O_4/rGO nanocomposite stability would thus be improved by the interaction between a negatively charged rGO and a positively charged CHIT. Lastly, the fabricated hybrid NC Pb (II) sensor offered potential to examine vegetable samples.

A highly sensitive electrochemical aptasensor for the detection of Pb^{2+} was fabricated by modifying the sensing platform Au@MoS_2/rGO NC and employing AuPd@Fe-MOFs as the electrocatalytic material by Wang and co-workers. The use of DNAzyme, which can precisely recognize Pb^{2+}, improved the aptasensor's

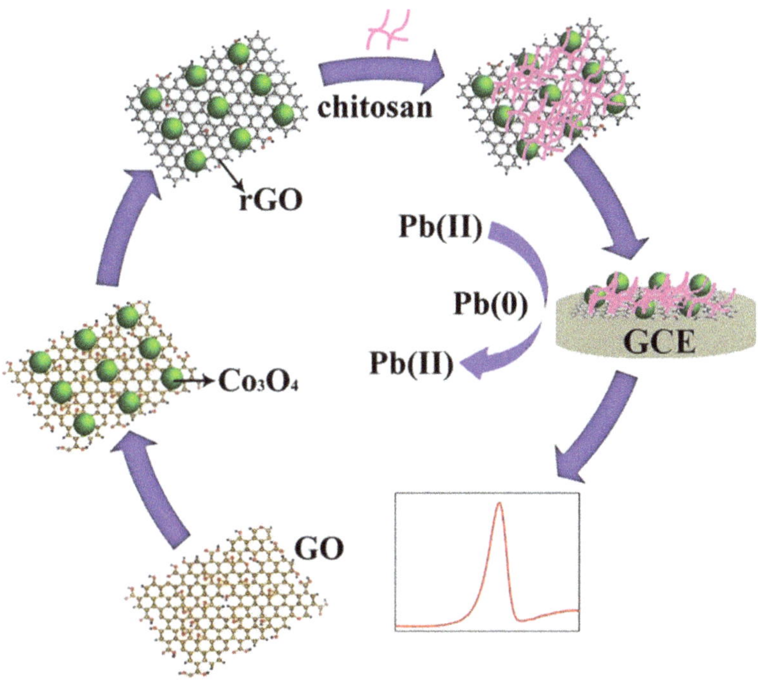

Figure 13.4. A scheme of Co_3O_4/RGO/CHIT/GCE production methods and electrochemical detection of Pb (II). (Reproduced with permission from [52]. Copyright 2017 Elsevier.)

selectivity, and the use of AuPd@Fe-MOFs to label Apt2 significantly increased the aptasensor's sensitivity. The prepared aptasensor demonstrated improved behaviour with a detection limit of 5.0 pmol l^{-1}–2.0 μmol l^{-1} and a lower LOD (0.07 pmol l^{-1}), along with good selectivity, strong chemical stability, and good repeatability. Additionally, the suggested strategy showed good performance for Pb^{2+} recognition for the analysis of real samples (water). The replacement of the aptamers in this analysis will also open the door for the detection of other metal ions, and the suggested method may have applications in environmental monitoring [53]. A novel greener method was reported involving the preparation of TiO_2 NPs employing gum Arabic (*Acacia senegal*) for the detection of Pb^{2+} by Sivan *et al.* A carbon paste prepared with the produced TiO_2 NPs have been used as an electrode for the electrochemical detection of Pb^{2+} ions. The prepared electrode showed increased electrocatalytic activity for the reduction of the Pb^{2+} ions when compared to the bare electrode. Using differential pulse voltammetry, the produced electrode's LOD, LOQ, and sensitivity were observed to be 506 ppb, 1.68 ppm, and 0.52 ± 0.01 μA μM^{-1}, respectively. The created electrode was verified for lead content determination in plastic toys [54]. By immobilizing the surfactant 1-dodecanoyl-3-phenylthiourea (DPT) across the GCE surface, a sensitive and focused voltammetric electrochemical determination for Pb ions was established by Shah *et al.* Using the electrode's capabilities to sequester Pb ions via complexation of its donor atoms with the target specific analyte, DPT increased the electrode's sensitivity. The increased electrocatalytic role of DPT in facilitating electron transfer between the Pb ions and GCE is suggested by the fact that the oxidation current response of electroreduced Pb ions at the DPTGCE revealed a nearly nine-fold increase in comparison to the unmodified electrode. The broad concentration–current linearity limit (11–45 μM) and LOD for Pb ion detection of 0.695 μg l^{-1} further validated the high sensitivity and robustness of the proposed approach. The ideal parameters for the finest detection performance of DPTGCE for the determination of Pb ions were determined for 120 s of deposition time with an accumulation potential of 1.1 V in a solution comprising HCl as the electrolyte background. Insignificant disturbance from coexisting interfering metal ions and good retrieval from spiked samples preserved the fabrication's selectivity and practicability for accurate many-metal ions detection in polluted water samples [55].

13.3.4 Simultaneous detection of heavy metals

A fabricated effective ASPE-PM/g-C_3N_4* nanonetwork (NNW) customized electrode was used to simultaneously monitor $Pb^{2+}$and Cd^{2+} ions electrochemically in environmental samples (figure 13.5). Owing to the porous nature (nitrogen rich) present in graphitic carbon nitride to attach the HMIs and the polymelamine's (PM) amine (NH^+) functional classes to produce complexes with HMIs, the created ASPE-PM/g-C_3N_4* NNW functions as a potent ligand. Since melamine is the only chemical employed in creating this NNW, the formed electrodes are inexpensive, simple to make, simple to use, and provide excellent long-term cyclic stability and reproducibility. The customized electrode demonstrated a remarkable anodic

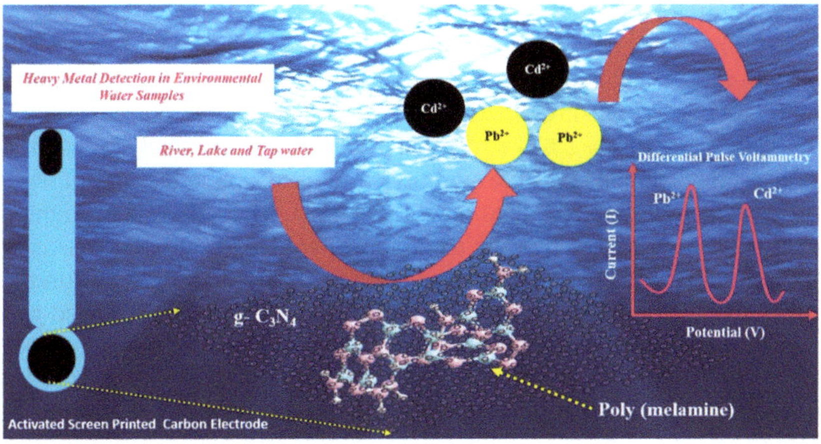

Figure 13.5. Interaction mechanism of ASPE-PM/g-C_3N_4* with HMIs in environmental samples. (Reproduced with permission from [56]. Copyright 2021 Elsevier.)

stripping voltammetric current response to Pb^{2+} ions and Cd^{2+} ions in the linear concentration limit from 0.1 to 1.0 μM, with an LOD of 0.008 and 0.02 μM for Pb^{2+} and Cd^{2+}, respectively. Furthermore, the detection of target HMIs in ambient environmental samples (water) underwent testing to confirm its viability in real-world scenarios. The outcomes showed that the suggested ASPE-PM/g-C_3N_4* NNW-customized electrode could be employed for the real-time detection of Pb^{2+} and Cd^{2+} in various environmental (tap, river, and lake) samples [56]. A straightforward and efficient electrode set-up was created for the instantaneous determination of Cd^{2+} and Pb^{2+}. Using this modified electrode, DPSV was employed to assess Cd^{2+} and Pb^{2+} in an aqueous solution. The ideal working circumstances were explained. Pb^{2+} and Cd^{2+} were simultaneously determined in water samples using a diacetyldioxime-customized carbon paste electrode, with an average recovery of 100% for Pb^{2+} and 101% for Cd^{2+}. The aforementioned findings suggest that the carbon paste electrode customized with diacetyldioxime is an extremely sensitive and repeatable chemical sensor for the concurrent determination of Cd^{2+} and Pb^{2+} in environmental water species [57].

For instantaneous and highly sensitive assessment of lead and cadmium ions utilizing anodic stripping voltammetric techniques, a poly methyl thymol blue (PMTB) film-customized electrode was proposed. Applying a voltage between 40 and 80 mV at a scan rate of 50 mV s^{-1} for 30 cycles in 0.1 M pH 7.0 PBS, comprising 5×10^{-4} M of methyl thymol blue monomer, the modified PMTB film was created on a paraffin-impregnated electrode (PIGE). Pb (II) was discovered to have a linear range in DPV of 0.54–77.7 $\mu g\ l^{-1}$ and Cd (II) of 1.05–75.3 $\mu g\ l^{-1}$, with LODs of 0.18 and 0.35 $\mu g\ l^{-1}$, respectively [58]. Hai *et al* successfully established a sensitive and reliable ASV method for the determination of Pb^{2+} and Cd^{2+} ions simultaneously utilizing a series of MWCNT electrodes modified with antimony oxide (Sb_2O_3/MWCNTs) paste electrodes. First, a paste electrode was successfully made using a composite of MWCNTs. Then, to enhance the ability to detect Cd^{2+} and Pb^{2+} ions, different amounts of Sb_2O_3 (1, 3, and 4 wt%, correspondingly) were added to the MWCNT bulk electrode. When using the linear anodic stripping voltammetry (LASV) technique, the customized electrode with a 3% Sb_2O_3 content demonstrates outstanding analytical sensing of Cd^{2+} and Pb^{2+}. The calculated sensitivity of Pb^{2+} and Cd^{2+} was interestingly 2.694 and 1.932 μAL μg^{-1}, respectively [59]. For the first time, Baghayeri *et al* described the synthesis and electrochemical use of a magnetic NC with glutathione functionalization (GSH@Fe_3O_4) to create a straightforward, reliable, and selective sensor for the detection of HMIs in real-world samples. The basis of the monitoring approach is the electrochemical preconcentration/reduction of metal ions onto a GSH@Fe_3O_4 customized magnetic GCE, followed by anodic stripping. This technique was tested on several water samples and is very sensitive for cadmium (II) and lead (II) (evaluated LODs of 0.182 and 0.172 $\mu g\ l^{-1}$, respectively), inexpensive, and convenient to use. The findings suggest that the GSH@Fe_3O_4 nanocomposite is suitable for real-world electrochemical metal ion detection applications [59]. Overall, table 13.1 summarizes and compares the electrochemical detection performance using various materials towards HMIs.

Table 13.1. Comparative performance of electrochemical determination of various HMIs.

Electrochemically active materials for sensors	Technique	HMIs	LOD	References
SnO_2/Nafion/Au electrode	CV	Cd^{2+}	0.5 ppm	[31]
SPCE-CB-AuNPs	SWV	Cd^{2+}	0.14 ppb	[36]
Heterocyclic ligands/nanocellulose /Co_3O_4/carbon-nanotube-nanocomposite	EIS	Cd^{2+}	1.5×10^{-13} M	[34]
PEDOT/GO/GCE	DPSV	Hg^{2+}	2.78 nM	[44]
Herbal silver nanoparticles (AgNPs)	CV	Hg^{2+}	8.43 μM	[46]
Cu-MOF	DPV	Hg^{2+}	0.0633 nM	[47]
BiM/GR	CV	Hg^{2+}	5.0 nM	[48]
Co_3O_4/rGO/CHIT/GCE	CV	Pb^{2+}	0.35 nM	[52]
AuPd@Fe-MOFs	CV	Pb^{2+}	0.07 pmol l^{-1}	[53]
TiO_2@gum Arabic-carbon paste electrode	DPV	Pb^{2+}	506 ppb	[54]
ASPE-PM/g-C_3N_4*	ASV	Pb^{2+} Cd^{2+}	0.008 μM, 0.02 μM	[56]
PMTB film-customized electrode	ASV	Pb^{2+} Cd^{2+}	0.18 $\mu g\ l^{-1}$, 0.35 $\mu g\ l^{-1}$	[58]
GSH@Fe_3O_4	ASV	Pb^{2+} Cd^{2+}	0.182 $\mu g\ l^{-1}$, 0.172 $\mu g\ l^{-1}$	[60]

13.3.5 Chromium pollution in the environment

Chromium (Cr) is one of the HMIs that is present broadly in our environment (soil, water, etc). In corrosion industries chromium is a commonly used material for anti-corrosive coating. In petrochemical applications chromium is used as an oxidative reagent and it is also use in other fields, including metal finishing, dyeing, ternary, textile, electroplating, paper making, and atomic power stations [61–63]. Generally, Cr ions exist in two oxidation states, the trivalent (Cr^{3+}) and hexavalent (Cr^{6+}) forms. Compared with hexavalent chromium, trivalent chromium has been considered a crucial nutrient for living species, and it has very small oral toxicity owing to very low bioavailability [64, 65]. In contrast, hexavalent chromium species cause significant local aggravation and have a corrosive nature in large concentration ranges, and are highly carcinogenic at low concentrations due to their oxidative solid properties. The hexavalent chromium ions are highly toxic and easily absorbed by the human body and cause various illnesses such as lung and nasopharyngeal cancer [66, 67].

13.3.5.1 Electrochemical detection of Cr^{6+} ions

Different types of classical analytical techniques have been used to detect chromium ions; however, these techniques required a sample pretreatment process with preconcentration and sample separation [68]. Classical analytical methods such as chromatography [69], electrophoresis [70], etc, are not suitable for performing repeated measurements, as they are laborious, have complex instrument operation protocols, and high cost instrument [71, 72]. To overcome these issues, electroanalytical tools provide attractive techniques for target metal detection. These techniques have high reliability, high sensitivity, and low-cost instruments and are mainly highly suitable for onsite analysis [73]. The electrochemical response of hexavalent chromium ions allows a direct redox electron transfer process with separation. Initially, the Cr^{6+} ions to form chromate are diffused on the electrode surface in the supporting electrolyte, followed by the reduced form of Cr^{3+} ions at a very low potential range of –0.05 V [74]. The electrochemical selective detection of both Cr^{6+} and Cr^{3+} ions is highly possible due to the different reduction potentials of both chromium ions. The selective detection of Cr^{6+} ions depends on the electrode material composition (the nature of the modified materials) and the background electrolyte [75].

13.3.5.2 CNT based materials for the detection of chromium ions

As we discussed above, there is a strong requirement to prepare portable sensors for the field analysis of Cr^{6+} ions. While a variety of chromatography techniques and absorption spectroscopic techniques coupled with mass spectral methods are used for Cr^{6+} ions detection, as we know, these kinds of instruments are highly expensive, need special equipment location, require maintaining a clean environment, are time-consuming processes, and require well-trained operators. The electroanalytical methods using nanomaterial-coated electrodes have shown outstanding performance for heavy metal analysis. Moreover, these techniques have plenty of advantages, including fast response times, easy operation protocols, and onsite detection [76, 77].

The Cr^{6+} ion detection and electrochemical techniques permit a direct redox electron transfer process without sample pretreatment, which is simple for the selective detection of Cr^{6+}ions owing to the dissimilar reduction potentials of Cr^{3+} and Cr^{6+} ions [78]. For this reason, electrochemical techniques are extensively applied to detect the Cr species, which are highly important for the rising demand for onsite detection with sensitivity [79]. The selective detection of Cr^{6+} ions in the presence of a high concentration of other metals including Cr^{3+}, Na^+, Ca^{2+}, K^+, and Mg^{2+} ions has been studied, along with their anti-interference capability with high sensitivity in different environments. A GCE was modified with carbon nanotubes and conducting polymers, improving the selective detection of Cr^{6+} ions. For instance, Rosaline *et al* developed SWCNTs with a 2-pyridine moiety through the sol–gel technique. The prepared materials were coated on the GCE surface using physical and electrochemical deposition methods [80]. The pyridine moiety can easily preconcentrate with Cr^{6+} ions in linear concentrations from 5 to 300 $\mu g\ l^{-1}$ with an LOD of 4.6 $\mu g\ l^{-1}$. Compared to earlier works, the sol–gel method prepared electrode materials showed better sensitivity and wide linear concentrations.

Moreover, the composite or conjugation between the polymer films with carbon nanotubes can overcome a wide range of analytical applications. The carbon-based materials have fewer electron transfer process and selectivity issues. Transition metal oxide based nanomaterials have recently received significant attention in chromium ion detection. Zirconium oxide (ZrO_2) can effectively be used as an electrocatalyst for the reduction of Cr^{6+} ions, as reported by Motaghedifard and their research group [76]. Initially, nanocomposites of polyaniline (PANI), MWCNTs, and zirconium oxysulfide (ZrO_2–SO_4^{2-}) were prepared, followed by drop casting on the GCE surface. The electrochemical performance of the prepared materials showed Cr^{6+} ion reduction at −0.27 V. After optimizing the experimental conditions, the developed electrode materials were tested for target ion detection by DPV. They found that the LOD for Cr^{6+} ions is 64.3 nmol l^{-1}. Finally, the fabricated electrode was effectively utilized to detect Cr^{6+} ions in industrial wastewater samples (from a barium chromate production company) and obtained satisfactory recovery values. The electrostatic communication between the negatively charged carboxylic acids and positively charged Cr^{6+} ion studies was performed by Ouyang and their research group [78]. Initially, SWCNTs were functionalized with a carboxylic acid group using a chemical oxidation method and these carboxylic acids could interact easily with bismuth cations, followed by an electroreduction process, subsequently the bismuth film was formed on the surface of the electrode. However, diethylene triaminepentaacetic acid was used as a metal chelating agent for the selective binding of Cr^{6+} ions and this metal complex reduction was observed at −106 mV. Further, the prepared electrode was tested for Cr^{6+} ion detection through cathodic stripping voltammetry techniques. After optimizing all the experimental parameters, the developed electrode was tested for Cr^{6+} ions in the linear concentration from 0 to 25 nM with an LOD of 0.036 nM. Finally, the selectivity and reproducibility of the developed electrode showed better results, and the realistic usage of the electrode was tested in different water samples and better recoveries were found.

Carbon nanotubes with other nanomaterials, such as gold and silver-based materials, can enhance the selectivity and sensitivity of trace Cr^{6+} ion detection. When compositing or mixing the carbon nanotubes with metal nanoparticles, the developed materials have shown excellent analytical parameters, including enhanced sensitivity and selectivity, wide linear concentrations with low detection limits, and so on. For example, Breslin and their research group developed oxygen-containing MWCNTs followed by deposition of AuNPs by a facile chemical reduction method [77]. The AuNP decorated MWCNT suspension was drop-cast on a gold working electrode. After optimizing all the experimental conditions, the prepared electrode was tested for detecting Cr^{6+} ions and achieved an LOD of 0.07 μM with a sensitivity of 0.28 $mA.mM^{-1}$. Finally, selectivity studies of the prepared electrode were tested in the presence of interfering species (high concentrations of Cl^{-}, NO_3^{-}, and Cu^{2+} ions). The prepared electrode was successfully applied to detect Cr^{6+} ions in various water samples and showed improved recovery results.

Similarly, SPCEs were customized with MWCNTs and quercetin materials for selective detection of Cr^{6+} ions in the presence of Cr^{3+} ions, as reported by Sadeghi and their research group [81]. The Cr^{6+} ions were selectively preconcentrated with quercetin to form a Cr^{6+}-quercetin metal complex on the working electrode surface. However, the Cr^{3+} ions did not effectively form an electrochemically active metal complex. Hence, the selective sensing of Cr^{6+} ions in the presence of an excess amount of Cr^{3+} ions was found through the selective preconcentration of Cr^{6+} ions in which the selective redox reactions will be occured without additional pretreatment process. However, the Cr^{3+}-quercetin complex was formed when using Cr^{6+} under a lower accumulation potential. After optimizing all the experimental parameters, the developed electrode was tested for selective sensing of Cr^{6+} ions in linear concentrations from 1 to 100 $\mu M\ l^{-1}$ with an LOD of 0.3 $\mu M\ l^{-1}$. An RSD of 3.0% was found at seven repeatable claculations of current at 50 $\mu M\ l^{-1}$ of Cr^{6+} ions.

Further, an enzyme-based altered electrode was established for the determination of Cr^{6+} ions, as reported by Silva and their research group. The modified electrode was prepared by immobilizing glucose oxidase (GOx) enzyme with MWCNTs and a polymer film of poly(brilliant green) materials through electrodeposition using deep eutectic solvent (DESs) as a transporter and enzyme reticence to detect bioactive trace heavy metals such as Hg^{2+}, Cd^{2+}, Pb^{2+}, and Cr^{6+} ions. The enzyme-immobilized modified electrodes provide a fast and direct electron transfer reaction which can improve the target HMI detection owing to the high selectivity of enzyme performance. The polymer films have a homogeneous surface morphology and enhanced electrochemical detection properties. In this enzyme-based electrode, they initially examined GOx activity through an amperometry test. After optimizing the baseline, different glucose concentrations were added to the electrochemical cell, and their obtained oxidation peak current was measured (i_0). Subsequently, known concentrations of trace HMIs were included to prevent the enzyme activity then the oxidation current was calculated (i_1). The loss of current is directly proportional to the concentration of inhibitor in the supporting electrolyte. They calculated the percentage of inhibition $I(\%) = i_0 - i_1/i_0 \times 100$, the influence of various glucose concentrations (0.25–1 mM), and i_{50}, the inhibitor concentration essential to obstruct half of the primary response in the substratum. Afterwards, the linear concentrations of Cr^{6+} ions were measured and obtained from 2.5 to 60 nM with an LOD of 2.44 nM. Finally, the fabricated electrode was tested for different heavy metals in different milk samples and showed improved recoveries. The developed GOx/PBGDES/MWCNT/GCE showed a nanomolar level of LOD and did not require any pretreatment process to rebuild a GOx enzyme activity after every measurement. The limitation of the enzyme-based electrodes is a very short-term stability of response.

The improved long-term stability and reduced leakage of membrane materials have many properties like high conductivity for electrode reproducibility in the sensor applications. For example, Filik and their research group prepared MWCNTs/neutral red, and AuNP composites were drop-cast on the carbon SPE surface. The prepared MWCNTs/NR/AuNPs/SCPE was employed for the instantaneous determination of Cr^{6+} and V^{5+} ions using linear sweep voltammetry techniques [82] and their detailed material preparation procedure is shown in figure 13.6(a). In optimized experimental conditions (figures 13.6(b) and (c)), the prepared electrode showed Cr^{6+} ion detection in linear concentrations from 0.4 to 80 μM with an LOD of 0.025 μM (figures 13.6(d) and (e)). The developed method does not require any preconcentration process or chelating agents.

Similarly, Rudnitskaya and their research group fabricated CNTs doped lignin polyepoxypropane copolymer for the potentiometric detection of Cr^{6+} ions [84].

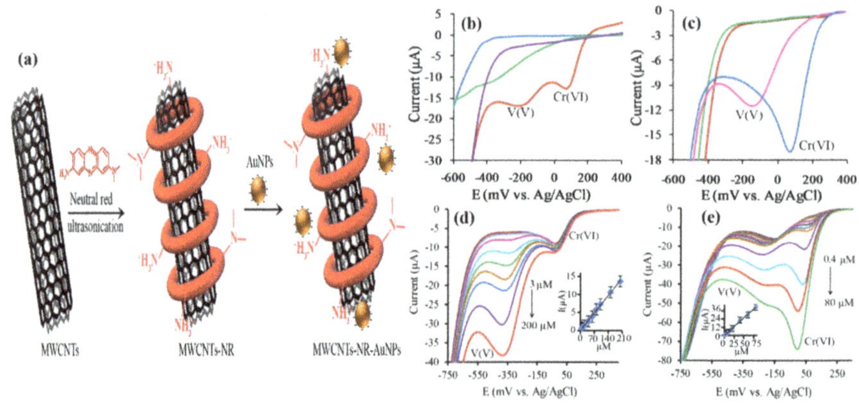

Figure 13.6. (a) Schematic illustration of neutral red interlinked absorption of AuNPs on MWCNTs. (b) LSV response of different modified electrodes for 60 μM of V^{5+} and 20 μM Cr^{6+} ions in 0.5 M sulphuric acid solution. (c) LSV response of 60 μM V^{4+} (green), 20 μM Cr^{3+} (red), 60 μM V^{5+} (pink), and 20 μM Cr^{6+} (blue). (d) LSV response of V^{5+} at MWCNTs-NR-AuNPs/SPCE in the presence of different V^{5+} concentrations (from 3 to 200 μM) at MWCNTs-NR-AuNPs/SPCE in the presence of 10 μM Cr^{6+} in 0.5 M sulphuric acid solution. (e) LSV response of Cr^{6+} at MWCNTs-NR-AuNPs/SPCE in the presence of different Cr^{6+} concentrations from 0.4 to 80 μM V^{5+} in 0.5 M sulphuric acid solution. (Reproduced with permission from [83]. Copyright 2020 Elsevier.)

Additionally, CNTs with copolymer membrane interactions of the potentiometric electrode were studied briefly. Initially, the MWCNTs (0.72 wt%) were added to lignin-based polyurethane materials because delocalized π-electrons may interact with CNTs and lignin in this scenario, enhancing the conductivity of the prepared material. In acetic conditions (pH = 2.0), the lignin-based electrodes detected Cr^{6+} ions selectively. However, in alkaline, transition, and alkaline metal cations, electrolyte conditions have shown the favourable potential of the lignin and propylene oxide copolymerization process with MWCNTs.

Additionally, the derivatives of phosphinic acids (PAs) have been used in the liquid–liquid descent and separation of different heavy metals. The derivatives of PA have high pH and thermal stability, organized with improved redevelopment abilities that could get a derivative of PA-appealing materials for heavy metal ions detection applications, as reported by Deep and their research group [85]. The chlorinated SWCNTs were functionalized with derivatives of PA at 80 °C. The complex of CNTs-PA was turned on borosilicate flakes to construct electrode materials for sensitive detection of Cr^{6+} ions. After optimized experimental parameters, including supporting electrolyte pH and preconcentration potentials, the fabricated electrode was tested for Cr^{6+} ion detection in linear concentrations from 0.01 to 10 ppb with an LOD of 0.01 ppb and a calculated sensitivity of 35 (±4) nA ppb^{-1}. The practical application of the fabricated electrode was tested for Cr^{6+} ion detection in industrial wastewater effluent and underground water samples and achieved better recovery results.

13.3.5.3 Graphene-based composite materials for the detection of chromium ions

Carbon-based materials have been used for the electrode materials for selective and sensitive determination of Cr^{6+} ions. In particular, graphene materials have several physio-chemical advantages, such as good conductivity, large electroactive surface area, rapid electron transfer, flexibility, optical transparency, and high thermal conducting properties, which can encourage the immediate improvement of electrode materials [86]. Other carbon forms such as graphene oxide, graphene, and reduced graphene oxide have been utilized for the fabrication of electrocatalyst materials for sensitive detection of Cr^{6+} ions [87]. Moreover, graphene-enabled nanomaterials such as metal, metal oxide NPs, quantum dots (QDs), biomolecules, and organic conducting polymers result in a different range of NCs [88]. Therefore, graphene-enabled nanomaterials can be used as electrode materials for the selective detection of Cr^{6+} ions. The highly selective determination of Cr^{6+} ions with suitable chelating or modified materials coated on the GCE surface was briefly studied using different nanomaterials. For example, Liu and their research group prepared rGO using a chemical reduction technique with ammonia as a reducing agent [89]. The fabricated rGO/GCE was successfully applied to detect the Cr^{6+} ions by differential pulse cathodic stripping voltammetry methods. The rGO/GCE has shown that the Cr^{6+} ions reduction potential was anodically shifted at 0.16 V, and the peak current was increased by ~8 times.

Additionally, the electrode/electrolyte interference charge transfer resistance of the rGO/GCE in the redox probe of $[Fe(CN)_6]^{3-/4-}$ was 75.02 Ω cm^{-2}, which is almost nine times lower in comparison to total GCE. In optimized experimental conditions, the fabricated electrode showed Cr^{6+} ion detection at an LOD of 0.15 $\mu M\ l^{-1}$. In this work, the rGO was prepared using a simple chemical reduction reaction using liquid ammonia as a reducing agent, which is highly efficient, eco-friendly, and less carcinogenic than other reducing agents such as hydrazine hydrate. Similarly, Sari and their research group developed a graphene and AuNP composite by ultrasonication techniques and coating on the GCE surface, followed by amperometry detection of trace Cr^{6+} ions. The CV and EIS results of the prepared graphene/AuNPs/GCE showed enhanced stability and electrocatalytic assets for reducing the Cr^{6+} ions. After optimizing the experimental conditions, the fabricated graphene/AuNPs/GCE was effectively utilized for detecting Cr^{6+} ions using chronoamperometry. It achieved wide linear concentrations from 0 to 20 μM with an LOD of 10 nM. The modified electrode showed excellent sensitivity, reproducibility, and good anti-interference ability for the selective detection of Cr^{6+} ions. Also, Liu and their research group synthesized rGO with AuNP nanocomposites using facile dual-step electrochemical deposition techniques [90]. Initially, the GO suspension was drop-cast on the GCE surface, followed by CV sweeping at a high potential range; as a result, the rGO films were uniformly deposited on the GCE surface. Subsequently, the AuNPs were electrochemically deposited on the surface of the rGO by similar CV parameters. The rGO and AuNP decorated GCE was effectively utilized for the determination of Cr^{6+} ions using square wave voltammetry techniques. After optimizing the experimental conditions (deposition time, supporting electrolytes, supporting electrolyte concentrations, etc) the prepared rGO/AuNPs/GCE used square wave voltammetric detection of Cr^{6+} ions in linear concentrations of 0.1–30 μM with an LOD of 0.046 μM and achieved high sensitivity of 0.7991 $\mu A\ \mu M^{-1}\ l^{-1}$. For 12 h of repetitive detection of Cr^{6+} ions the current response was reported as ~92.98%, which suggests that the prepared electrode had outstanding stability for Cr^{6+} ion detection. Additionally, the anti-interference ability of the rGO/AuNPs/GCE was tested in the presence of interfering HMIs such as Mg^{2+} and Mn^{2+} ions, the electrode selectively detected Cr^{6+} ions with a current deviation of <2%, which suggested that the modified electrode has better selectivity for Cr^{6+} ion detection. Compared with the preparation of other modified electrode materials, the rGO and AuNP material was electrodeposited directly on the working electrode surface by a facile electrodeposition technique and this kind of electrode fabrication process has several advantages, such as being an eco-friendly method, not requiring harmful chemicals and solvents, having no need for separation or purification steps, and, mainly, the direct deposition of nanosized materials on the electrode surface without external processes.

Recently, gold nanocubes (AuNcs) and their composite materials have improved the selective detection of chromium ions. Ternary nanocomposites of rGO, nickel sulphide, and AuNcs were synthesized using a hydrothermal process [91]. The obtained rGO/NiS/AuNc nanocomposites were characterized well and confirmed with different physicochemical techniques, and the prepared trinary nanocomposite showed a flower-like structural morphology. Afterwards, the rGO/NiS/AuNc nanocomposites were dispersed in ethanol, followed by drop casting on the GCE surface.

The fabricated rGO/NiS/AuNcs/GCE was used for the square wave voltammetric detection of Cr^{6+} ions in acetate medium (pH 5), following which the analytical parameters which are affected by Cr^{6+} ions were optimized. Finally, the square wave voltammetry determination of Cr^{6+} ions was tested and achieved linear concentrations from 34.4 to 269.2 μM with an LOD of 0.09 μg l^{-1} with a preconcentration potential of −50 mV and 30 s as the preconcentration time. The developed electrode showed better long-term stability, reproducibility, and superior sensitivity with a very short-time signal response due to the highly active surfaces on rGO and NiS permitting rapid electron transfer between the electrolyte/electrode boundary.

Similarly, single-step preparation of an NC with Au–PdNPs and electrochemically reduced graphene oxide (ErGO) was carried out as electrode materials for the sensitive determination of Cr^{6+} ions in environmental water samples and was reported by Hu and their research group [92]. NCs of Au–PdNPs/ErGO containing a bimetallic form of Au–Pd nanomaterials and ErGO were produced by electrochemical reduction of $HAuCl_4$, GO, and H_2PdCl_4 materials using CV techniques; as a result, the nanosized composites were deposited on the GCE surface. After optimizing the experimental conditions, the prepared electrode was tested for selective detection of Cr^{6+} ions and achieved a lower LOD owing to a synergistic interaction between the Au–Pd nanoalloys and rGO film. Based on the voltammograms, the Au–PdNPs/ErGO/GCE showed enhanced catalytic current response compared to pristine PdNPs/GCE, AuNPs/GCE, and ErGO/GCE. Additionally, the prepared Au–PdNPs/ErGO/GCE was effectively employed for the determination of the Cr^{6+} ions and achieved two linear concentrations of 0.05–5 μM and 5–1000 μM with an LOD of 0.013 μM. Finally, the prepared Au–PdNPs/ErGO/GCE was tested for Cr^{6+} ion detection in different water samples and 95% recovery results were found.

Modified electrode materials with high sensitivity and the achieve the lowest detection limits are a key part in the electrochemical sensing of HMIs. In recent years, 2D materials have played an essential role in the electrochemical sensing of Cr^{6+} ions and their detailed experimental procedure is shown in figure 13.7(a). The versatile 2D semiconducting material of g-C_3N_4 was decorated with silver molybdate (AgM), with Nafion film coated on the GCE surface (g-C_3N_4/AgM/Nf/GCE), developed by Karthika and their research group [93]. A facile thermal polycondensation technique synthesized the 2D nanomaterial of g-C_3N_4. The g-C_3N_4/AgM nanocomposites were prepared using ultrasonic treatment followed by hydrothermal treatment for 12–72 h. Afterwards, the prepared nanocomposite was dissolved in ethanol (3 mg ml^{-1}), and 5 ml of Nafion polymer was dispersed and then drop-cast on the GCE surface. After optimizing all the experimental parameters, such as different supporting electrolytes and their concentrations, different scan rates, etc, the amperometric detection of Cr^{6+} ions was achieved. The linear concentration ranged from 0.1 to 0.7 μM with an LOD of 0.0016 μM and calculated sensitivity of 65.8 $\mu A\ mM^{-1}\ cm^{-2}$ (figures 13.7(b) and (c)). Finally, the prepared electrodes were utilized for the determination of the Cr^{6+} ions in different real water and industrial effluent water samples by a standard addition method and recovery results of 89.4%–100.2% were found.

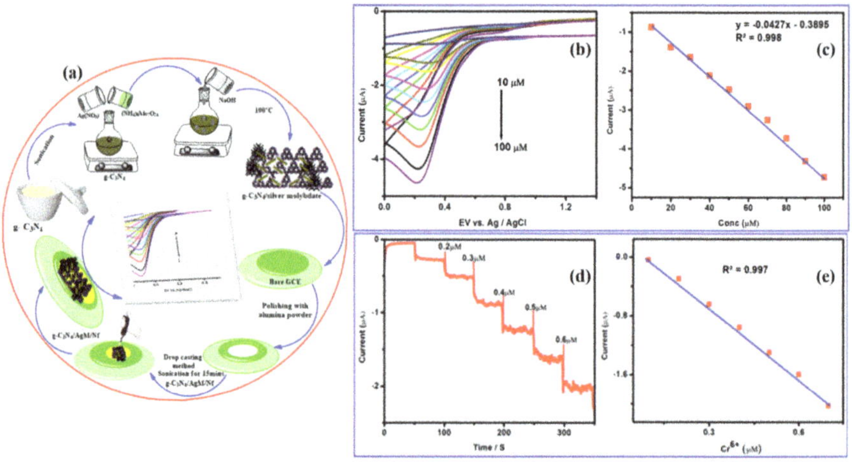

Figure 13.7. (a) Stepwise preparation of g-C_3N_4/AgM/Nf/GCE for Cr^{6+} detection. (b) CV curves of g-C_3N_4/AgM/Nf/GCE in 50 μM of 0.1 M HCl containing varied Cr^{6+} ion concentrations from 10 to 100 μM and the linear plot for cathodic and anodic peak current responses of Cr^{6+} ions versus concentration. (c) Amperometric current response of the g-C_3N_4/AgM/Nf/GCE in 0.1 M HCl versus different Cr^{6+} ion concentrations from 0.1 to 0.7 μM and a calibration plot for the linear dependence of peak current versus concentrations of Cr^{6+} ions. (Reproduced with permission from [93]. Copyright 2020 Elsevier.)

GO-based materials have been employed widely as electrode materials for the detection of Cr^{6+} ions to improve the sensitivity and high selectivity due to rapid electron transfer kinetics. For example, Singh and their research group developed a green preparation method for calcium oxide (CaO) nanoparticles composited with GO [94]. Initially, the biogenic CaO NPs were obtained from Lala clam shells. The prepared nanocomposites were characterized and confirmed by various analytical studies. Afterwards, the prepared nanocomposites were applied to remove Cr^{6+} ions effectively. The electrochemical interaction between the target Cr^{6+} ions and the prepared nanocomposite was examined by amperometric techniques. The CaO nanomaterials were drenched on the GO surface to promote a high active surface area and enhanced absorption capacity (38.04 mg g^{-1}). The unmodified electrode for the detection of Cr^{6+} ions showed a poor current response due to a slower electron transfer process. The synergistic and electrocatalytic effect of the prepared CaO/GO/GCE showed Cr^{6+} ion reduction at 0.2 V. The DPV response of the obtained linear concentrations ranged from 2 to 18 μM with a detection limit of 0.02 μM.

Further, the developed electrode was applied for the amperometric detection of Cr^{6+} ions studied under optimized investigational circumstances (0.2 M acetate buffer (pH 3.6) as the supporting electrolyte, applied potential, etc). 10 μM of Cr^{6+} ions were added by successive addition at a bias potential of 0.28 V; as a result, a stepwise increase of cathodic current response was recorded. The achieved linear concentrations ranged from 10 to 80 μM with a lower LOD of 0.02 μM. The selectivity studies of the developed electrode were carried out in the presence of 10 μM of Fe^{3+}, and 100 μM of Cr^{3+} ions and the modified electrode detected selective sensing of 10 μM Cr^{6+} ions. Additionally, the CaO/GO/GCE showed outstanding long-term stability and reproducibility (up to 1 month), and the electrode was verified for real water samples and achieved 102% recovery results.

Similarly, Jayaraman and their research group prepared GO covalently functionalized with thymine (T) and carbohydrazide (C) through the epoxide ring opening method followed by a reduction method [95]. The GO was synthesized by a well-known modified Hummer technique. The obtained GO was functionalized with T-GO-C through a nucleophilic attack of nitrogen with an epoxide bond cleavage reaction. The prepared T-GO-C (1 mg l^{-1} ml) suspension of 5 μl^{-1} was drop-cast on the GCE surface, followed by square wave voltammetric simultaneous detection of Hg^{2+} and Cr^{6+} ions. Under optimized experimental conditions, the preconcentration potential (from −0.3 to −1.5 V), incubation time (120 s), deposition time (120 s), and different supporting electrolytes (pH-4.5) were briefly examined. The fabricated T-GO-C/GCE showed linear concentrations of 5–600 ppb for both Hg^{2+} and Cr^{6+} ions and the LODs were calculated to be 1 and 20 ppb for the Hg^{2+} and Cr^{6+} ions, respectively, and these obtained LOD values are in line with the United States Environmental Protection Agency (US-EPA) permitted limit of heavy metal ions in drinking water. Moreover, the selectivity studies of the T-GO-C/GCE in the presence of ten-fold excess concentrations of Pb^{2+}, Cd^{2+}, Zn^{2+}, Cu^{2+}, Cr^{3+}, and As^{3+} were examined; as a result, the electrode showed a predominant current response for Hg^{2+} and Cr^{6+} ions. However, above ten-fold excess concentrations of the above-mentioned heavy metals minimized the current signals due to non-specific interactions with the active binding sites. Under similar experimental conditions, the prepared T-GO-C/GCE showed excellent reproducibility and reusability for target metal detection. Finally, the feasibility of the prepared electrode was verified in various water samples by the standard addition technique and obtained 93%–98% recovery results. Further, the obtained outcomes were validated with ICP-OES methods.

Photoactive materials such as TiO_2 materials often have a low application potential in the visible light region due to their high energy band gap. Hence, the need to modify such materials with different kinds of conduction nanomaterials has become a hot research field. Photoelectrochemical detection and photoactive electrode materials with a huge proton to electron transfer proficiency are significant factors in achieving high detection sensitivity. Recently, Soleymani and their research group developed nanostructured TiO_2 with graphene quantum dots (GQDs) and pyrocatechol violet nanocomposites, and these nanocomposites significantly improved the photocurrent under visible light as well as enhanced TiO_2 dispersion [96]. The photocurrent was enhanced up to five times compared with TiO_2 with pyrocatechol violet. Hence, these results suggested a synergistic effect between the pyrocatechol violet and GQDs. Founded on this scenario, the enhancement of photoelectrode and photocurrent to prepare a photoelectrochemical current off the sensor for sensing of Cr^{6+} ions with an LOD of 0.04 μM. These LOD values are lower than the WHO-permitted level of Cr^{6+} ions (0.96 μM) in drinking water. Therefore, these prepared photoactive nanocomposites are delivered a new kind of TiO_2 as electrode materials, which is also a valuable material for manufacturing photoelectrochemical sensing devices.

Graphene-based materials have been modified on a gold surface for the sensitive detection of Cr^{6+} ions. For example, Wu and their research group fabricated an ion-impregnated type of electrode using graphene–chitosan nanocomposites using a facile electrodeposition method [97]. The electrodeposited chitosan–graphene on the gold electrode for the sensitive and selective determination of Cr^{6+} ions was examined by DPV methods. After optimizing the conditions, the mass production of the fabricated ion-impregnated graphene–chitosan composite electrode has simple fabrication steps. CV, EIS, and DPV techniques were use to examine the electrochemical performance of the fabricated electrode and the obtained linear concentration range of 1 nM to 10 μM with an LOD of 0.64 nM. The constructed electrode showed excellent selectivity for Cr^{6+} ion detection in the presence of other interfering species (Cu^{2+}, Cr^{3+}, and MnO_4^-, etc) and 85% of reproducible current was achieved after 9 cycles for 14 days. Finally, the constructed electrode was utilized for real-world (tap and river water) sample analysis and satisfactory recovery results were obtained.

Similarly, Punrat and their research group constructed a conducting polymer of polyaniline (PANI) and graphene quantum dot (GQD) modified carbon screen-printed electrodes for a stopped-flow injunction technique coupled with voltammetric techniques for the detection of Cr^{6+} ions (figure 13.8) [98]. A two-step process was used to synthesize the modified materials; initially, the GQDs were synthesized by a top-down method using citric acid as a precursor; subsequently, a monomer of aniline was mixed with the reaction mixture; as a result, the GQD/PANI nanocomposites were formed. Next, synthesized GQD/PANI composites were coated (drop-cast) on the working carbon screen-printed electrode surface for the analysis of Cr^{6+} ions. Under optimized experimental parameters, the developed electrode was tested for selective detection of Cr^{6+} ions in linear concentrations from 0.1 to 10 mg l^{-1} with an LOD of 0.097 mg l^{-1}. However, the GQD/PANI/SPCE was constantly examined in the sample of 0.5 ml for more than 90 samples every hour. Finally, the established technique was effectively utilized to detect Cr^{6+} ions in mineral water samples and achieved better recovery results. The GQD and

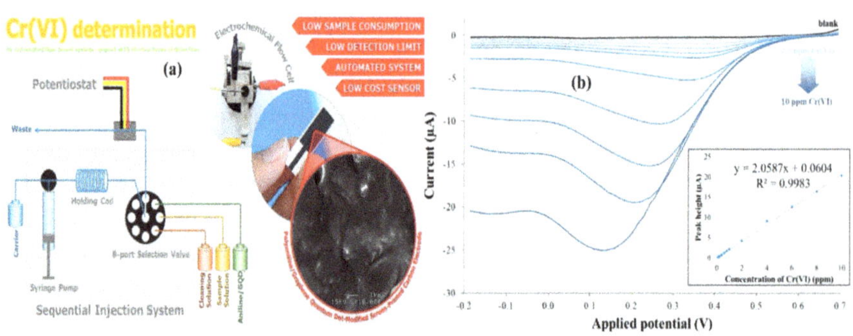

Figure 13.8. Schematic illustration and LSV response of Cr^{6+} ion detection in the Auto-Pret system coupled with electrochemical detection for online modification of PANI/GQDs-modified SPCEs and the determination of Cr^{6+} ion concentrations between 0.1 and 10 mg l^{-1} in 1 M HCl by stopped-flow analysis. (Reproduced with permission from [98]. Copyright 2016 Elsevier.)

Table 13.2. Comparison studies of carbon-based nanomaterials enabled modified electrodes for Cr^{6+} ions detection.

Modified materials	Technique	Concentration range (μM)	LOD (μM)	Sensitivity ($\mu A\ \mu M^{-1}$)	References
Gr/AuNPs/GCE	Amp	0–20	0.01	—	[97]
rGO/NiS/AuNcs/GCE	SWV	34.4–269.2	1.73×10^{-3}	3.38	[91]
AuNPs-Chit-GR/AuE	CV	5–800	—	—	[101]
AuNPs/rGO/GCE	SWV	0.1–30	0.046	0.79	[102]
AuNPs/PANI-co-PoT/GO/AuE	SWV	5–500	0.02	4.2	[103]
$g\text{-}C_3N_4$/AgM/Nf/GCE	Amp	0.1–0.7	0.0016	65.8	[93]
Carbon/SPCE	Amp	3–10 000	1	0.08	[104]
MWCNTs-PEI/GCE	Amp	0.002–20	0.0006	—	[105]
Au–PdNPs/ErGO/GCE	DPV	0.05–1000	0.013	—	[92]
QH_2/MWCNTs/SPCE	DPV	1–200	0.3	0.43	[81]
Bi/SWCNTs/GCE	CSV	0–25 000	36	—	[78]
MWCNTs-NR-AuNPs/SPCE	LSV	0.4–80	0.02	0.51	[82]
PBG_{DES}150-MWCNT/GCE	Amp	0.0025–0.060	0.02	1.15	[106]
PET/AuNPs/MPTS/AuNPs/GCE	SWV	2×10^{-4}–0.02	5.6×10^{-5}	—	[107]

peroxydisulfate materials for electrochemiluminescence-based sensors for the selective determination of Cr^{6+} ions were described by Chen and their research group [99, 100]. The Cr^{6+} ions were detected by quenching the strong cathodic electrochemiluminescence signal of the developed GQD and peroxydisulfate system. After optimizing the experimental conditions, the fabricated sensors showed Cr^{6+} ion detection in a linear concentration of 50–60 mM with a detection limit of 20 nM. Moreover, the prepared electrochemiluminescence system showed sensitive and selective determination of Cr^{6+} ions in the presence of other interfering species. Finally, the real-world application of the developed sensor to detect Cr^{6+} ions in various environmental water samples showed good retrieval outcomes.

The summarized version of the different kinds of carbon-based nanomaterials for the electrochemical determination of Cr^{6+} ions and their analytical behaviour in terms of concentration range, LOD, sensitivity, and stabilities were related and shown in table 13.2.

13.4 Conclusion and future scope

Sensitive electrochemical methods such as cyclic voltammetry, differential pulse voltammetry, impedimetry, and so on, can be employed as authoritative tools for the simultaneous detection of various HMIs (e.g. Pb^{2+}, Cd^{2+}, Hg^{2+}, and Cr^{6+}).

Many advanced nanomaterials and modified nanocomposites have achieved substantial results and progress with their extraordinarily selective and sensitive electrochemical properties in the study of target HMIs. This chapter shows the risks of HMI pollutants for humans and their environment, emphasizing Cd, Hg, Cr, and Pb, which are the topic of numerous investigations. The review also assesses the carbon-based nanomaterials (CNTs, graphene, graphene oxide, and carbon nitrides) employed as electrode materials for the electrochemical detection of Cr^{6+} ions. In light of the primary outcomes obtained in recent years, electrochemical methods have been demonstrated to be the most appropriate technique to detect HMIs with lower detection limits, ultra-high sensitivity, selectivity, reproducibility, etc. It seems that research on the application of advanced materials and their hybrid nanocomposites to detect harmful HMI pollutants is still authentic and not exhausted. Efforts could be directed toward developing synthesis methods/conditions, portable sensor systems for remote monitoring, and the search for economical materials to design nanocomposite-based electrochemical sensors.

Abbreviations

HMI	heavy metal ion
AAS	atomic absorption spectroscopy
ENAA	experimental neutron activation analysis
ICPMS	inductively coupled plasma mass spectroscopy
XFS	x-ray fluorescence spectroscopy
CV	cyclic voltammetry
CA	chronoamperometry
DPV	differential pulse voltammetry
ASV	anodic stripping voltammetry
LOD	limit of detection
DNA	deoxyribonucleic acid
PSA	polysulfoaminoathraquinone
LOQ	limit of quantitation
PAN	polyacrylonitrile
PMTB	poly methyl thymol blue
NC	nanocomposite
NP	nanoparticle
PGE	pencil graphite electrode
SPCE	screen printed carbon electrode
AuNP	gold nanoparticle
MMA-DMA	methyl methacrylate and decyl methacrylate
CNT	carbon nanotube
SEM	scanning electron microscopy
XRD	x-ray diffraction
FT-IR	Fourier transform-infrared
NF	Nafion
GCE	glassy carbon electrode
rGO	reduced graphene oxide
AuNP	gold nanoparticle
TPP	tetraphenylporphyrin

SPE	screen-printed electrode
DOP	dioctyl phthalate
RSD	relative standard deviation
MWCNT	multi-walled carbon nanotube
PIGE	paraffin-impegrenated electrode
NNW	nanonetwork
EG	ethyl green
DPT	1-dodecanoyl-3-phenylthiourea
MOF	metal–organic framework
BiM/GR	bismuth molybdate/graphene

Acknowledgements

The author BC thanks the Centre of Material Chemistry, Department of Science and Humanities, and the management of Karpagam Academy of Higher Education, Coimbatore, Tamil Nadu, India, for research support. The authors ME and IM thank the research support and grant from Vinnova-Swelifes, Medtech4Healths [2020-04733], Vetenskapsrådet [2020-04096], and Novo Nordisk Foundation grant NNF20CC003550. The author PA gratefully acknowledges the FONDECYT/ANID (2022) Post-doctoral Project No. 3220176, Government of Chile. The author KR gratefully acknowledges the FONDECYT/ANID (2021) Post-doctoral Project No. 3200232, Government of Chile. This work was supported by the Basic Science Research Program and Engineering and Academic Research (R&D) Program through the National Research Foundation of Korea (NRF) and funded by the Ministry of Education (2022R1I1A3072535 and RS202300249778).

References

[1] Tekaya N, Saiapina O, Ouada H B, Lagarde F, Ouada H B and Jaffrezic-Renault N 2013 Ultra-sensitive conductometric detection of heavy metals based on inhibition of alkaline phosphatase activity from *Arthrospira platensis Bioelectrochemistry* **90** 24–9

[2] Turdean G L 2011 Design and development of biosensors for the detection of heavy metal toxicity *Int. J. Electrochem.* **2011** 343125

[3] Gumpu M B, Sethuraman S, Krishnan U M and Rayappan J B B 2015 A review on detection of heavy metal ions in water—an electrochemical approach *Sensors Actuators* B **213** 515–33

[4] Preethi S, Arulraj A, Mangalaraja R V, Ravichandran V and Subramanian N 2022 Biomass-derived carbon-based materials for microbicidal applications *Biomass-Derived Carbon Materials* (New York: Wiley) pp 63–92

[5] Salmain M, Vessieres A, Jaouen G and Butler I S 1991 Fourier transform infrared spectroscopic method for the quantitative trace analysis of transition-metal carbonyl-labeled bioligands *Anal. Chem.* **63** 2323–9

[6] Dewald H D, Watkins J, Elder R and Heineman W R 1986 Development of extended x-ray absorption fine structure spectroelectrochemistry and its application to structural studies of transition-metal ions in aqueous solution *Anal. Chem.* **58** 2968–75

[7] Mohamed R, Abdel-Lateef A, Mahmoud H and Helal A 2012 Determination of trace elements in water and sediment samples from *Ismaelia canal* using ion chromatography and atomic absorption spectroscopy *Chem. Speciation Bioavailability* **24** 31–8

[8] Zhu R and Kok W T 1998 Determination of trace metal ions by capillary electrophoresis with fluorescence detection based on post-column complexation with 8-hydroxyquinoline-5-sulphonic acid *Anal. Chim. Acta* **371** 269–77

[9] Jones P, Williams T and Ebdon L 1990 Development of a novel multi-element detection system for trace metal determination based on chemiluminescence after separation by ion chromatography *Anal. Chim. Acta* **237** 291–8

[10] Tormen L, Gil R A, Frescura V L, Martinez L D and Curtius A J 2012 The use of electrothermal vaporizer coupled to the inductively coupled plasma mass spectrometry for the determination of arsenic, selenium and transition metals in biological samples treated with formic acid *Anal. Chim. Acta* **717** 21–7

[11] Shah A, Sultan S, Zahid A, Aftab S, Nisar J, Nayab S, Qureshi R, Khan G S, Hussain H and Ozkan S A 2017 Highly sensitive and selective electrochemical sensor for the trace level detection of mercury and cadmium *Electrochim. Acta* **258** 1397–403

[12] Padmanaban A, Bharathkumar S, Dhanasekaran T, Manigandan R, Pandian M S, Ramasamy P, Kathirvelu D and Valdes H 2022 Investigation of photo-/electrocatalytic activity of hydrothermal synthesized novel copper ion-modulated bifunctional $NiTe_2$ nanoflakes *Surf. Interfaces* **32** 102124

[13] Annamalai P, Thangavelu D, Ramadoss M, Subramani S, Muthu S P, Perumalsamy R, Ranganathan S and Hector V 2021 Electrochemical sensing of tyrosine and removal of toxic dye using self-assembled three-dimensional $CuBi_2O_4$/rGO microsphere composite *Colloid Interface Sci. Commun.* **45** 100523

[14] Padmanaban A, Murugadoss G, Venkatesh N, Hazra S, Kumar M R, Tamilselvi R and Sakthivel P 2021 Electrochemical determination of harmful catechol and rapid decolorization of textile dyes using ceria and tin doped ZnO nanoparticles *J. Environ. Chem. Eng.* **9** 105976

[15] Padmanaban A, Padmanathan N, Dhanasekaran T, Manigandan R, Srinandhini S, Sivaprakash P, Arumugam S and Narayanan V 2020 Hexagonal phase Pt-doped cobalt telluride magnetic semiconductor nanoflakes for electrochemical sensing of dopamine *J. Electroanal. Chem.* **877** 114658

[16] Padmanaban A, Dhanasekaran T, Manigandan R, Kumar S P, Gnanamoorthy G, Stephen A and Narayanan V 2017 Facile solvothermal decomposition synthesis of single phase $ZnBi_{38}O_{60}$ nanobundles for sensitive detection of 4-nitrophenol *New J. Chem.* **41** 7020–7

[17] Li M, Gou H, Al-Ogaidi I and Wu N 2013 *Nanostructured Sensors for Detection of Heavy Metals: A Review ACS Sustainable Chem. & Engg.* 1 713–23

[18] Tarley C R T, Santos V S, Baêta B E L, Pereira A C and Kubota L T 2009 Simultaneous determination of zinc, cadmium and lead in environmental water samples by potentiometric stripping analysis (PSA) using multiwalled carbon nanotube electrode *J. Hazard. Mater.* **169** 256–62

[19] de la Escosura-Muñiz A, Ambrosi A and Merkoçi A 2008 Electrochemical analysis with nanoparticle-based biosystems *TrAC, Trends Anal. Chem.* **27** 568–84

[20] Sall M L, Fall B, Diédhiou I, Dièye E H, Lo M, Diaw A K D, Gningue-Sall D, Raouafi N and Fall M 2020 Toxicity and electrochemical detection of lead, cadmium and nitrite ions by organic conducting polymers: a review *Chem. Africa* **3** 499–512

[21] Wu Y, Liu L, Zhan S, Wang F and Zhou P 2012 Ultrasensitive aptamer biosensor for arsenic (III) detection in aqueous solution based on surfactant-induced aggregation of gold nanoparticles *Analyst* **137** 4171–8
[22] Wang G, Chen X, Liu S, Wong C and Chu S 2016 Mechanical chameleon through dynamic real-time plasmonic tuning *ACS Nano* **10** 1788–94
[23] Anandhakumar S, Mathiyarasu J, Phani K L N and Yegnaraman V 2011 Simultaneous determination of cadmium and lead using PEDOT/PSS modified glassy carbon electrode *Am. J. Anal. Chem.* **2** 470
[24] Yang H, Li J, Lu X, Xi G and Yan Y 2013 Reliable synthesis of bismuth nanoparticles for heavy metal detection *Mater. Res. Bull.* **48** 4718–22
[25] Liu G, Lin Y, Tu Y and Ren Z 2005 Ultrasensitive voltammetric detection of trace heavy metal ions using carbon nanotube nanoelectrode array *Analyst* **130** 1098–101
[26] Gu P-Y, Deng C and Tang L 2011 Experimental study on damage identification based on operational strain modal shape *Zhendong yu Chongji (J. Vib. Shock)* **30** 175–8
[27] Aragay G, Pons J and Merkoçi A 2011 Enhanced electrochemical detection of heavy metals at heated graphite nanoparticle-based screen-printed electrodes *J. Mater. Chem.* **21** 4326–31
[28] Tajik S, Beitollahi H, Nejad F G, Dourandish Z, Khalilzadeh M A, Jang H W, Venditti R A, Varma R S and Shokouhimehr M 2021 Recent developments in polymer nanocomposite-based electrochemical sensors for detecting environmental pollutants *Ind. Eng. Chem. Res.* **60** 1112–36
[29] Khan A, Khan A A P, Rahman M M, Asiri A M, Alfaifi S Y and Taib L A 2018 Toward facile preparation and design of mulberry-shaped poly (2-methylaniline)-$Ce_2(WO_4)_3$@CNT nanocomposite and its application for electrochemical Cd^{2+} ion detection for environment remediation *Polym.-Plast. Technol. Eng.* **57** 335–45
[30] Ramachandran R, Chen T-W, Chen S-M, Baskar T, Kannan R, Elumalai P, Raja P, Jeyapragasam T and Dinakaran K 2019 A review of the advanced developments of electrochemical sensors for the detection of toxic and bioactive molecules *Inorg. Chem. Front.* **6** 3418–39
[31] Bhanjana G, Dilbaghi N, Kumar R, Umar A and Kumar S 2015 SnO_2 quantum dots as novel platform for electrochemical sensing of cadmium *Electrochim. Acta* **169** 97–102
[32] Sreekanth S, Alodhayb A, Assaifan A K, Alzahrani K E, Muthuramamoorthy M, Alkhammash H I, Pandiaraj S, Alswieleh A M, Van Le Q and Mangaiyarkarasi R 2021 Multi-walled carbon nanotube-based nanobiosensor for the detection of cadmium in water *Environ. Res.* **197** 111148
[33] Fakude C T, Arotiba O A and Mabuba N 2020 Electrochemical aptasensing of cadmium (II) on a carbon black-gold nano-platform *J. Electroanal. Chem.* **858** 113796
[34] Magar H S, Abu-El Magd E E, Hassan R Y and Fahim A M 2022 Rapid impedimetric detection of cadmium ions using nanocellulose/ligand/nanocomposite (CNT/Co_3O_4) *Microchem. J.* **182** 107885
[35] Martín-Yerga D, González-García M B and Costa-García A 2012 Use of nanohybrid materials as electrochemical transducers for mercury sensors *Sensors Actuators* B **165** 143–50
[36] Chen K, Lu G, Chang J, Mao S, Yu K, Cui S and Chen J 2012 Hg (II) ion detection using thermally reduced graphene oxide decorated with functionalized gold nanoparticles *Anal. Chem.* **84** 4057–62

[37] Fernández E, Vidal L, Martín-Yerga D, del Carmen Blanco M, Canals A and Costa-García A 2015 Screen-printed electrode based electrochemical detector coupled with ionic liquid dispersive liquid–liquid microextraction and microvolume back-extraction for determination of mercury in water samples *Talanta* **135** 34–40

[38] Fernández E, Vidal L, Costa-García A and Canals A 2016 Mercury determination in urine samples by gold nanostructured screen-printed carbon electrodes after vortex-assisted ionic liquid dispersive liquid–liquid microextraction *Anal. Chim. Acta* **915** 49–55

[39] Cinti S, Santella F, Moscone D and Arduini F 2016 Hg^{2+} detection using a disposable and miniaturized screen-printed electrode modified with nanocomposite carbon black and gold nanoparticles *Environ. Sci. Pollut. Res.* **23** 8192–9

[40] Bui M-P N, Brockgreitens J, Ahmed S and Abbas A 2016 Dual detection of nitrate and mercury in water using disposable electrochemical sensors *Biosens. Bioelectron.* **85** 280–6

[41] Razmi H, Musevi S J and Mohammad-Rezaei R 2016 Solid phase extraction of mercury (II) using soluble eggshell membrane protein doped with reduced graphene oxide, and its quantitation by anodic stripping voltammetry *Microchim. Acta* **183** 555–62

[42] Palanisamy S, Thangavelu K, Chen S-M, Velusamy V, Chang M-H, Chen T-W, Al-Hemaid F M, Ali M A and Ramaraj S K 2017 Synthesis and characterization of polypyrrole decorated graphene/β-cyclodextrin composite for low level electrochemical detection of mercury (II) in water *Sensors Actuators* B **243** 888–94

[43] Zhang J, Zhu Z, Di J, Long Y, Li W and Tu Y 2015 A sensitive sensor for trace Hg^{2+} determination based on ultrathin g-C_3N_4 modified glassy carbon electrode *Electrochim. Acta* **186** 192–200

[44] Martín-Yerga D and Costa-García A 2017 Recent advances in the electrochemical detection of mercury *Curr. Opin. Electrochem.* **3** 91–6

[45] Zuo Y, Xu J, Zhu X, Duan X, Lu L, Gao Y, Xing H, Yang T, Ye G and Yu Y 2016 Poly (3,4-ethylenedioxythiophene) nanorods/graphene oxide nanocomposite as a new electrode material for the selective electrochemical detection of mercury (II) *Synth. Met.* **220** 14–9

[46] Eksin E, Erdem A, Fafal T and Kıvçak B 2019 Eco-friendly sensors developed by herbal based silver nanoparticles for electrochemical detection of mercury (II) ion *Electroanalysis* **31** 1075–82

[47] Singh S, Numan A, Zhan Y, Singh V, Van Hung T and Nam N D 2020 A novel highly efficient and ultrasensitive electrochemical detection of toxic mercury (II) ions in canned tuna fish and tap water based on a copper metal-organic framework *J. Hazard. Mater.* **399** 123042

[48] Sakthi Priya T, Chen T-W, Chen S-M and Kokulnathan T 2022 Bismuth molybdate/graphene nanocomposite for electrochemical detection of mercury *ACS Appl. Nano Mater.* **5** 12518–26

[49] Huang M-R, Ding Y-B and Li X-G 2014 Combinatorial screening of potentiometric Pb (II) sensors from polysulfoaminoanthraquinone solid ionophore *ACS Comb. Sci.* **16** 128–38

[50] Shen L, Chen Z, Li Y, He S, Xie S, Xu X, Liang Z, Meng X, Li Q and Zhu Z 2008 Electrochemical DNAzyme sensor for lead based on amplification of DNA–Au Bio-Bar codes *Anal. Chem.* **80** 6323–8

[51] Tarley C R T, Basaglia A M, Segatelli M G, Prete M C, Suquila F A C and de Oliveira L L G 2017 Preparation and application of nanocomposite based on imprinted poly (methacrylic acid)-PAN/MWCNT as a new electrochemical selective sensing platform of Pb^{2+} in water samples *J. Electroanal. Chem.* **80** 144–21

[52] Zuo Y, Xu J, Jiang F, Duan X, Lu L, Xing H, Yang T, Zhang Y, Ye G and Yu Y 2017 Voltammetric sensing of Pb (II) using a glassy carbon electrode modified with composites consisting of Co_3O_4 nanoparticles, reduced graphene oxide and chitosan *J. Electroanal. Chem.* **801** 146–52
[53] Wang Y, Zhao G, Zhang G, Zhang Y, Wang H, Cao W, Li T and Wei Q 2020 An electrochemical aptasensor based on gold-modified MoS_2/rGO nanocomposite and gold-palladium-modified Fe-MOFs for sensitive detection of lead ions *Sensors Actuators* B **319** 128313
[54] Sivan S K, Shankar S S, Kandambath Padinjareveetil A, Pilankatta R, Kumar V S, Mathew B, George B, Makvandi P, Černík M and Padil V V 2020 Fabrication of a greener TiO_2@ gum Arabic-carbon paste electrode for the electrochemical detection of Pb^{2+} ions in plastic toys *ACS Omega* **5** 25390–9
[55] Shah A, Zahid A, Khan A, Iftikhar F J, Nisar J, Fernandez C, Akhter M S, Almutawah A A and Kraatz H-B 2019 Development of a highly sensitive electrochemical sensing platform for the trace level detection of lead ions *J. Electrochem. Soc.* **166** B3136
[56] Eswaran M, Tsai P-C, Wu M-T and Ponnusamy V K 2021 Novel nano-engineered environmental sensor based on polymelamine/graphitic-carbon nitride nanohybrid material for sensitive and simultaneous monitoring of toxic heavy metals *J. Hazard. Mater.* **418** 126267
[57] Hu C, Wu K, Dai X and Hu S 2003 Simultaneous determination of lead (II) and cadmium (II) at a diacetyldioxime modified carbon paste electrode by differential pulse stripping voltammetry *Talanta* **60** 17–24
[58] Manikandan R and Narayanan S S 2019 Simultaneous anodic stripping voltammetric determination of Pb (II) and Cd (II) using poly methyl thymol blue film-modified electrode *Ionics* **25** 6083–92
[59] Le Hai T, Hung L C, Phuong T T B, Ha B T T, Nguyen B-S, Hai T D and Nguyen V-H 2020 Multiwall carbon nanotube modified by antimony oxide (Sb_2O_3/MWCNTs) paste electrode for the simultaneous electrochemical detection of cadmium and lead ions *Microchem. J.* **153** 104456
[60] Baghayeri M, Amiri A, Maleki B, Alizadeh Z and Reiser O 2018 A simple approach for simultaneous detection of cadmium (II) and lead (II) based on glutathione coated magnetic nanoparticles as a highly selective electrochemical probe *Sensors Actuators* B **273** 1442–50
[61] Liu S, Chen M, Cao X, Li G, Zhang D, Li M, Meng N, Yin J and Yan B 2020 Chromium (VI) removal from water using cetylpyridinium chloride (CPC)-modified montmorillonite *Sep. Purif. Technol.* **241** 116732
[62] Saif M, Kamal R and Hafez H S 2019 $BaZrO_3:Eu^{3+}$ nanophosphor: a potential fluorescent sensor for highly selective and sensitive detection of chromium ions from tannery leather and wastewater *J. Alloys Compd.* **803** 658–63
[63] Arellano-Sánchez M G, Devouge-Boyer C, Hubert-Roux M, Afonso C and Mignot M 2022 Chromium determination in leather and other matrices: a review *Crit. Rev. Anal. Chem.* **52** 1537–56
[64] Garcia-Miranda Ferrari A, Crapnell R D, Adarakatti P S, Suma B P and Banks C E 2022 Electroanalytical overview: the detection of chromium *Sensors Actuators Rep.* **4** 100116
[65] Ambika S, Kumar M, Pisharody L, Malhotra M, Kumar G, Sreedharan V, Singh L, Nidheesh P V and Bhatnagar A 2022 Modified biochar as a green adsorbent for removal of

hexavalent chromium from various environmental matrices: mechanisms, methods, and prospects *Chem. Eng. J.* **439** 135716
[66] Wise J P, Young J L, Cai J and Cai L 2022 Current understanding of hexavalent chromium [Cr(VI)] neurotoxicity and new perspectives *Environ. Int.* **158** 106877
[67] Shrivastava P, Jain V K and Nagpal S 2022 Nanoparticle intervention for heavy metal detection: a review *Environ. Nanotechnol., Monitor. Manag.* **17** 100667
[68] Jin Q, Dai M, Zhan X, Wang S and He Z 2022 Carbon nanotubes and graphene composites used in Cr (VI) detection techniques: a review *J. Alloys Compd.* **922** 166268
[69] Tirez K, Brusten W, Cluyts A, Patyn J and De Brucker N 2003 Determination of hexavalent chromium by species specific isotope dilution mass spectrometry and ion chromatography–1,5-diphenylcarbazide spectrophotometry *J. Anal. At. Spectrom.* **18** 922–32
[70] Memon R, Memon A A, Sirajuddin, Balouch A, Shah M R, Mujeeb ur R, Sherazi S T H, Memon S S, Memon and K 2021 Highly selective nanomolar level colorimetric sensing of Cr^{3+} through biosynthesized gold nanoparticles in the presence of Cr^{6+} *Optik* **248** 168188
[71] Wu Y, Midinov B and White R J 2019 Electrochemical aptamer-based sensor for real-time monitoring of insulin *ACS Sens.* **4** 498–503
[72] Liuzhu Z, Sekar S, Chen J, Lee S, Kim D Y and Manikandan R 2022 A polyrutin/AgNPs coated GCE for simultaneous anodic stripping voltammetric determination of Pb(II) and Cd(II)ions in environmental samples *Colloids Surf.* A **648** 129082
[73] Sekar S, Huijun J, Liuzhu Z, Jin C, Lee S, Kim D Y and Manikandan R 2022 Copper phthalocyanine conjugated graphitic carbon nitride nanosheets as an efficient electrocatalyst for simultaneous detection of natural antioxidants *Electrochim. Acta* **413** 140150
[74] Jorge E O, Rocha M M, Fonseca I T E and Neto M M M 2010 Studies on the stripping voltammetric determination and speciation of chromium at a rotating-disc bismuth film electrode *Talanta* **81** 556–64
[75] Tsai M-C and Chen P-Y 2008 Voltammetric study and electrochemical detection of hexavalent chromium at gold nanoparticle-electrodeposited indium tinoxide (ITO) electrodes in acidic media *Talanta* **76** 533–9
[76] Motaghedifard M H, Pourmortazavi S M and Mirsadeghi S 2021 Selective and sensitive detection of Cr(VI) pollution in waste water via polyaniline/sulfated zirconium dioxide/multi walled carbon nanotubes nanocomposite based electrochemical sensor *Sensors Actuators* B **327** 128882
[77] Breslin C B, Branagan D and Garry L M 2019 Electrochemical detection of Cr(VI) with carbon nanotubes decorated with gold nanoparticles *J. Appl. Electrochem.* **49** 195–205
[78] Ouyang R, Zhang W, Zhou S, Xue Z-L, Xu L, Gu Y and Miao Y 2013 Improved Bi film wrapped single walled carbon nanotubes for ultrasensitive electrochemical detection of trace Cr(VI) *Electrochim. Acta* **113** 686–93
[79] Jin W and Yan K 2015 Recent advances in electrochemical detection of toxic Cr(VI) *RSC Adv.* **5** 37440–50
[80] Rosolina S M, Bragg S A, Ouyang R, Chambers J Q and Xue Z-L 2016 Highly sensitive detection of hexavalent chromium utilizing a sol–gel/carbon nanotube modified electrode *J. Electroanal. Chem.* **781** 120–5
[81] Sadeghi S and Garmroodi A 2013 A highly sensitive and selective electrochemical sensor for determination of Cr(VI) in the presence of Cr(III) using modified multi-walled carbon nanotubes/quercetin screen-printed electrode *Mater. Sci. Eng.* C **33** 4972–7

[82] Filik H and Aslıhan Avan A 2020 Neutral red interlinked gold nanoparticles/multiwalled carbon nanotubes modified electrochemical sensor for simultaneous speciation and detection of chromium (VI) and vanadium (V) in water samples *Microchem. J.* **158** 105242
[83] Filik H and Avan A A 2020 Neutral red interlinked gold nanoparticles/multiwalled carbon nanotubes modified electrochemical sensor for simultaneous speciation and detection of chromium (VI) and vanadium (V) in water samples *Microchem. J.* **158** 105242
[84] Rudnitskaya A, Evtuguin D V, Costa L C, Graça M P F, Fernandes A J S, Correia M R P, Gomes M T S R and Oliveira J A B P 2013 Potentiometric chemical sensors from lignin–poly(propylene oxide) copolymers doped by carbon nanotubes *Analyst* **138** 501–8
[85] Deep A, Sharma A L, Tuteja S K and Paul A K 2014 Phosphinic acid functionalized carbon nanotubes for sensitive and selective sensing of chromium(VI) *J. Hazard. Mater.* **278** 559–65
[86] Das T, Sharma B K, Katiyar A K and Ahn J-H 2018 Graphene-based flexible and wearable electronics *J. Semiconduct.* **39** 011007
[87] Carvalho A P G, Alegria E, Fantoni A, Ferraria A M, do Rego A M B and Ribeiro A P C 2022 Effect of graphene vs. reduced graphene oxide in gold nanoparticles for optical biosensors: a comparative study *Biosensors* **12** 163
[88] Sainz-Urruela C, Vera-López S, San Andrés M P and Díez-Pascual A M 2021 Graphene-based sensors for the detection of bioactive compounds: a review *Int. J. Mol. Sci.* **22** 3316
[89] Liu C, He C, Xie T and Yang J 2015 Reduction of graphite oxide using ammonia solution and detection Cr (VI) with graphene-modified electrode *Fuller. Nanotub. Carbon Nanostruct.* **23** 125–30
[90] Sari T K, Takahashi F, Jin J, Zein R and Munaf E 2018 Electrochemical determination of chromium(VI) in river water with gold nanoparticles-graphene nanocomposites modified electrodes *Anal. Sci.* **34** 155–60
[91] Sahoo S and Satpati A K 2021 Fabrication of rGO/NiS/AuNCs ternary nanocomposite modified electrode for electrochemical sensing of Cr(VI) at ultra-trace level *Surf. Interfaces* **24** 101096
[92] Hu J, Liu Y, Gao G and Zou X 2018 One-step synthesis of AuPdNPs/ERGO nanocomposite for enhanced electrochemical sensing of Cr(VI) in water *J. Electrochem. Soc.* **165** B893
[93] Karthika A, Nikhil S, Suganthi A and Rajarajan M 2020 A facile sonochemical approach based on graphene carbon nitride doped silver molybdate immobilized nafion for selective and sensitive electrochemical detection of chromium (VI) in real sample *Adv. Powder Technol.* **31** 1879–90
[94] Singh S, Naik T S S K, Anil A G, Khasnabis S, Nath B U B, Kumar V, Garg V K, Subramanian S, Singh J and Ramamurthy P C 2022 A novel CaO nanocomposite cross linked graphene oxide for Cr(VI) removal and sensing from wastewater *Chemosphere* **301** 134714
[95] Jayaraman N, Palani Y, Jonnalagadda R R and Shanmugam E 2022 Covalently dual functionalized graphene oxide-based multiplex electrochemical sensor for Hg(II) and Cr (VI) detection *Sensors Actuators* B **367** 132165
[96] Bakhshandeh F, Saha S, Sakib S, Zhitomirsky I and Soleymani L 2022 TiO_2 nanoparticles co-sensitized with graphene quantum dots and pyrocatechol violet for photoelectrochemical detection of Cr(VI) *J. Electrochem. Soc.* **169** 057520

[97] Wu S, Dai X, Cheng T and Li S 2018 Highly sensitive and selective ion-imprinted polymers based on one-step electrodeposition of chitosan–graphene nanocomposites for the determination of Cr(VI) *Carbohydr. Polym.* **195** 199–206

[98] Punrat E, Maksuk C, Chuanuwatanakul S, Wonsawat W and Chailapakul O 2016 Polyaniline/graphene quantum dot-modified screen-printed carbon electrode for the rapid determination of Cr (VI) using stopped-flow analysis coupled with voltammetric technique *Talanta* **150** 198–205

[99] Nagarajan L, Arulraj A, Senguttuvan G and Kumaraguru K 2020 Solar assisted reduced graphene oxide as adsorbent for carbondioxide and its kinetic studies *Physica E: Low-dimesional Sys. and Nanostruc* **116** 113739

[100] Chen Y, Dong Y, Wu H, Chen C, Chi Y and Chen G 2015 Electrochemiluminescence sensor for hexavalent chromium based on the graphene quantum dots/peroxodisulfate system *Electrochim. Acta* **151** 552–7

[101] Murugan C S, Rajendran S, Felix R, Velmurugan V S and Grace A 2014 Graphene/gold nanocomposites-based thin films as an enhanced sensing platform for voltammetric detection of Cr(VI) Ions *J. Nanotechnol.* **2014** 7

[102] Santhosh C, Saranya M, Ramachandran R, Felix S, Velmurugan V and Nirmala Grace A 2014 Graphene/gold nanocomposites-based thin films as an enhanced sensing platform for voltammetric detection of Cr (VI) ions *J. Nanotechnol.* **2014** 304526

[103] Hussein M A, Ganash A A and Alqarni S A 2019 Electrochemical sensor-based gold nanoparticle/poly(aniline-co-o-toluidine)/graphene oxide nanocomposite modified electrode for hexavalent chromium detection: a real test sample *Polym.-Plast. Technol. Mater.* **58** 1423–36

[104] Miscoria S A, Jacq C, Maeder T and Martín Negri R 2014 Screen-printed electrodes for electroanalytical sensing, of chromium VI in strong acid media *Sensors Actuators* B **195** 294–302

[105] Peng Y, Wang Y, Li Y and He X 2016 Polyethylenimine functionalized multi-walled carbon nanotubes for electrochemical detection of chromium(VI) *Electroanalysis* **28** 2029–36

[106] da Silva W, Ghica M E and Brett C M A 2020 Biotoxic trace metal ion detection by enzymatic inhibition of a glucose biosensor based on a poly(brilliant green)–deep eutectic solvent/carbon nanotube modified electrode *Talanta* **208** 120427

[107] Ouyang R, Bragg S A, Chambers J Q and Xue Z-L 2012 Flower-like self-assembly of gold nanoparticles for highly sensitive electrochemical detection of chromium(VI) *Anal. Chim. Acta* **722** 1–7

www.ingramcontent.com/pod-product-compliance
Ingram Content Group UK Ltd.
Pitfield, Milton Keynes, MK11 3LW, UK
UKHW051042120926
8556IPUK00030B/723